Large Carnivore Conservation and Management

Large carnivores include iconic species such as bears, wolves and big cats. Their habitats are increasingly being shared with humans, and there is a growing number of examples of human-carnivore coexistence as well as conflict. Next to population dynamics of large carnivores, there are considerable attitude shifts towards these species worldwide, with multiple implications.

This book argues and demonstrates why human dimensions of relationships to large carnivores are crucial for their successful conservation and management. It provides an overview of theoretical and methodological perspectives, heterogeneity in stakeholder perceptions and behaviour as well as developments in decision making, stakeholder involvement, policy and governance informed by human dimensions of large carnivore conservation and management. The scope is international, with detailed examples and case studies from Europe, North and South America, Central and South Asia, as well as debates of the challenges faced by urbanization, agricultural expansion, national parks and protected areas. The main species covered include bears, wolves, lynx and leopards.

The book provides a novel perspective for advanced students, researchers and professionals in ecology and conservation, wildlife management, human-wildlife interactions, environmental education and environmental social science.

Tasos Hovardas is the Human Dimensions Expert of CALLISTO – Wildlife and Nature Conservation Society, and through CALLISTO, he is providing consultancy services to the EU Platform on Coexistence between People and Large Carnivores. Tasos acts as Editor-in-Chief of *Society & Natural Resources* (2017–2020), together with Prof. Linda Prokopy (2017–2020). He is an Editorial Board Member of the *Journal of Research in Science Teaching* (2017–2020) and a member of the Research in Science & Technology Education Group at the University of Cyprus.

Earthscan Studies in Natural Resource Management

For more information on books in the Earthscan Studies in Natural Resource Management series, please visit the series page on the Routledge website: www.routledge.com/books/series/ECNRM/

Large Carnivore Conservation and Management

Human Dimensions

Edited by
Tasos Hovardas

LONDON AND NEW YORK

First published 2018
by Routledge

2 Park Square, Milton Park, Abingdon, Oxfordshire OX14 4RN
52 Vanderbilt Avenue, New York, NY 10017

Routledge is an imprint of the Taylor & Francis Group, an informa business

First issued in paperback 2020

British Library Cataloguing-in-Publication Data
A catalogue record for this book is available from the British Library

Library of Congress Cataloging-in-Publication Data
Names: Hovardas, Tasos, editor.
Title: Large carnivore conservation and management : human dimensions / edited by Tasos Hovardas.
Description: Abingdon, Oxon ; New York, NY : Routledge, 2018. | Series: Earthscan studies in natural resource management | Includes bibliographical references and index.
Identifiers: LCCN 2018001463 | ISBN 9781138039995 (hardback) | ISBN 9781315175454 (ebook)
Subjects: LCSH: Carnivora—Conservation. | Predatory animals— Conservation. | Wildlife conservation. | Conservation biology. | Human-animal relationships.
Classification: LCC QL737.C2 L33587 2018 | DDC 599.7—dc23
LC record available at https://lccn.loc.gov/2018001463

ISBN: 978-1-138-03999-5 (hbk)
ISBN: 978-0-367-60588-9 (pbk)

Typeset in Bembo
by Apex CoVantage, LLC

Contents

Contributors

Justine Shanti Alexander works to protect wildlife and wild spaces in high altitude areas of Asia. She works for the Snow Leopard Trust as their Regional Ecologist and is involved in research, conservation and capacity building initiatives supporting national teams in China and Mongolia.

Anna Bendz is a Researcher and Senior Lecturer at the department of Political Science, University of Gothenburg, Sweden. Apart from wolf policy and opinion, her research interests concern drinking water risk management and welfare policy/welfare public opinion.

Yash Veer Bhatnagar has been involved in snow leopard research and conservation for nearly 25 years and assists governments and conservation organizations in creating and implementing snow leopard conservation plans. He is a senior scientist at the Snow Leopard Trust and the Nature Conservation Foundation.

Michal Bojda has worked as a field biologist and educator with Friends of the Earth Czech Republic since 2008. He focuses on field monitoring of large carnivores, educational programmes about large carnivores for schools and communication with different stakeholders such as farmers and hunters.

Serena Cinque has a PhD in public administration at the University of Gothenburg, Sweden. Her research interests focus on implementation of political decisions in natural resource management particularly regarding the exercise of administrative discretion by regulatory agencies.

Fredrik Dalerum is a Research Fellow at the Research Unit of Biodiversity, University of Oviedo, Spain. He is also affiliated with University of Pretoria (South Africa) and Stockholm University (Sweden). He is a terrestrial ecologist who focuses his research on the biology and conservation of mammalian carnivores, functional aspects of biodiversity and general issues of conservation and environmental sustainability.

Martina Dušková works as a Czech-English translator. She has been participating in the large carnivore monitoring programme in the West Carpathians since 2014.

Adam Eagle is a lawyer and an associate at the international law firm Clifford Chance LLP. He has provided legal advice to numerous wildlife conservation charities, specifically in the areas of species reintroductions and international treaties on wildlife conservation. He has acted as project manager on a number of rewilding projects, and has a specific interest in landscape scale ecosystem restoration.

Göran Ericsson is the chaired professor in Wildlife Ecology in the Department of Wildlife, Fish, and Environmental Studies at the Swedish University of Agricultural Sciences (SLU) in Umeå. He earned a PhD from SLU, and B.S and M.S. degrees from Uppsala University, Sweden. In 2001–2002, he was a Senior Fulbright Fellow at the University of Wisconsin-Madison (USA). He is one of the European forerunners of research and teaching in Human Dimensions of Wildlife Conservation.

Laurent Garde has a PhD in Ecology and a Master's degree in Human Sciences. He is Deputy Director of the CERPAM (Centre d'Etudes et de Realisations pastorales Alpes Méditerranée), the Agency for studies, support and advisory to grazing livestock breeders in Provence and the Southern Alps. Since 2004, he is member of the French National Wolf Group of experts. His personal work focuses on the direct and indirect impacts of predation and related constraints on livestock grazing systems, mostly in Mediterranean and Alpine areas.

Sunetro Ghosal researches human-large carnivore relationships to help mitigate conflicts. He teaches courses on environment and development at St. Xavier's Autonomous College, Mumbai, and Norwegian University of Life Sciences, Ås (Aas), and is currently serving as editor of Stawa and Ladakh Studies.

Meredith L. Gore is an Associate Professor in the Department of Fisheries & Wildlife at Michigan State University (East Lansing, Michigan, USA) and Jefferson Science Fellow with the US Department of State. She is a conservation social scientist whose interdisciplinary research explores relationships between human behavior and the environment. Most recently, she has helped lead development of the interdisciplinary framework, conservation criminology, for understanding environmental risks. She is a MSU Global Research Fellow and Past President of the Society for Conservation Biology's Social Science Working Group.

Juha Hiedanpää acts as Research Professor at the Natural Resources Institute Finland (Luke). He is an environmental economist, environmental policy scientist and a specialist of wildlife and forest biodiversity policy and governance issues.

Tasos Hovardas is the Human Dimensions Expert of CALLISTO – Wildlife and Nature Conservation Society, and through CALLISTO, he is providing consultancy services to the EU Platform on Coexistence between People and large Carnivores. Tasos acts as Editor-in-Chief of *Society & Natural Resources* (2017–2020), together with Prof. Linda Prokopy (2017–2020). He is an Editorial Board Member of the *Journal of Research in Science Teaching* (2017–2020) and a member of the Research in Science & Technology Education Group at the University of Cyprus.

Djuro Huber is Professor emeritus at the Biology Department of the Veterinary Faculty in Zagreb. Since 1981, he has studied brown bears in Croatia. In 1996, he started a multidisciplinary project on large carnivores in Croatia, which also included wolves and lynxes. He is member of several national and international organizations like International Union for Conservation of Nature (IUCN), Species Survival Commisions (SSCs) for Bears, Canids and Veterinary Medicine; International Association for Bear Research and Management; Large Carnivore Initiative for Europe and Wildlife Disease Association.

Catherine Jampel is a PhD candidate in the Graduate School of Geography at Clark University (Worcester, Massachusetts, USA). Her master's degree is in Geography and Women's Studies, from The Pennsylvania State University (State College, Pennsylvania, USA). Her bachelor's degree is in History and Literature from Harvard College (Cambridge, Massachusetts, USA).

Klemen Jerina is a wildlife researcher working mainly on applied issues of ecology and management of ungulates and large carnivores in south-central Europe. He is Professor of Wildlife Ecology at the Forestry Department of University of Ljubljana in Slovenia.

Örjan Johansson is a scientist at the Snow Leopard Trust and Swedish University of Agricultural Sciences. He works mainly with snow leopard ecology and behaviour. His skills include capture and collaring of animals using safe, ethical and efficient methods.

Petr Kovařík works as a zoologist and ecologist at Palacký University in Olomouc and in Nature Conservation Agency of the Czech Republic. He focuses mainly on ecology and ornithology, but is also passionate about other wildlife topics including carnivores, bats and insects.

Olve Krange, PhD, is a Sociologist at the Norwegian Institute for Nature Research – NINA. Together with Ketil Skogen, he has studied the conflicts over large carnivore management in Norway for almost two decades. They have worked together on several studies of large carnivore management.

Miha Krofel is wildlife researcher and Assistant Professor at University of Ljubljana in Slovenia. His research focus is on large carnivore ecology, management and conservation in Eurasia and Africa, and to a lesser degree on ungulates, lizards, and scavenger communities. He dedicates most of his studies on bears to improve understanding of their basic ecology and effects of bear management measures, especially regarding brown bears in Dinaric Mountains and the Alps.

Miroslav Kutal has coordinated activities of Friends of the Earth Czech Republic in the field of large carnivore conservation and monitoring since 2002. Since 2015, he has also worked at Mendel University in Brno, focusing his research on the ecology of the grey wolf and the Eurasian lynx as they are recolonizing Central Europe, and the transfer of scientific results to practical conservation measures. He is a member of the Large Carnivore Initiative for Europe, IUCN/SSC specialist group.

Leona Kutalová worked as a coordinator of volunteers participating in field monitoring of large carnivores with Friends of the Earth Czech Republic. She has concentrated on investigating attitudes of people towards large carnivores in the West Carpathians as part of her studies at Palacký University Olomouc.

Nicolas Lescureux has a PhD in Ethnoecology and has been Researcher at the National Center for Scientific Research (CNRS) since 2014. For over 15 years, he has been studying the impacts of large carnivores on human knowledge, perceptions and practices as well as the impacts of human practices – notably livestock breeding and hunting – on large carnivore ecology and behaviour in several countries (France, Kyrgyzstan, Former Yugoslav Republic of Macedonia, Poland, Bulgaria, Brazil).

Steven Lipscombe currently works in conservation for the North Pennines AONB Partnership while pursuing independent ecology projects; he has also worked all over the world on independent eco-based farming and building schemes. He recently earned a Masters degree in conservation ecology from Newcastle University, converting from a career as an industrial and research chemist.

José Vicente López-Bao is a Research Fellow at the Research Unit in Biodiversity (UO-CSIC-PA), Oviedo University, Spain. His research interests include large carnivore conservation and management in human–dominated landscapes.

Michelle L. Lute is the Wildlife Coexistence Campaigner for WildEarth Guardians, working to promote human-wildlife coexistence throughout the American West. She is an interdisciplinary conservation scientist and animal ethologist whose work has spanned issues from water to wolves, on public and private lands, from Madagascar to Michigan.

William S. Lynn is a research scientist in the George Perkins Marsh Institute at Clark University (Worcester, Massachusetts, USA), a research fellow at New Knowledge Organization, and former Director of the Masters in Animals and Public Policy (MAPP) program at Tufts University. The focus of his work is the ethics and politics of sustainability, with a special eye for human-animal relations. Schooled in ethics, geography and political theory, his interdisciplinary approach examines why and how we ought to care for nature and society.

Katrina Marsden works for adelphi, a leading environmental think tank and public policy consultancy, as Senior Project Manager responsible for the field of biodiversity and nature conservation. Her professional focus over the last 12 years has been on the analysis of the effects of a range of policies, especially land use and nature, on biodiversity and the involvement of different stakeholders in policy making. She originally studied Earth Sciences and Environmental Change and Management at Oxford University.

Michel Meuret has a PhD in Animal Ecology Sciences and is Research Director at the French National Institute for Agricultural Research (INRA) since 2005, which concentrates on grazing practices and animal feeding behaviour at pasture. Most of his research has been conducted with herders and their family farmers, focusing on herding practices on rangeland. Since 2016, he has been the Coordinator of the research network COADAPHT, focusing on co-adaptation processes between predators (e.g., wolves), humans and other animals, to better anticipate human-predator conflicts.

Charudutt Mishra serves as the Science and Conservation Director of the Snow Leopard Trust and Executive Director of Snow Leopard Network, and is Senior Scientist at the Nature Conservation Foundation. He is involved in multi-disciplinary research, conservation, and policy initiatives in snow leopard landscapes of High Asia.

Muhammad Ali Nawaz is a wildlife ecologist involved in research, conservation and policy issues related to large carnivores in Pakistan for two decades. He is based at Quaid-i-Azam University and leads the Pakistan Program of the Snow Leopard Trust.

Andrés Ordiz has worked in the last 20 years mostly in Spain and Scandinavia, where he is a researcher with the Scandinavian Brown Bear Research Project (SBBRP, since 2004) and the Scandinavian Wolf Research Project (SKANDULV, since 2014). His main motivation is that scientific research is the base for conservation-oriented management of large carnivores. Main topics of his research are behavioral reactions of brown bears to human activities and interactions between apex predators (bears and wolves).

Jani Pellikka, PhD, works as Senior Research Scientist at the Natural Resources Institute Finland (Luke), as an Adjunct Professor at the University of Helsinki and as an entrepreneur. He focuses on human–wildlife-related knowledge production, wildlife-related nature wellbeing, wildlife management, governance and large carnivore conflicts at his academic activities.

Vincenzo Penteriani is a permanent researcher at the Spanish Council of Scientific Research (CSIC, Pyrenean Institute of Ecology). He is currently working on brown bears in the Cantabrian Mountains (northwestern Spain; www.cantabrianbrownbear.org). His professional interests include brown bear ecology and behaviour in human-modified landscapes, large carnivore attacks on humans and animal visual communication.

Mari Pohja-Mykrä is an ecologist specializing in biodiversity research with a PhD in Environmental Science. She has over 15 years' experience in human-wildlife conflicts and wildlife management, and she is working as a Senior Researcher at the Ruralia Institute, University of Helsinki, Finland. Her particular research interests are in interdisciplinary approaches in conservation conflicts, human dimensions of natural resource and wildlife conflicts, and human-environment interactions.

Shawn J. Riley is the Parrish Storrs Lovejoy Professor of Wildlife Management in the Department of Fisheries and Wildlife at Michigan State University (East Lansing, Michigan, USA), and a scientist in the Partnership for Ecosystem Research and Management, a long-term collaboration between MSU and wildlife agencies in Michigan. He earned a PhD from Cornell University (Ithaca, New York, USA), and B.S. and M.S. degrees from Montana State University (Bozeman, Montana, USA). In 2009–2010, he was a Senior Fulbright Fellow at the Swedish University of Agricultural Science in Umeå. Shawn is coauthor of the The Wildlife Society-Johns Hopkins book, *Human Dimensions of Wildlife Management.*

Gustaf Samelius is the Assistant Director of Science at the Snow Leopard Trust and is working on research and community-based conservation of snow leopards and mountain ecosystems in Central Asia. His interests include both applied and theoretical conservation biology.

Camilla Sandström is Professor in Political Science at Umeå University, Sweden. Her research focuses on environmental politics and policy, in particular new modes of governance such as co-management and public-private partnerships. She has published numerous peer-reviewed papers and has recently co-edited two books published by Routledge: *Indigenous Rights in Modern Landscapes: Nordic Conservation Regimes in Global Context*, and *Forest Governance and Management Across Time: Developing a New Forest Social Contract.*

Francisco J. Santiago-Ávila holds Masters' degrees in environmental public policy as well as environmental management (MPP/MEM) from Duke University (Durham, North Carolina, USA), after which he worked in environmental projects for international organizations such as the Inter-American Development Bank and World Bank. As a PhD student at the University of Wisconsin-Madison (USA), his research has revolved around the integration and application of environmental and animal ethics to coexistence with wildlife, and the evaluation of the effectiveness of lethal and non-lethal methods to prevent conflicts with large carnivores. His main objective is to reform wildlife management by embedding in it the much needed acknowledgement of moral standing for individual nonhuman animals.

Koustubh Sharma is based in Bishkek, Kyrgyzstan, where he manages the International Snow Leopard Secretariat of GSLEP, an inter-governmental conservation effort involving all snow leopard range countries. He studies snow leopards and is a Senior Scientist at the Snow Leopard Trust and Nature Conservation Foundation.

Annelie Sjölander-Lindqvist is Associate Professor in Social Anthropology, Lecturer in Human Ecology and Researcher at the Gothenburg Research Institute and School of Global Studies, University of Gothenburg, Sweden. Her research concerns conflicting values in natural resource management, including the limits and possibilities of deliberative measures, the analysis of local identity, science and politics in environmental contests and the link between place attachment, landscape dynamics and resource management.

Ketil Skogen, PhD, is a Sociologist at the Norwegian Institute for Nature Research – NINA. He has studied the conflicts over wolves and wolf management in Norway for almost two decades, with a particular view to situating the controversies within societal power structures and processes of social and cultural change in rural areas.

Kulbhushansingh R. Suryawanshi serves as the Program Director of the Snow Leopard Trust in India and Scientist at Nature Conservation Foundation. He works on research and conservation of wildlife in the high altitude regions of the Himalaya and other Central Asian mountains.

Adrian Treves earned his PhD at Harvard University (Cambridge, Massachusetts, USA) in 1997 and is now a Professor of Environmental Studies at the University of Wisconsin-Madison (USA), and founder of the Carnivore Coexistence Lab. His research focuses on agroecosystems where crop and livestock production overlap carnivore habitat. He and his students work to understand and manage the balance between human needs and carnivore conservation. With his students, he investigates conservation and ecology of large carnivores, as well as the attitudes and behaviors of the people who live alongside those carnivores.

Erwin van Maanen (BSc Hons, MSc) is a landscape and conservation ecologist with a track record in advocating and promoting the new conservation science of rewilding in many parts of the world. He is particularly keen to advance the endeavor of safeguarding and restoring large-scale ecosystems with more of the original trophic interactions, including more ecological balance through the key interactions between mammalian herbivores and carnivores. He is current chairman of the Rewilding Foundation and a member of the recently established IUCN Rewilding Taskforce.

Chris White is an environmental economist at the US-based multinational engineering firm AECOM. His role is to provide specialist economic input and modelling into a wide range of sustainable development, environmental policy and climate change projects. His primary area of work focuses on ecosystem services, in particular working with businesses and governments to account for their environmental impacts, quantifying environmental values in monetary terms and assessing the use of market-based instruments in public policy.

Alejandra Zarzo-Arias is mainly interested in the ecology and conservation biology of large carnivores. Her PhD research is focusing on the small and endangered Cantabrian brown bear population, as an example of a large carnivore in human-modified landscapes. Principally, she is studying bear behaviour and habitat requirements, potential expansion range in Asturias and temporal patterns of bear damages.

Preface

An ancient Greek myth portrays Zeus taking the form of Goddess Artemis to seduce Callisto, companion of the goddess. Callisto, her name meaning "the most beautiful one", was left pregnant by Zeus and she was expelled from Artemis after having broken the followers' vow of chastity. Callisto wandered in the woods to give birth to her child, Arcas. Goddess Hera, the jealous wife of Zeus, took revenge by transforming Callisto into a bear and separating her from her child.

Figure 0.1 Callisto turning into a bear.

(image created by Iordanis Stylidis, Associate Professor at the University of Thessaly, Department of Architecture)

Since mother and son could not have remained together, Zeus sent Hermes to take Arcas to Maia, Hermes' mother, to raise the son of Callisto. The years passed, and as young Arcas was hunting in the woods, his mother come across him and rushed to take him in her arms. Arcas did not recognise his mother and was about to kill her with his spear. To avoid this matricide, Zeus moved Callisto and Arcas in the sky and turned them into star constellations: Ursa Major (Callisto) and Ursa Minor (Arcas).

There are various versions of the myth and there are many more myths and narratives all around the world, where bears and wolves and other large carnivores are featured. All these mythologies depict a transient world of humans and nonhumans, a transient world of agency. Imagine the scene: Callisto turning into a bear (Figure 0.1), capturing the permeability between the human and nonhuman, the moment of exchange.

The present volume is devoted to current versions of a narrative that has initiated many years ago. Our wish was to engage in the discussion which takes human dimensions as a necessary component of large carnivore conservation and management, and to bring this discussion forward, with an aim to shed light on theoretical and methodological perspectives, heterogeneity in perceptions and challenges for decision-making and stakeholder involvement.

We are indebted to all who have contributed to this volume, either directly or indirectly, providing their position and concern. We hope that they will find an interesting story to tell further among the lines that follow.

Tasos Hovardas

Editor

CALLISTO – Wildlife and Nature Conservation Society,

Human Dimensions Expert

Society & Natural Resources, Editor-in-Chief (2017–2020)

Journal of Research in Science Teaching, Editorial

Board Member (2017–2020)

Research in Science and Technology Education Group,

University of Cyprus

Part I

Theoretical and methodological perspectives

1 Addressing human dimensions in large carnivore conservation and management

Insights from environmental social science and social psychology

Tasos Hovardas

Introduction

The need to incorporate human dimensions in large carnivore conservation and management has long been acknowledged. Next to trends of population dynamics of large carnivore species worldwide, we can observe noticeable changes in attitudes towards these species (e.g., Ericsson et al., in this volume; Kutal et al., in this volume). In a very coarse and broad trajectory, we may distinguish a first period of research aiming at the examination of stakeholder attitudes towards large carnivores, and a subsequent period concentrating on stakeholder analysis, consultation, and engagement (e.g., Mishra et al., in this volume; Sandström et al., in this volume). All these records reflect the various initiatives taken at international, national, or regional and local levels in reaction to dynamics in large carnivore populations. These have stimulated social science research using quantitative, qualitative (see Jampel, in this volume), and mixed methods, have unravelled an increasing heterogeneity of stakeholder positions, and have highlighted implications for large carnivore conservation and management. Although the local context and stakeholder synthesis has always been decisive (see Ghosal, in this volume; Sjölander-Lindqvist et al., in this volume), some incidents and outcomes appear again and again, which indicates that they do not surface due to local circumstances only. Our intention in this chapter is to shed light on several aspects of that kind by taking into account insights from environmental social science and social psychology.

One first point to underline is that stakeholder perceptions are not formed just by their interaction with large carnivores, but by stakeholder interaction, as well. In this regard, tension and conflict among stakeholder groups as well as agreement and collaboration, whenever this may be possible, are anticipated to shape their positions. Stakeholder groups are constantly addressing one another formally or informally and at different venues: Stakeholder interaction is promoted through an interplay of arguments, where each stakeholder group is engaged in offensive (i.e., attempting to attack a point presented by another stakeholder group) and defensive acts (i.e., intending to support its own argumentation against counter-arguments) (Davies and Harré, 1990). In this conceptualization, which we will call discursive positioning, discursive practices gain their meaning in the confrontation of social actors in antagonistic camps, where rival speech acts enable one's own positioning (see Hovardas, 2017, for a detailed theoretical and methodological approach to discursive positioning). Although discursive positioning may prove quite important in investigating stakeholder interaction, it has not received proper attention in social research targeting large carnivore issues.

Discursive positioning may be detected across many topics in large carnivore conservation and management, which address ongoing or future developments (e.g., Lipscombe et al., in this volume). Lethal control is such a topic (e.g., Lute and Gore, in this volume). It may be favoured in order to keep a large carnivore population within a range but also for legitimizing local knowledge and practices (e.g., Pohja-Mykrä, in this volume) and re-establishing a symbolic border between human settlements and wildlife, which large carnivores are assumed to have overridden (e.g., Lescureux et al., in this volume). However, lethal methods will not be readily endorsed by all engaged actors, for instance, due to ethical consideration of non-human animals (e.g., Santiago-Ávila et al., in this volume), but also because they may not be more effective than non-lethal methods in damage prevention (Treves et al., 2016). Moreover, human-caused mortality may have a substantial impact on population dynamics of large carnivore species. For instance, lethal control of wolf populations is usually equated to just removing individuals, but the consequences of that option may expand well beyond having a numerical effect. This may be due to a complex interplay among population, reproduction, and dispersal indices, e.g., when probability of finding mates is impacted at low densities during the dispersal phase, giving rise to an Allee effect (Hurford et al., 2006). Density-dependent mechanisms for brown bear populations (e.g., sexually selected infanticide or female reproductive suppression) may also initiate non-linear effects and may not allow for simple extrapolation of current trends to predict future population trajectories (Cano et al., 2016). This complexity may expand to transboundary issues of large carnivore conservation and management (e.g., see Bischof et al., 2016; see also Penteriani et al., in this volume).

There have been numerous episodes in large carnivore conservation and management which have set new challenges for environmental social science and social psychology. In most cases, it is acknowledged that natural data alone cannot suffice for decision making. This makes interdisciplinary perspectives indispensable. Examining how natural and social data may give rise to new place-based mappings is a demanding task. The problem here is not just to merge two different types of data and present their overlap, but to investigate how different inputs may produce new meanings and elicit new motivations and practices. All this novel information and knowledge is expected to fuel and refuel discursive positioning of stakeholders, which is increasingly transferred to inclusionary schemes for consultation and engagement (e.g., Hovardas, in this volume; Hovardas and Marsden, in this volume). The experience available, in this regard, indicates that stakeholders need to recognize and respect other positions in order to be respected and recognized in the first place. And they have to do so, while disagreement and tension will most probably remain. Stakeholders need to interact in disagreement and explore possible points of convergence. They need to come to terms even if total consensus will never be achievable (e.g., Jacobsen and Linnell, 2016). In the following sections, we will concentrate on theoretical and methodological approaches to examine stakeholder perceptions and how these may inform their positioning and interaction.

"Risk" vs. "danger"

Damage caused to livestock has been an ongoing source of tension among stakeholder groups in large carnivore conservation and management. A closely associated topic is fear of large carnivores (e.g., Johansson et al., 2016), which may also relate to fear for human safety (Pohja-Mykrä and Kurki, 2014). When approaching stakeholder interaction on this matter from the vantage point of discursive positioning, we can observe how pro-carnivore groups have attempted to compare these emotions and fear with actual records of damage caused to livestock or attacks to humans. In the first case, claims for damage show that large carnivore numbers do not always correlate with damage caused (e.g., Bautista et al., 2017). In the latter case, comparisons of various types of threats to human safety indicate that actual attacks and threat stemming from large carnivores is almost absent (Hoffmann et al., 2017). The main point here is that local people may be taken to overestimate livestock loss or danger for human safety based on actual incidents. Therefore, local complaints are often dismissed by pro-carnivore groups as overreaction.

To take this discussion one step further, we need to concentrate on risk theory and how several important concepts have been elaborated upon within this theoretical frame. A crucial contribution of risk theory has been the definition of "risk" as opposed to "danger" (e.g., Luhmann, 1990; Beck, 1997). The main difference here is if actors are capable of anticipating the sequence of events that may expose them to a threat and if they can act proactively to confront that threat. If such an anticipation is possible and can be attempted, then actors are said to take a "risk" (i.e., take an informed decision after weighting the possibility of suffering any harm as compared to any benefit that may be derived). Therefore, taking action within the frame of risk perception rests upon a risk calculus given by a balance between perceived costs and benefits (e.g., Rasborg, 2012). However, if actors do not have the means to prepare themselves for the threat and can only deal with it retroactively – namely, if they can respond only after they have been harmed – then they are exposed to a "danger".

How can we apply this distinction of risk theory between "risk" and "danger" in large carnivore conservation and management? For many local producers in regions with large carnivores, the call for coexistence equates with a call for accepting a risk of damage caused by large carnivores. Such a tolerance has been frequently reported and has also been incorporated in formal agreements (see Hovardas, in this volume; Hovardas and Marsden, in this volume). Damage prevention methods like the use of electric fences and livestock guarding dogs may assist in achieving this balance. However, since no damage prevention method can provide absolute safety, some probability of damage remains. Indeed, some local people, who would not be ready to accept a compromise of this kind, will still perceive exposure to damage or threat by large carnivores as a "danger" and not a "risk". For these local residents, illegal killing may be an eligible alternative (Rauset et al., 2016; see also Pohja-Mykrä, in this volume). Independent of the intent behind their introduction in stakeholder argumentation and counter-argumentation, actual records point towards this non-dismissible chance of damage to livestock or the possibility of attack to humans, no matter how unlikely it may be. These instances will always provide a justification for local people who would never accept the "risk" of coexisting with large carnivores.

Although compensation cannot guarantee easement of conflict (e.g., Fernández-Gil et al., 2016), there are suggestions to link compensation systems to damage prevention methods and consider the latter as a prerequisite for the former (see for instance Bautista et al., 2017). One problem among many in this direction is to align compensation systems with the real loss encountered by local producers, which does not really equate with the monetary equivalent of depredated livestock or other damage. For instance, livestock cannot be treated in the same the way non-living assets are usually compensated, since producers will build on their current stock to schedule their future investment and production. This means that livestock cannot be conceptualized in a static fashion but needs to be approached within the dynamics inherent in their nature and one's farming holding. In addition, there may be collateral damage, which is often not accounted for; for example, accumulative loss in the case of pregnant livestock and milk production, not to mention workload devoted for putting the damaged asset in place or restoring any part of the damage after it has occurred. What is more, many different socio-cultural characteristics of rural areas that are explicitly or implicitly linked to large carnivore conservation and management, such as pluriactivity (e.g., Giourga and Loumou, 2006), may remain unnoticed or be ignored. These additional aspects add to the reluctance of local people to accept an involuntary exposure to the threat of large carnivores and may also apply to instances such as illegal killing that are still observed (Tosi et al., 2015).

A substantial change in compensation of damage caused by large carnivores has been introduced with conservation performance payments in Sweden (Persson et al., 2015; Skonhoft, 2017; see also Hovardas and Marsden, in this volume). In contrast to *ex-post* compensation, where a fixed payment follows after a damage has been documented, *ex-ante* compensation rewards reproduction of large carnivores (e.g., wolverines in Sweden). Such an *ex-ante* compensation system with conservation performance payments has been in place in the Sámi reindeer herding area in Sweden since 1996, where there is a considerable likelihood of reindeer loss due to vulnerability of reindeer to large carnivore attacks (reindeer are rarely confined in corrals) and dependence of large carnivores on reindeer as a main feeding source (especially during the winter). The monetary amount paid for each individual animal (SEK = 200,000; around 20,000 Euros or 23,000 USD) has been set to balance total estimated livestock damage caused by the animal in its lifetime. Payments are first directed to Sámi villages, and then each village is responsible for their allocation to individual reindeer herders or any other exploitation of the reward, e.g., if it will be invested for community expenses. This *ex-ante* compensation scheme triggers an incentive structure that re-organizes the risk calculus for reindeer herders so as to allow them to respond truly proactively. In that regard, it turns "danger" into a "risk": Since beneficiaries have been rewarded a monetary sum for successful wolverine reproductions, they need to take additional action to protect their livestock and retain as much of the benefit as possible. Indeed, it has been underlined that conservation performance payments may provide more incentives for an optimal uptake of damage prevention methods (Zabel et al., 2011). Overall, there are multiple indications that *ex-ante* compensation may be more effective than *ex-post* compensation (Skonhoft, 2017).

Polyphasic representational fields

Cognitive polyphasia has been developed as a concept within the theory of social representations (Moscovici, 1961/2008)[1] to describe a state where different forms of knowledge, rationality, reasoning, or practice (e.g., scientific vs. non-scientific; expert vs. lay) may be voiced or performed by the same social group to address a certain social object and this will lead to what has been termed "polyphasic representational fields" (Jovchelovitch, 2008). Cognitive polyphasia has been often used as a term to describe a "state" of social representations; namely, a state of heterogeneity, drawing upon different and at times inconsistent or even contradictory modes of reasoning and acting. However, cognitive polyphasia has always implied a type of agency enacted by groups who resort to this multiplicity of forms of knowledge, reasoning, and practice. This means that apart from being a state denoting representational fields of diverse forms of reasoning and practice, cognitive polyphasia should also be seen as a process – and indeed, as a discursive one, since it would give social groups new opportunities for discursive positioning. Polyphasic representational fields are multivalent, meaning that they may allow for integrating or addressing a multiplicity of dimensions of social objects, or even for creating new dimensions of these objects, which would highlight a productive character for cognitive polyphasia. Since social groups with similar or dissimilar perspectives and representations of social objects may position themselves towards each other, cognitive polyphasia, inter-group relations, and dynamics of social representations need to be jointly observed (e.g., Jovchelovitch, 2008;

Marková, 2008; see also Batel et al., 2016). Cognitive polyphasia can be seen as the result of ongoing inter-group interaction, which drives dynamics of social representations.

Cognitive polyphasia is exemplified when social groups attempt to: (1) appropriate different types of knowledge or modes of thinking and acting, (2) strategically re-interpret them, and (3) integrate them in their argumentation in order to serve their own ends (e.g., Wagner, 2007). The social representation of a social object for a social group is configured according to what Bauer and Gaskell (2008) have termed the "project" of the group, meaning a central demand, quest, or mission to be pursued by group members. Aspects that may promote this project are highly likely to be elaborated upon by social groups and adapted to strengthen their argumentation lines. For instance, adopting some elements of the dominant environmentalist discourse[2] or recognizing the normative character of environmental law may be a prerequisite for local people before they voice their concern or formulate an alternative position (e.g., Krange and Skogen, 2007; Hovardas and Korfiatis, 2012; Mauro and Castro, 2012). In a similar vein, there were instances where large carnivores were presented by local people to threaten local biodiversity, redirecting a keyword of the environmentalist discourse against wolves and bears (López-Facal and Jiménez-Aleixandre, 2009; Von Essen, 2017). In addition, hunters and rural breeders have been reported to claim selected connotations of the terms "ecology" and "ecological", since they believed that their relationship with nature is truly authentic, especially when compared to some environmental non-governmental organizations which were criticized as "bureaucratic" and detached from the local context (e.g., Hovardas, 2005; Theodorakea, 2014).

Previous research in Greek protected areas has revealed how the wolf-reintroduction narrative, voiced by local interviewees, may present the same elements over different regions (Hovardas, 2010a, 2012, 2015, Hovardas and Korfiatis, 2008). The wolf-reintroduction narrative can provide an exemplary case of how a polyphasic representational field can be established along certain themata (e.g., overarching sets of concepts arranged in bipolar pairs; see Liu, 2004) with reference to which local residents represent the wolf and which they also employ to position themselves against environmental non-governmental organizations. According the wolf-reintroduction narrative, so-called "ecologists" (meaning members of environmental non-governmental organizations) were supposed to breed wolves in captivity and then release them secretly in the wild. To substantiate their claims, interviewees referred to instances when environmental organizations released large carnivores after recovery from injury or instances of relocation of "problem" animals, even if these referred to species other than the wolf. Interviewees also claimed that the supposed hybridization of wolves induced a significant alteration in wolf appearance and behaviour: They highlighted that hybrid wolves revealed intermediate characteristics between wild wolves and dogs. Moreover, they said that these hybrid wolves presented unexpected high levels of tolerance towards human presence, which had not been observed in past. This behavioural characteristic made hybrid wolves more dangerous to rural life than wild wolves, since they did not fear humans and were much more likely to cause damage to livestock. Overall, the wolf-reintroduction narrative pictured "ecologists" as an out-group that promoted their interests at the expense of local communities. Knowledge from hybridization (relating to changes in animal populations) was integrated with ecologists' perceived role to express rural resentment.

There were three different themata in this representational field ("wild vs. domesticated", "fearful vs. fearless", "harmless vs. dangerous"), which were used jointly to provide a sequence of premises and contrast wild wolves to hybrid wolves (Table 1.1): The wild wolf is fearful of humans and, therefore, almost harmless; the hybrid wolf is partly domesticated and, therefore, fearless of humans and dangerous; namely, capable of causing severe damage to livestock. The constructive potential of cognitive polyphasia here refers to appropriating hybridization and drawing on ecologists' perceived role in order to produce a "hybrid wolf" with new characteristics that were not to be found neither in wild animals (which would fear humans) nor in domesticated animals (which would not constitute a threat).[3] An interesting point to underline is that intermediate characteristics in the representational field of the hybrid wolf (i.e., a hybrid with phenotypic characteristics between wild wolves and dogs) were associated with extreme possibility of harm (i.e., the "hybrid wolf" was more dangerous than the "wild wolf").

The themata of Table 1.1, or versions of them, have been reported in many different regions in the world (see, for instance, Buller, 2004; Skogen et al., 2008; Von Essen, 2017), while an analogous discursive scheme has been also been reported in some cases for bears (e.g., López-Facal and Jiménez-Aleixandre, 2009). The implications for stakeholder interaction may be manifold. First, the narrative may be taken as an attempt by local residents in rural areas to gain justification for their resistance against large carnivore expansion, which is

Table 1.1 Themata in the representational field of "wild wolves" and "hybrid wolves"

Themata	"Wild wolves"	"Hybrid wolves"
"Wild vs. domesticated"	Wild wolves have some distinguishing features that allow for their recognition	Hybrid wolves have different characteristics from wild wolves in terms of appearance and behavior
"Fearful vs. fearless"	Wild wolves have a natural fear for humans	Hybrid wolves are not deterred by human presence since they have lost their natural fear for humans
"Harmless vs. dangerous"	Wild wolves would not risk coming close to humans for causing damage to livestock	Hybrid wolves are much more prone to causing damage to livestock as compared to wild wolves

Data sources: Hovardas, 2010a, 2012, 2015; Hovardas and Korfiatis, 2008.

thought to be imposed upon them by an urban pro-carnivore elite not belonging to the local context and not entitled to dominate in that context. Since "ecologists" are to be found behind the threat of hybrid wolves, "ecologists" are to blame for that threat, too. Second, the domesticated hybrid wolf may not deserve the protection usually granted to genuine wild species. In line with such an assumption, lethal control of wolf populations may be supported to retain the natural fear of humans (Von Essen, 2017). Within the frame of discursive positioning, the debate will be probably advanced and become more complicated, especially under the light of recent verification of wolf-dog hybridization (Pacheco et al., 2017; Torres et al., 2017).

Social influence mechanisms

Cognitive polyphasia and dynamics of social representations would indicate that inter-group relations leave their mark on stakeholder interaction and may also facilitate changes in stakeholder perspectives or action. But how do ideas and practices change? If all social influence was to be sought in majorities as a source, then all individuals within a society would have to conform to majority positions. Then, how could we explain social change? This was the challenge formulated by Moscovici (1976) in his work on minority influence (see also Martin and Hewstone, 2010). Minority influence describes the influence minorities may exert on majorities.[4] This may occur at a latent level: Conversion of members of the majority to the position of the minority may proceed with a delay and may not be explicitly admitted by majority members. Conversion is promoted through a concentration of recipients on the content of the minority position, which may eventuate in a position change of the majority. Social influence is therefore different when the source is a majority or a minority: In the former case, the majority position may be adopted but this may happen due to identification with the source (majority) and without involving much processing of the content of the message. In the latter case, however, engagement in social comparison is not expected, since an identification with the source, in this case (minority), would not be desirable or favoured; instead, the target audience is highly likely to engage in an evaluation of the argumentation put forward by the minority.

Local communities which have to coexist with large carnivores are found within a complex web of social influence effects. Several social groups within these local communities – for instance, local producers engaged in primary sector activities and local hunters – seem to comprise a salient majority who are quite reluctant to celebrate the return of large carnivores and their increase in abundance and distribution. However, local communities also host a number of experts working in numerous projects concerning large carnivore conservation and management; for instance, scientists, wildlife conservation professionals, members of environmental non-governmental organizations, and people involved in ecotourism (see Ghosal, in this volume). Many of these scientists and professionals have a prolonged stay or even reside within local communities.

They comprise a local minority that largely endorses the comeback of large carnivores. This local minority is usually the recipient of harsh criticism for all initiatives that start outside local communities (e.g., state initiatives for nature or wildlife conservation, environmental legislation, etc.) and which are perceived by local people as being imposed upon them in a top–down fashion. In this larger scale, the local community is positioned as a minority (inferior position, having less power) against the state and central governments.

Overall, local communities in regions with large carnivores are integrated in a double system of social influence mechanisms, where they can be presented either as majorities opposed to local pro-carnivore minorities or as minorities, themselves, opposed to the state and central government (Figure 1.1). The complexity of social influence effects is augmented by the fact that majority influence at the large scale (i.e., majority influence exerted by the state and central government on the local community) is aligned to minority influence exerted at the local scale (i.e., minority influence exerted by the pro-carnivore local minority on the local community). This double system of social influence mechanisms has been previously proposed for protected areas and Natura 2000 sites (Hovardas, 2010b), and seems to have mediated many relevant developments in the last few decades; for example, the diffusion of the environmentalist discourse in rural communities (being mainly a result of minority influence exerted by local minorities on local communities; see also Mauro and Castro, 2012), and the need for inclusionary schemes for environmental and protected area governance (being mainly a result of minority influence exerted by local communities on the state and central government). The double system of social influence mechanisms may also explain the difficulty of administrative

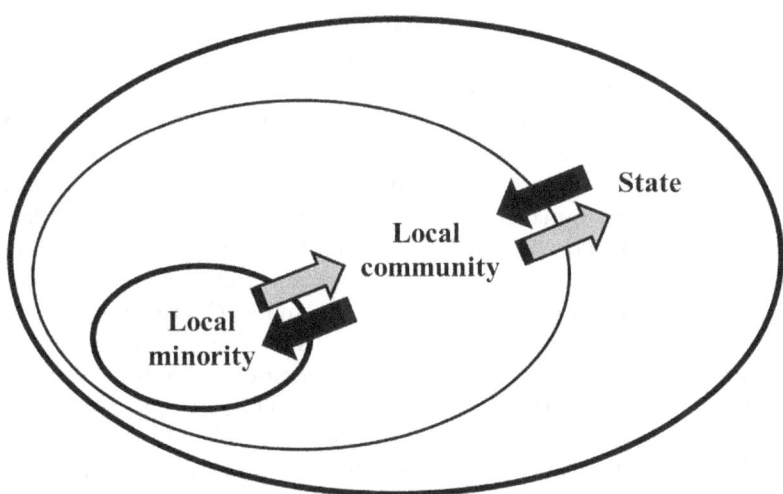

Figure 1.1 Double system of social influence mechanisms. Black arrows depict social influence exerted by majorities on minorities; grey arrows depict minority influence.

personnel of regional authorities working at the local level, who may be considered among the local minority, when they strive to strike a balance between advocating for local demands and their responsibility to enforce rules and regulations (Mauro and Castro, 2012; see also Sjölander et al., in this volume).

What are the implications of social influence effects for stakeholder consultation and engagement in large carnivore conservation and management? Social groups in local communities face a series of motives and counter-motives when offered the opportunity to take part in inclusionary schemes. On the one hand, participation is in itself an instance of recognition for local groups and an opportunity to voice their position and make an impact on future events. Previous research has shown that the negotiation style chosen by minorities will be decisive for their final impact (see for instance Mugny, 1975): Minorities will need to be both consistent (e.g., defending their position with certainty, confidence and commitment) and flexible (e.g., showing readiness to negotiate under certain conditions), since having a more rigid style that excludes compromise may stigmatize the minority position as unproductive and the minority message may remain largely unaccounted for.[5] Timeline and scheduling of stakeholder meetings may also play a role. Recent research has shown that majority members were more receptive to the minority position when they expected interaction to go on in the future, increasing the odds and intensity of minority influence (San Martin et al., 2015).

On the other hand, stakeholder interaction in formal or informal governance schemes will necessarily advance in parallel with standard in-group relations and inter-group confrontation in society overall. This may create and maintain a tension between outcomes in decision-making schemes and wider stakeholder interaction. Spokespersons or representatives of local groups will need to position themselves in these schemes, but they will also be held accountable for that positioning from members of their own group. Any compromise, no matter how well articulated, prepared, and democratically supported, will first need to take a shape in negotiations and then be announced to the peer group. It is likely that spokespersons or representatives, in their attempt to reach a compromise, may need to deviate from expectations of other in-group members. If this is taken to be threatening to the image of the group, then these representatives may be isolated and stigmatized as in-group deviants to protect the identity of the group (Pérez and Mugny, 1987). The "black-sheep" stigma describes this derogation and rejection of deviant in-group members, which may be even stronger when compared to out-group members (Marques et al., 2001). Recent research has exemplified that the "black-sheep" effect may be triggered for elements in the central core of social presentations (Zouhri and Rateau, 2015). When core assumptions of a social group will be considered challenged by in-group members, then deviants will be stigmatized. All these effects may provide background and explanation for the slow and insubstantial diffusion of any positive results of decision-making and deliberation schemes among in-group members who have not taken part themselves in these processes.

Implications for science communication, science education, environmental education, and outreach

Social influence effects also need to be considered in contrast to calls for providing scientific knowledge to local groups and raising their environmental awareness. According to the "deficit" model of science communication (e.g., Castro and Batel, 2008; Brossard and Lewenstein, 2009; Wibeck, 2014), lay people would need additional scientific knowledge to grasp the full array of consequences and implications of their practices, and then they are expected to change their behaviour and align it with the new acquired knowledge. Many initiatives in science communication and science education, as well as environmental education and outreach, have been based on this "deficit" model, and still are. Such a conceptualization, however, retains a skewed role for the addressee or learner, generally, as a passive receptor of information or knowledge and fails to recognize the active role of all targeted audiences in deciphering and interpreting that information and knowledge as well as their various sociocultural implications (see a relevant discussion by López-Facal and Jiménez-Aleixandre, 2009). Polyphasic representational fields and social influence mechanisms may provide valuable insight in this regard: New information or knowledge will be integrated in representational fields of the social groups targeted and it will be fine-tuned so as to serve their own needs and desires. Such a provision will not necessarily lead to behaviour change or change towards the desired direction, since representational fields do not have "gaps" or "deficits". Instead, they provide a fully-fledged and coherent representation of social groups for social objects, which guide inter-group interaction. Leaving the "deficit" model behind may add numerous challenges to stakeholder involvement, especially when approaching the latter as a co-creation process within a frame of social learning (see Hovardas, in this volume).

Notes

1 The theory of social representations has originated in contrast to experimental social psychology long practiced in North America, and has involved a strong refocusing from individual actors on social groups (e.g., Moscovici, 1972). A social representation is defined as a system of values, ideas, and practices that a social group employs to address a social object (Moscovici, 1973). The elements of a representation may be distinguished in central ones, which form a salient core that largely defines the social object under reference, and in peripheral elements, which function to adapt the representation to the context (Abric, 1996). Two basic procedures have been proposed for monitoring or reconstructing the genesis of a social representation: When a social group is confronted with a new, unfamiliar, or threatening development, phenomenon, or event, then it attempts to anchor this new development in a pre-existing system of meanings to render it familiar and less threatening (anchoring). Moreover, the social group may also portray the new phenomenon by means of an image that may add concrete visual content to account for its vagueness (objectification) (Jodelet, 2008). For an overview of the theory of social presentations in environmental social science, see Buijs et al. (2012).

2 With the term "environmentalist discourse", we address the multifarious meanings and practices related to environmental and wildlife conservation and management as well as natural resource management, which surfaced in the 1970s and have gradually been applauded by large segments of most societies worldwide. Although denoting a heterogeneous field, all different versions of the environmentalist discourse include the necessity of a proper consideration of the relationship between society and nature. Of course, "proper" may have quite different connotations – and this is where all relevant discussions start.

3 The constructive character of cognitive polyphasia needs to be contrasted to responses expected due to cognitive dissonance (Festinger, 1957). In the latter case, "corrective" action taken at the individual level is anticipated to address an inconsistency between different attitudes or attitudes and behaviour. In the case of cognitive polyphasia, however, polyphasic representational fields are arranged so as to account for a social group's "project" and promote inter-group interaction.

4 Minorities are defined with reference to a quantitative criterion of number (minorities having fewer members than majorities) and/or to a qualitative criterion of countering hegemony (minorities challenging established norms) (Gardikiotis, 2011).

5 The incorporation of aspects of the environmentalist discourse in the argumentation forwarded by local communities may establish a flexible negotiation style (when local communities are approached as minorities in opposition to states or central governments); see previous section on polyphasic representational fields.

References

Abric, J.-C. (1996) 'Nature and function of the core system of social representations', *International Journal of Psychology*, vol 31, pp1443–1443.

Batel, S., Castro, P., Devine-Wright, P., and Howarth, C. (2016) 'Developing a critical agenda to understand pro-environmental actions: Contributions from social representations and social practices theories', *WIREs Climate Change*, vol 7, no 5, pp727–745.

Bauer, M. W., and Gaskell, G. (2008) 'Social representations theory: A progressive research programme for social psychology', *Journal for the Theory of Social Behavior*, vol 38, no 4, pp335–353.

Bautista, C., Naves, J., Revilla, E., Fernández, N., Albrecht, J., Scharf, A. K., Rigg, R., Karamanlidis, A. A., Jerina, K., Huber, D., Palazón, S., Kont, R., Ciucci, P., Groff, C., Dutsov, A., Seijas, J., Quenette, P.-I., Olszańska, A., Shkvyria, M., Adamec, M., Ozolins, J., Jonozovič, M., and Selva, N. (2017) 'Patterns and correlates of claims for brown bear damage on a continental scale', *Journal of Applied Ecology*, vol 54, no 1, pp282–292.

Beck, U. (1997) *The Reinvention of Politics: Towards a New Modernity*, Polity Press, Cambridge.

Bischof, R., Brøseth, H., and Gimenez, O. (2016) 'Wildlife in a politically divided world: Insularism inflates estimates of brown bear abundance', *Conservation Letters*, vol 9, no 2, pp122–130.

Brossard, D., and Lewenstein, B. (2009) 'A critical appraisal of models of public understanding of science: Using practice to inform theory', in L. Kahlor and P. Stout (eds) *Communicating Science: New Agendas in Communication*, Routledge, New York.

Buijs, A., Hovardas, T., Figari, H., Castro, P., Devine-Wright, P., Fischer, A., Mouro, C., and Selge, S. (2012) 'Understanding people's ideas on natural resource management: Research on social representations of nature', *Society & Natural Resources*, vol 25, no 11, pp1167–1181.

Buller, H. (2004) 'Where the wild things are: The evolving iconography of rural fauna', *Journal of Rural Studies*, vol 20, pp131–141.

Cano, I. M., Taboada, F. G., Naves, J., Fernández-Gil, A., and Wiegand, T. (2016) 'Decline and recovery of a large carnivore: Environmental change and long-term trends in an endangered brown bear population', *Proceedings of the Royal Society B*, vol 283, no 1843, 20161832.

Castro, P., and Batel, S. (2008) 'Social representation, change and resistance: On the difficulties of generalizing new norms', *Culture & Psychology*, vol 14, pp475–497.

Davies, B., and Harré, R. (1990) 'Positioning: The discursive production of selves', *Journal for the Theory of Social Behaviour*, vol 20, no 1, pp43–63.

Ericsson, G., Sandström, C., and Riley, S. J. (in this volume) 'Rural-urban heterogeneity in attitudes towards large carnivores in Sweden, 1976–2014', in T. Hovardas (ed) *Large Carnivore Conservation and Management: Human Dimensions*, Routledge, London.

Fernández-Gil, A., Naves, J., Ordiz, A., Quevedo, M., Revilla, E., and Delibes, M. (2016) 'Conflict misleads large carnivore management and conservation: Brown bears and wolves in Spain', *PLoS ONE*, vol 11, no 3, ppe0151541.

Festinger, L. (1957) *A Theory of Cognitive Dissonance*, Stanford University Press, Stanford, CA.

Gardikiotis, A. (2011) 'Minority influence', *Social and Personality Compass*, vol 5, pp679–693.

Ghosal, S. (in this volume) 'Heterogeneity in perceptions of large carnivores: Insights from Sanjay Gandhi National Park, Mumbai, and Ladakh', in T. Hovardas (ed) *Large Carnivore Conservation and Management: Human Dimensions*, Routledge, London.

Giourga, C., and Loumou, A. (2006) 'Assessing the impact of pluriactivity on sustainable agriculture: A case study in rural areas of Beotia in Greece', *Environmental Management*, vol 37, no 6, pp753–763.

Hoffmann, C. F., Montgomery, R. A., and Jepson, P. R. (2017) 'Examining the effect of billboards in shaping the great wolf debate of the American West', *Human Dimensions of Wildlife*, vol 2, no 3, pp267–281.

Hovardas, T. (2005) *Social Representations on Ecotourism: Scheduling Interventions in Protected Areas*, PhD thesis, Aristotle University of Thessaloniki.

Hovardas, T. (2010a) *Stakeholder analysis*, LIFE EXTRA – Improving the conditions for large carnivore conservation – A transfer of best practices (LIFE07NAT/IT/000502), Report of Action A5.

Hovardas, T. (2010b) 'The contribution of social science research to the management of the Dadia Forest Reserve: nature's face in society's mirror', in C. Catsadorakis and Kälander (eds) *The Dadia-Lefkimi-Soufli Forest National Park, Greece: Biodiversity, Management, and Conservation*, WWF-International, Athens, www.wwf.gr/images/pdfs/Hovardas.pdf, accessed 19 December 2017.

Hovardas, T. (2012) *Follow up surveys of stakeholder attitudes*, LIFE EXTRA – Improving the conditions for large carnivore conservation – A transfer of best practices (LIFE07NAT/IT/000502), Report of Action E3.

Hovardas, T. (2015) *Questionnaire development and administration*, Deliverable 2, Monitoring of knowledge and attitudes of stakeholders in national park management, Management Authority of Rodopi Mountain Range National Park (in Greek).

Hovardas, T. (2017) '"Battlefields" of blue flags and sea horses: Acts of fencing and defencing place in a gold mining controversy', *Journal of Environmental Psychology*, vol 53, no pp100–111.

Hovardas, T. (in this volume) 'A methodology for stakeholder analysis, consultation and engagement in large carnivore conservation and management', in T. Hovardas (ed) *Large Carnivore Conservation and Management: Human Dimensions*, Routledge, London.

Hovardas, T., and Korfiatis, K. (2008) Report of local environmental knowledge in the National Park of Northern Pindos. EU Community Initiative Programme INTERREG

IIIB Archimed, East Mediterranean Network for the Sustainable Development of Protected Aeas – East-Med-Net.

Hovardas, T., and Korfiatis, K. J. (2012) 'Adolescents' beliefs about the wolf: Investigating the potential of human – Wolf coexistence in the European south', *Society & Natural Resources*, vol 25, no 12, pp1277–1292.

Hovardas, T., and Marsden, K. (in this volume) 'Good practice in large carnivore conservation and management: Insights from the EU Platform on Coexistence between People and Large Carnivores', in T. Hovardas (ed) *Large Carnivore Conservation and Management: Human Dimensions*, Routledge, London.

Hurford, A., Hebblewhite, M., and Lewis, M. A. (2006) 'A spatially explicit model for an Allee effect: Why wolves recolonize so slowly in Greater Yellowstone', *Theoretical Population Biology*, vol 70, pp244–254.

Jacobsen, K. S., and Linnell, J. D. C. (2016) 'Perceptions of environmental justice and the conflict surrounding large carnivore management in Norway – Implications for conflict management', *Biological Conservation*, vol 203, pp197–206.

Jampel, C. (in this volume) 'Situated, reflexive research in practice: Applying feminist methodology to a study of human-bear conflict', in T. Hovardas (ed) *Large Carnivore Conservation and Management: Human Dimensions*, Routledge, London.

Jodelet, D. (2008) 'Social representations: The beautiful invention', *Journal for the Theory of Social Behaviour*, vol 38, no 4, pp411–430.

Johansson, M., Ferreira, I. A., Støen, O.-G., Frank, J., and Flykt, A. (2016) 'Targeting human fear of large carnivores – Many ideas but few known effects', *Biological Conservation*, vol 201, pp261–269.

Jovchelovitch, S. (2008) 'The rehabilitation of common sense: Social representations, science and cognitive polyphasia', *Journal for the Theory of Social Behavior*, vol 38, no 4, pp431–448.

Krange, O., and Skogen, K. (2007) 'Reflexive tradition – Young working-class hunters between wolves and modernity', *Young*, vol 15, no 3, pp215–233.

Kutal, M., Kovařík, P., Kutalová, L., Bojda, M., and Dušková, M. (in this volume) 'Attitudes towards large carnivore species in the West Carpathians: Shifts in public perception and media content after the return of the wolf and the bear', in T. Hovardas (ed) *Large Carnivore Conservation and Management: Human Dimensions*, Routledge, London.

Lescureux, N., Garde, L., and Meuret, M. (in this volume) 'Considering wolves as active agents in understanding stakeholder perceptions and developing management strategies', in T. Hovardas (ed) *Large Carnivore Conservation and Management: Human Dimensions*, Routledge, London.

Lipscombe, S., White, C., Eagle, A., and van Maanen, E. (in this volume) 'A community divided: Local perspectives on the reintroduction of Eurasian lynx (*Lynx lynx*) to the UK', in T. Hovardas (ed) *Large Carnivore Conservation and Management: Human Dimensions*, Routledge, London.

Liu, L. (2004) 'Sensitising concept, themata and shareness: A dialogical perspective of social representations', *Journal for the Theory of Social Behaviour*, vol 34, no 3, pp249–264.

López-Facal, R., and Jiménez-Aleixandre, M. P. (2009) 'Identities, social representations and critical thinking', *Cultural Studies of Science Education*, vol 4, pp689–695.

Luhmann, N. (1990) *Risiko und Gefahr: Soziologische Aufklärung 5. Konstruktivistische Perspektiven*, Westdeutscher Verlag, Opladen.

Lute, M. L., and Gore, M. L. (in this volume) 'Challenging the false dichotomy of Us vs Them: Heterogeneity in stakeholder identities regarding carnivores', in T. Hovardas (ed) *Large Carnivore Conservation and Management: Human Dimensions*, Routledge, London.

Marková, I. (2008) 'The epistemological significance of the theory of social representations', *Journal for the Theory of Social Behavior*, vol 38, no 4, pp461–487.

Marques, J. M., Abrams, D., Paez, D., and Hogg, M. A. (2001) 'Social categorization, social identification, and rejection of deviant group members', in M. A. Hogg and S. Tindale (eds) *Blackwell Handbook of Social Psychology: Group Processes*, Blackwell, Malden, MA.

Martin, R., and Hewstone, M. (eds) (2010) *Minority Influence and Innovation – Antecedents, Processes and Consequences*, Psychology Press, Hove and New York.

Mauro, C., and Castro, P. (2012) 'Cognitive polyphasia in the reception of legal innovations for biodiversity conservation', *Papers on Social Representations*, vol 21, pp3.1–3.21.

Mishra, C., Alexander, J. S., Bhatnagar, Y. V., Johansson, O., Sharma, K., Suryawanshi, K., Nawaz, M. A., and Samelius, G. (in this volume) 'Science, society and snow leopards: Bridging the divides through collaborations and best practice convergence', in T. Hovardas (ed) *Large Carnivore Conservation and Management: Human Dimensions*, Routledge, London.

Moscovici, S. (1961/2008) *Psychoanalysis: Its Image and Its Public* (D. Macey, Trans.), Polity Press, Cambridge.

Moscovici, S. (1972) 'Society and theory in social psychology', in J. Israel and H. Tajfel (eds) *The Context of Social Psychology: A Critical Assessment*, Academic Press, London.

Moscovici, S. (1973) 'Introduction', in C. Herzlich (ed) *Health and Illness: A Social Psychological Analysis*, Academic Press, London.

Moscovici, S. (1976) *Social Influence and Social Change*, Academic, London.

Mugny, G. (1975) 'Negotiations, image of the other and the process of minority influence', *European Journal of Social Psychology*, vol 5, pp209–228.

Pacheco, C., López-Bao, J. V., García, E. J., Lema, F. J., Llaneza, L., Palacios, V., and Godinho, R. (2017) 'Spatial assessment of wolf-dog hybridization in a single breeding period', *Scientific Reports*, vol 7, 42475.

Penteriani, V., Huber, D., Jerina, K., Krofel, M., López-Bao, J. V., Ordiz, A., Zarzo-Arias, A., and Dalerum, F. (in this volume) 'Trans-boundary and trans-regional management of a large carnivore: Managing brown bears across national and regional borders in Europe', in T. Hovardas (ed) *Large Carnivore Conservation and Management: Human Dimensions*, Routledge, London.

Pérez, J. A., and Mugny, G. (1987) 'Paradoxical effects of categorization in minority influence: When being an outgroup is an advantage' *European Journal of Social Psychology*, vol 17, pp157–169.

Persson, J., Rauset, G. R., and Chapron, G. (2015) 'Paying for an endangered predator leads to population recovery', *Conservation Letters*, vol 8, pp345–350.

Pohja-Mykrä, M. (in this volume) 'Socio-political illegal acts as a challenge for wolf conservation and management: Implications for legitimizing traditional hunting practices', in T. Hovardas (ed) *Large Carnivore Conservation and Management: Human Dimensions*, Routledge, London.

Pohja-Mykrä, M., and Kurki, S. (2014) 'Strong community support for illegal killing challenges wolf management', *European Journal of Wildlife Research*, vol 60, no 5, pp759–770.

Rasborg, K. (2012) '"(World) risk society" or "new rationalities of risk"? A critical discussion of Ulrich Beck's theory of reflexive modernity', *Thesis Eleven*, vol 108, no 1, pp3–25.

Rauset, G. R., Andrén, H., Swenson, J. E., Samelius, G., Segerström, P., Zedrosser, A., Persson, J. (2016) 'National parks in Northern Sweden as refuges for illegal killing of large carnivores', *Conservation Letters*, vol 9, no 5, pp334–341.

San Martin, A., Swaab, R. I., Sinaceur, M., Vasiljevic, D. (2015) 'The double-edged impact of future expectations in groups: Minority influence depends on minorities' and majorities' expectations to interact again', *Organizational Behavior and Human Decision Processes*, vol 128, pp49–60.

Sandström, C., Sjölander-Lindqvist, A., Pellikka, J., Hiedanpää, J., Krange, O., and Skogen, K. (in this volume) 'Between politics and management: Governing large carnivores in Fennoscandia', in T. Hovardas (ed) *Large Carnivore Conservation and Management: Human Dimensions*, Routledge, London.

Santiago-Ávila, F. J., Lynn, W. S., and Treves, A. (in this volume) 'Inappropriate consideration of animal interests in predator management: Towards a comprehensive moral code', in T. Hovardas (ed) *Large Carnivore Conservation and Management: Human Dimensions*, Routledge, London.

Sjölander-Lindqvist, A., Bendz, A., Cinque, S., and Sandström, C. (in this volume) 'Research amidst the contentious issue of wolf presence: Exploration of reference frames and social, cultural, and political dimensions', in T. Hovardas (ed) *Large Carnivore Conservation and Management: Human Dimensions*, Routledge, London.

Skogen, K., Mauz, I., and Krange, O. (2008) 'Cry wolf: Narratives of wolf recovery in France and Norway', *Rural Sociology*, vol 73, no 1, pp105–133.

Skonhoft, A. (2017) 'The silence of the lambs: Payment for carnivore conservation and livestock farming under strategic behavior', *Environmental and Resource Economics*, vol 67, no 4, pp905–923.

Theodorakea, I. (2014) *Who let the wolves out? Perceptions about the presence of the Wolf in Central Greece*, Master's thesis, Swedish University of Agricultural Sciences, Uppsala.

Torres, R. T., Ferreira, E., Rocha, R. G., and Fonseca, C. (2017) 'Hybridization between wolf and domestic dog: First evidence from an endangered population in central Portugal', *Mammalian Biology*, vol 86, pp70–74.

Tosi, G., Chirichella, R., Zibordi, F., Mustoni, A., Giovannini, R., Groff, C., Zanin, M., Apollonio, M. (2015) 'Brown bear reintroduction in the Southern Alps: To what extent are expectations being met?' *Journal for Nature Conservation*, vol 26, pp9–19.

Treves, A., Krofel, M., and McManus, J. (2016) 'Predator control should not be a shot in the dark', *Frontiers in Ecology and the Environment*, vol 14, pp380–388.

Von Essen, E. (2017) 'Whose discourse is it anyway? Understanding resistance through the rise of "barstool biology" in nature conservation', *Environmental Communication*, vol 11, no 4, pp470–489.

Wagner, W. (2007) 'Vernacular science knowledge: Its role in everyday life communication', *Public Understanding of Science*, vol 16, no 1, pp7–22.

Wibeck, V. (2014) 'Social representations of climate change in Swedish lay focus groups: Local or distant, gradual or catastrophic?', *Public Understanding of Science*, vol 23, no 2, pp204–219.

Zabel, A., Pittel, K., Bostedt, G., and Engel, S. (2011) 'Comparing conventional and new policy approaches for carnivore conservation: Theoretical results and applications to tiger preservation' *Environmental and Resource Economics*, vol 48, no 2, pp287–311.

Zouhri, B., and Rateau, P. (2015) 'Social representation and social identity in the black sheep effect', *European Journal of Social Psychology*, vol 45, no 6, pp669–677.

2 Research amidst the contentious issue of wolf presence

Exploration of reference frames and social, cultural, and political dimensions

Annelie Sjölander-Lindqvist, Anna Bendz, Serena Cinque, and Camilla Sandström

Introduction

Around the world, resolving human-wildlife conflicts remains a challenge for wildlife management. The return and recovery of large carnivores may be locally undesired – though of course also highly anticipated. In the case of the Scandinavian wolf, the species is both considered an impediment to rural livelihoods and survival, and is valued as an inextricable part of the fauna (Sjölander-Lindqvist, 2008, 2009, 2015). These understandings pit conservationists against private property owners, and farmers and hunters against wolves as well as against policy and policy work (Treves and Karanth, 2003; Decker et al., 2012). Unsurprisingly, such controversies are loaded with emotive human responses, issues of social and political trust, conflicting values and norms, clashing knowledge claims, and politicized arenas of interaction (Knight, 2003; Clark and Rutherford, 2014; Sjölander-Lindqvist et al., 2015). Understanding these complex problems, which involve uncertainty regarding future prospects for species recovery and human-wildlife coexistence, conflicting environmental goals and values, and disputes over the burdens and benefits of conservation initiatives, requires a research approach that can contribute to a renewed and broadened understanding of contemporary aspects of the cultural, social, and political dimensions of large carnivore presence.

Undertaking such research is itself beset with complexity. The researcher is stepping into a highly politicized context characterized by multiple competing interests and values, polarization among actors, political and social distrust, as well as ideological attributes and factions (Knight, 2000, 2003; Sjölander-Lindqvist et al., 2008; Cinque et al., 2012). Scientific exploration and analysis encounter tension and challenges when attempting to study politicized phenomena. Part of the research task is to seek in-depth understanding of the issue's cultural layers and implications, while staying true to scientific imperatives and maintaining distance. Simultaneously, the researcher must create a trustful conversational space and balance the entanglements arising from the socio-cultural and political embeddedness of the connections between individuals, groups, organizations, and institutions (Bernard, 1994). Therefore, to provide a 'generous, comparative but nevertheless critical understanding of human being and knowing in the world we all inhabit' (Ingold, 2008, p. 69) and maintain an exploratory stance, research should explore various dimensions, including the roles of scale, institutions, value orientations, and different actors' (competing) understandings and interpretations of the world. To advance understanding, various analytical aspects must be addressed, and the relationships between components taken into account (Sobo and de Munck, 1998). This undoubtedly creates opportunities to apply interdisciplinary research designs.

Using the example of a controversial case of national conservation policy implementation, this chapter discusses three topical aspects of complex wildlife recovery activities: (1) the local setting of the introduction of management regimes, illustrating how different perspectives and values regarding the landscape and local traditions shape local responses, in this case, to the political decision that the wolf should be allowed to survive and recover; (2) the mediating position of the authorities with the task of implementing political decisions; and (3) the role of public opinion when it comes to understanding the opportunities for policy makers to take various measures to enhance their legitimacy. The context of these aspects has both tangible and intangible dimensions. Collectively shared and transmitted memories of past generations (e.g., stories of how wolves could cause starvation, by attacking and killing a crofter's only cow) – as well as social and cultural constructions of the landscape, place, and identity – linger beside the implementation of political strategies for the enhancement of biodiversity; for example, protective measures and a legislative framework for preserving the wolf. To illustrate the usefulness of different theoretical and methodological perspectives, we will highlight the findings of qualitative and quantitative studies of the Swedish wolf controversy. We started with local conflicts over the presence of large carnivores, arising after the introduction of the first coherent large carnivore policy by the Swedish parliament in 2001. Ensuing conflicts and difficulties in attaining local consent to politics and policy implementation led to an amended policy in 2009 incorporating additional deliberative measures to increase locally approved decisions regarding wolf management.

The studies, running from 2003 to date, have been interdisciplinary in design, including theoretical perspectives from social anthropology, public administration, political science, and human ecology. When studying contemporary controversies, the scientist 'is at the front lines of battle' (Scott et al., 1990, p. 490). Actors' expectations of the research undertaken enter into the process, which may lead the analyst, wittingly or not, to 'become a partisan participant in the debate' (Scott et al., 1990, p. 491). To prevent unbalanced analysis, asymmetrical and biased exploration must be avoided. Because ideas, values, and norms are social, cultural, psychological, and political phenomena, both individual and collective dimensions must be interrogated and disciplinarily and methodologically triangulated. Through participant observations and semi-structured interviews undertaken in situ or over the phone (with people living in wolf areas, administrators, managers, politicians, and stakeholder representatives), survey data, and document analysis, we investigated the immanent norms and values regarding the disputed presence of wolves in Swedish rural areas (Sjölander-Lindqvist, 2008, 2009; Sjölander-Lindvist et al., 2015; Sjölander-Lindqvist and Cinque, 2014). Along with interrogating local perspectives, our research has also explored wildlife management and public opinion in a policy context staged in response to intense conflicts between stakeholder groups and reduced trust in the authorities (Cinque, 2008, 2015; Sandström et al., 2009).

We begin by presenting an overview of three research areas connected to various applicable perspectives and methodological approaches: local perspectives, the authorities' mediating position, and matters of public opinion. Building on this overview, we then discuss how environmental controversies must be approached as continuously reflecting social, cultural, and political norms and values. According to Sjölander-Lindqvist et al. (2015), it is increasingly important to expand knowledge of and insight into the multiple and complex value sets shaping and legitimizing understanding and action at different levels. This, we argue, requires a research approach that interrogates the multiplicity of complex structures.

Local perspectives and stakeholder positioning

In one sense, the well-established discourse on sustainable development can be understood as an arena in which the agency and authority of particular actors are established by means of political directives and agreements. These may, as in the case of Swedish wolf policy, be conceived as having material and social impacts that intrude on local lives, resulting in conflicting perspectives on how best to manage natural and cultural heritage resources. These controversies engender debate on what are regarded as optimal measures for halting biodiversity loss and securing endangered species' recovery while maintaining local livelihood opportunities (cf. Tsing, 2001; Sjölander-Lindqvist, 2008, 2009, 2015; Sjölander-Lindvist et al., 2015; Sjölander-Lindqvist et al., 2008).

On one 'side', we find the view that farming and hunting are affected by the presence of wolves (Sjölander-Lindqvist, 2008, 2009, 2015). In effect, interviewed farmers and hunters say that their relationships with the local environment are being disturbed by the recovery of the wolf population. Farmers and hunters sceptical of the wolf presence hold that wolves are affecting opportunities for recreational activities, such as hunting, fishing, horseback riding, orienteering, and berry- and mushroom-picking. The survival of small-scale farming is also considered to be at stake. Farmers who bring their cattle to summer pastures understand that they are perpetuating traditions of the past, supporting both rich cultural heritage and biodiversity as this pastoral practice helps maintain an open landscape and diversified flora. They claim, however, that the presence of large carnivores makes it difficult to uphold their traditions because their livestock are exposed to the risk of predatory attacks. Consequently, wolf presence is seen as endangering biological and cultural diversity when livestock breeders and farmers give up small-scale agriculture and husbandry because of the burdens of wolf presence. Similarly, hunters claim that reduced opportunities to hunt without exposing hunting dogs to prowling wolves will leave the Swedish forests empty of hunters. Hence, the effects of the presence of wolves – attacks on livestock and hunting dogs, potential attacks on people as growing wolf populations no longer have sufficient prey, and declining game stocks – will, according to these informants, increase the marginalization of rural people and the depopulation of the countryside. Concerned about the reality – or possibility – of attacks on private property, wolf sceptics say that wolves' intrusive behaviour and residence near villages and farmsteads must be controlled and restricted. Concomitant network building and mobilization of opinion have taken place, and there have been complaints and protests concerning Swedish wolf policy and management. Local hunters, pet and livestock owners, and farmers say they have learned that the authorities simply disregard local worries and knowledge. This has encouraged cohesion and cooperation in efforts to protect local values. In such situations, members of affected communities often lose trust in authorities and harbour feelings of suspicion and hostility towards the state because they believe that the authorities ignore local concerns (Sjölander-Lindqvist, 2004, 2008, 2009; Feldman and Khademian, 2007).

On the other 'side', wolf protectionists maintain that wolves have a right to exist in the Swedish countryside and that action must be taken to restore a threatened ecosystem. Grounding this position in the environmentalist discourse advocating action to restore ecosystems understood as under threat, they promote fauna diversity and believe that ongoing action must be taken to support the survival of wolves in the Swedish fauna. In support of this, people with pro-wolf attitudes have joined nature conservation and environmental organizations. Similar pro-wolf arguments have also been evident among the authorities handling governmental predator policy (Cinque, 2008). Administrators and pro-wolf groups share a belief that the anti-wolf groups' opinions as to the causes and effects of wolf presence in the countryside are inconsistent, and that these opinions must be addressed to resolve the persistence of unfounded fears of wolves. These parties advocate scientific knowledge as central to both decision making and alleviating worries about the effects of large carnivore presence (Sjölander-Lindqvist, 2008). Backed by a governmental inquiry (Government Official Report SOU, 1999, p. 146), it has been assumed that increased public dissemination of science-informed knowledge will increase concerned stakeholders' consent to current policy (cf. Limoges, 1993). In this case, science has taken a central position in science-informed management practice. Since the 1980s, wolves' population growth, genetics, behaviour, and effects on large herbivore populations have been investigated to support evidence-based wolf population management.

The context of the controversy also includes positioning in the debate. In this case, this occurs when sceptics maintain that society should learn from the experience of past generations, claiming that wolves can severely harm local communities.[1] Protectionists counter by pointing out that nobody has been injured or killed by a wolf in modern times, and the sceptics' fears are therefore groundless. People opposing the public wolf policy respond that local interests and knowledge gained over years and generations of residing in an area must be acknowledged and reflected in decisions regarding the community environs of rural Sweden. While protectionists generally hold the opinion that wolves have migrated to Sweden, wolf sceptics believe otherwise. Explanations from authorities and the biological science community that the Scandinavian wolf has migrated to Sweden from Russia, Finland, and the Baltic States, passing the Finnish-Swedish border, are not credible, according to interviewed wolf sceptic informants. Instead, they believe that wolves do not really belong in the Swedish countryside, but have been reintroduced by unidentified people and groups, adding to the debate on whether Sweden should strive to protect wolves' living conditions (Sjölander-Lindqvist, 2008).

Although both sides – to varying extents – support systematic wildlife observation (e.g., using radio telemetry, GPS tracking, DNA sampling of carcasses and droppings, and snow-tracking management), such observation may contribute to local conceptions of the wolf as extraneous to nature; for example, when scientific knowledge (e.g., Latour and Woolgar, 1986) signals that society must interfere in nature to preserve its assets. The incorporation of data gathered by evidence-producing methods into policy making and wildlife management lends the wolf exceptionality, and safeguarding the living conditions of the wolves and promoting their survival have, from this perspective, taken precedence over local living environments and traditions, inadvertently affecting local identities (Sjölander-Lindqvist, 2008).

The mediating position of county administrative board managers and field staff

Our second area of attention concerns the realities of policy work. Ideally, decision making in organizations should be logical: goals and alternatives should be made explicit, and decision makers should calculate the effects of alternative strategies and evaluate them against the goals set (Simon, 1987). However, instead of actors pursuing 'rational' ways of establishing accounts of alternative actions, rationality is instead mediated by personal contexts, feelings, beliefs, and – not least – experiences of different practices, all of which are essential to the decision-making process (Klein, 1998; McPhee and Zaug, 2001; Meier and O'Toole, 2006; Zinn, 2008). From a theoretical perspective, leadership in public authorities is likely to be experience-informed and intuitive when public managers conflate policy goals with personal preferences and necessities arising from previous experience. This forces a mediating position on the manager that can be expected to have extensive impact on the management of controversial and emotion-laden issues, and on the manager's personal life and well-being (Sjölander-Lindqvist and Cinque, 2013; Cinque, 2015).

In Sweden, the regulatory framework for wolf protection comprises an extensive administrative package including surveys, damage prevention, stakeholder compensation, controlled hunting, delegation of decision making, predator research programmes, enforcement of hunting legislation, and consultation with interest organizations. While the Swedish Environmental Protection Agency (SEPA) is responsible for implementing national legislation uniformly by providing coordination and guidance, the County Administrative Boards (CABs) apply regulations at the local level by adapting and interpreting them. This demands great ability to lead and coordinate collaborative and consensus-seeking processes (Huxham, 2000). In addition to planning, monitoring, and evaluating policy (Conley and Moote, 2003), CAB managers and field staff are also tasked with establishing and facilitating communication between the authorities and stakeholder and interest organizations. CAB managers, for example, have the responsibility to design and implement the quota-regulated wolf hunt, to be implemented through a collaborative management approach to increase the opportunity for local communities to take an active part in wildlife management. Another example is their work to compensate farmers and hunters for loss of private property (mainly farm and companion animals) due to large carnivore attacks. In a case of suspected harm, a formal investigation process to assess the damage and its cause is carried out by field staff under CAB supervision. Expected to be impartial, enforcing rules and regulations, the field staff should act supportively and empathetically so as to promote local support for national large carnivore protection policies. Interacting directly with concerned property owners at attack sites, as frontline actors in policy implementation, they are to orient themselves according to both organizational and social expectations (Sjölander-Lindqvist and Cinque, 2013). A third example refers to the implementation of collaborative approaches, suggested to be necessary to reduce local conflict and yield consensus on the contested matters of large carnivore presence (Sandström et al., 2009). In leading the collaborative process forward, CAB managers must maintain balance between the broad representation of interests and the need to quickly accomplish collaborative goals through pragmatic management.

In their everyday work, CAB managers are expected to harmonize the multiple interactions of scale and level, maintain administrative rules, and achieve the specific targets of policy decisions. They have to cope with innumerable relationships and interactions. These multiple tasks demand great ability to design and coordinate various processes (Feldman and Khademian, 2002). At the same time, CAB managers try to avoid and manage possible conflicts with both local stakeholders and SEPA. In interviews, they describe their work as well structured and never ad hoc, implying that they follow a rational decision-making process. However, managers recognize that they find themselves in an exposed situation in which discretionary latitude allows them to adapt rules to local conditions. For example, national legislation prescribes prevention as a prerequisite for obtaining damage compensation. Furthermore, the rules specify that the compensation funds should first be used for preventive measures. However, as many CAB managers believe that the compensation system is the main instrument to build trust and increase acceptance, they may decide to pay for preventable damage, as well. This illustrates the existence of a responsive working situation in which experiences and personal beliefs serve as a reference point for judgement and decision making, which they need to balance against national policy requirements (Sjölander-Lindqvist and Cinque, 2013). They lament that their work exposes them to contradictory expectations and demands from different parties. For example, while farmers or hunters exposed to large carnivore attacks demand that managers enforce targeted removal, conservationists expect the officers to apply protective rules and directives. This situation contributes to an ambiguous situation: on one hand, public authorities should be attentive to local circumstances, to build support for the government's large carnivore policy; on the other hand, the bureaucracy should enforce the rules. The solution to this dilemma is administrative discretion, which the CAB managers apply to navigate various claims and expectations, to reduce or defuse conflicts personally and bureaucratically, and to implement what they call a 'highly charged policy decision' (Cinque, 2008, 2015). As Lipsky (1980) observed, this decision entails various and ambiguous objectives, generating both unrealistically high expectations and opposition from part of the public. In this context, the role of the authorities will be crucial in interpreting, adapting, and partially modifying the content of the decision in order to achieve sustainable results.

According to the results of our empirical investigations, the encounter between citizens and public managers is a highly fluid interpretative zone of face-to-face interaction characterized by varied social, cultural, political, and economic dimensions (cf. Maynard-Moody and Musheno, 2003). This transforms the traditional perspective on bureaucracy as typically rational and hierarchical in structure, steered by written rules, graded authority, strict discipline and control, and expert qualification (Simon, 1947). When performing their tasks, public managers instead find themselves in a highly emotion-laden judgment situation in which essential parts of the decisions are formed by affective and cognitive factors (cf. Feldman and Khademian, 2007; cf. Maynard-Moody and Musheno, 2003; Sjölander-Lindqvist and Cinque, 2013).

Matters of public opinion

Along with local perspectives and bureaucratic circumstances, public attitudes, preferences, and opinions should be considered for a fuller understanding of the complexities of the politics and policies of large carnivore management. Otherwise, it will be difficult to understand the challenges that decision makers and authorities face when formulating and implementing policies, and to follow public responses to, for example, policy development and management measures. For democracy to work and gain legitimacy for its policies, decision makers must be responsive to citizen expectations and value orientations (e.g., Page and Shapiro, 1983; Soroka and Wlezien, 2010). In this case, democratic responsiveness is particularly challenging because of the polarization of public opinion when it comes to carnivore management, and the strong involved values that may lead to open conflict between groups. Politicians may have difficulties properly representing public opinion in such a situation. This, in turn, risks undermining the legitimacy of carnivore policy.

Looking at research into public attitudes towards carnivores, a recent study based on three large-scale surveys conducted in 2004, 2009, and 2014 demonstrated that, in general, the Swedish population is positive towards wolf presence in Sweden (Eriksson, 2016). In addition, 65–70% of respondents either supported current wolf policy objectives targeting a specific number of wolves, or wanted to increase the number of wolves still further. However, respondents living in rural areas, particularly within or near wolf territories, were more sceptical of or even negative towards the current wolf policy. Indeed, support for a more restrictive wolf policy has grown over time from 30% in 2004 to 35% in 2009 and 2014. The data also indicate ongoing attitude polarization based on geography, with attitudes towards wolves being rather stable in urban areas but increasingly negative in rural areas. This result aligns with the expectations of one of the classic cleavages suggested by Lipset and Rokkan (1967) concerning the centre/periphery, in this case expressed as a division between urban and rural areas.

It is possible to detect similar patterns in other countries (Dressel et al., 2015). A meta-analysis of 105 surveys of attitudes towards bears and wolves across Europe found that respondents were more positive towards bears than wolves. The analysis also found that the proportion of people with positive attitudes towards wolves decreased as the duration of their coexistence with them increased. In general, farmers and hunters had less positive attitudes towards wolves than did others. This rather well-known pattern was initially related to two wildlife value orientations; i.e., the wildlife benefits/existence and wildlife rights/use orientations (Fulton et al., 1996), precursors to the more developed concept of wildlife value orientations (Teel and Manfredo, 2009), which now include domination (also referred to as 'utilitarianism' or 'materialism') and mutualism (Teel et al., 2010). People holding a domination value orientation generally believe that wildlife should be managed for human benefit and are more likely to rank human well-being over wildlife in their attitudes and behaviours. They also tend to justify management measures in utilitarian terms and to rate lethal actions as acceptable. In contrast, people with a mutualist wildlife value orientation hold an egalitarian ideology that encompasses human–animal relationships. Wildlife is seen as part of an extended family, entitled to rights and care. They are less likely to support lethal wildlife management measures, and more likely to engage in actions to increase the welfare of and humanize wildlife, that is, to 'view wildlife in human terms' (Teel et al., 2010, p. 109).

Previous research also demonstrates that public opinion on carnivore attitudes can be divided according to socio-demographic and political factors. In

Sweden, the left-right ideological dimension is pronounced, not only regarding, for example, attitudes to taxes or welfare, but also when it comes to attitudes to wolves, with those who sympathize with left parties or the Greens being significantly less likely to approve of controlling the wolf population by hunting. In addition, socio-demographic factors such as age and education also seem to be of importance in explaining attitudes towards wolf hunting, with younger and more educated respondents being less willing to accept control of the wolf population by hunting (Sjölander-Lindqvist et al., 2008; Cinque et al., 2012; Eriksson, 2016).

An alternative value structure used in opinion research contrasts a green/alternative/libertarian (GAL) perspective with a traditional/authoritarian/nationalist (TAN) perspective (e.g., Hooghe et al., 2002). Although this configuration transcends the left-right political continuum, it may also be relevant to attitudes towards large carnivores. For instance, GAL advocates may lean towards a more conservationist view or, as defined by Teel et al. (2010), towards a mutualist approach to nature. Indeed, that positioning may be independent of any broader ideological positioning or socio-demographic characteristics, such as place of residence. In sum, public opinion may be polarized on several dimensions, which makes it challenging to formulate a policy perceived as legitimate by all. For decision makers, it is vital to pay attention to existing cleavages in public opinion and try to balance opposing preferences. It is important to recall that public opinion, as the general assemblage of varying opinions at the aggregated level, might not reflect the dilemmas existing at the local level or any substantial variations between local communities.

A complex and demanding research field

The Swedish wolf conflict has been researched since the early 2000s, when the first steps were taken by social science to explore the dimensions of this environmental controversy. It has been learned that to manage the associated socio-ecological conflicts without undermining the viability or welfare of wildlife and humans, local concerns about rural community livelihoods and vulnerable property must be balanced by international concerns about saving threatened species. Sensitivity to psychological and sociocultural factors – such as perceptions, triggered emotions, values, attitudes, and norms – pertaining to large carnivores must be examined and understood within a wider context of social, cultural, and political dynamics (Sjölander-Lindqvist, 2009; Sjölander-Lindqvist et al., 2015; Redpath et al., 2017). Social science perspectives are therefore crucial for understanding the individual and collective dimensions associated with the presence of disputed wild animals, dimensions shaping the legitimacy of policy development, implementation, and management (Lundmark et al., 2014; Sjölander-Lindqvist et al., 2015).

It is equally important to consider the methodological complexities associated with describing and understanding 'the multiplicity of conceptual complex structures' (Geertz, 1973, p. 9). Brilliantly described by the anthropologist

Clifford Geertz, the 'thing to ask (is) what is getting said' and what the 'import is' of the occurrence of the realities explored when entering the field of investigation (p. 10). Entering a contentious field, we may find authority officials asserting certain arguments and favouring certain coveted decisions. By doing this, they may construe particular actors (often local residents) as having 'indecorous' concerns and opinions about the issues at hand (cf. Cinque, 2008; Cox, 2010; Sjölander-Lindqvist and Cinque, 2014). Even deliberative arrangements, introduced to reduce conflicts and enhance legitimacy, may end up increasing polarization when representatives of engaged actors go on expressing their diverging beliefs (cf. Duit et al., 2009; Lundmark and Matti, 2015) or when relevant knowledge is sidelined in favour of certain interpretations of mandates and certain expectation of roles (Hallgren and Westberg, 2015). At the local level, state policy may be conceived as supporting wolf recovery over human well-being, prompting feelings of injustice and marginalization; for example, when farmers have to change their practices and routines due to wolf presence (Sjölander-Lindqvist, 2008). This highly politicized field context puts additional pressure on the research context.

Although single-method and disciplinary approaches have proven valuable in achieving insight into certain aspects of the Swedish wolf controversy, we argue that it is beneficial to aim for a comprehensive and exhaustive research design that can describe the complex relationship between the individual and collective dimensions, including the researcher's role as an impartial investigator. To create and maintain legitimacy in conducting their research, researchers should be able to present valid and reliable approaches to detecting value orientations at different scales. Given the complexity and dynamics of stakeholder attitudes and interactions regarding the wolf issue, researchers must be continuously aware of the major trends in the controversy and systematically update their records elaborating on both the global and the local, universals and particulars. Apart from scale, interrelatedness and consistency may also be issues for various theoretical and methodological approaches. It may be justifiable to integrate quantitative methods and statistical analyses for large datasets and sample sizes with finer-scaled, qualitative approaches that concentrate on specific social strata or groups. According to Bunge (1998):

> the fragmentation of the social studies into dozens of weakly related or even mutually independent disciplines . . . is artificial because . . . every one of the major subsystems of society – the biological, economic, political, and cultural ones – is intimately connected.
>
> (p. 452)

If research is incapable of giving a fair and objective picture of the attitudes of larger and smaller groups, social trust in research into wildlife conflicts may be threatened and researchers may be accused of taking sides. In addition, it is important to acknowledge that not all options for solving human-wildlife conflict are legitimate or even feasible from a political perspective. That is, how local interests or public opinion conceptualize an issue and how political decision makers are able and willing to respond to these conceptualizations do set limits on what is possible. Taking measures not considered legitimate by the majority may harm the democratic process. On the other hand, polls of larger segments of society may not be able to capture local aspects or uncover disagreements between local people and the general public at the regional or national scales (Ericsson et al., 2006). Such mismatch is especially significant when there is a clash between opinions emerging from different frames of reference; for example, how the cost of large carnivore management should be distributed, which arguably excessively burdens rural communities (Sjölander-Lindqvist, 2008, 2009).

Research legitimacy, i.e., achieving and maintaining acceptance of research activity and juggling divergent expectations, is therefore crucial. When researchers explore local perspectives, informants may expect the inquiries to bring about changed laws and bureaucratic procedures when the identified difficulties faced by hunters and farmers in their daily lives are presented to the decision makers. From the public manager's perspective, researchers might be expected to inform the farmers and hunters about how policy works. In such situations, when actors at different levels have different expectations, the researcher assumes the task of making findings understandable to decision makers (Crumley, 2001). The researcher adopts both an explanatory and troubleshooting role, and by conveying why society should support the further spread and recovery of the wolf population, hostility and resentment towards public managers may be decreased. Because wolf policy is said to increase the marginalization of rural residents and the depopulation of rural areas (Sjölander-Lindqvist, 2008, 2009), public administration may say that wolf-sceptical residents, stakeholders, and stakeholder organizations are refusing to consider the possible merits of alternative points of view (Cinque, 2008). Immersed in facilitating and legitimizing policy decisions, the manager is situated in the nexus between local conditions, personal opinions, and policy requirements, affecting his or her personal life (Cinque, 2008; Sjölander-Lindqvist and Cinque, 2013). Lipsky (1980) emphasized that when 'street-level bureaucrats' implement regulations, they tend to conflate organizational goals with personal preferences and perceived necessities (Lipsky, 1980; cf. Winter, 2007; Vinzant and Crothers, 1998). The field of policy implementation is therefore a socially negotiated encounter shaped by its actors (Fineman, 1998). Here, the researcher assumes a moderating role with the expectation of being an interlocutor both at the national level of administration, interrogating decisions that managers themselves may disagree with or find unfair, and at the local level, shedding light on the circumstances engendered by certain political decisions.

Ensuring continued re-entry to the field may therefore be problematic for researchers, challenging their integrity, knowledge, detachment, and objectivity; any scientific report may, depending on the actor's position, be regarded as 'choosing sides' or as aligned with a particular group or policy. This means that local stakeholders may misunderstand the research performed as yet another governmental initiative, or wildlife managers may believe that the research and resulting reports may negatively affect policy implementation. Critically, this refers to a boundary tension between research autonomy, knowledge integrity, authority over the research process, and the interests of the subjects of inquiry.

Those examples reflect the distinct social, cultural, and political spheres in which the researcher participates. Informed by cultural value frameworks developed over years of 'dwelling' – to apply the terminology of anthropologist Tim Ingold (1993) – local residents can be said to find themselves in a 'configurative complex of things' (Casey, 1996, p. 25). Besides relating to the natural and social worlds in ways that reproduce both collective memories (Schama, 1995) and the meanings that rural traditions bring to their practitioners, people's understandings and experiences of the contemporary world are also informed by other actors' endeavours to impose different sets of values (Scott, 1998). For example, whereas local traditions such as small-scale farming or hunting aid in reproducing collective memories and maintaining local identity (Sjölander-Lindqvist, 2008, 2009, 2015), public managers employ consensual and experience-driven heuristic strategies to meet organizational and social expectations (Sjölander-Lindqvist and Cinque, 2013). According to policy, these managers must bring about the broad involvement of local actors without favouring any particular interest, and must also maintain bureaucratic procedures to ensure efficacy and efficiency. For example, if a wolf causes harm to farm or companion animals, the property owner is entitled to state compensation. At the same time as the damage is assessed by regional administration staff, the inspection constitutes an opportunity for those materially affected by the presence of protected predators to meet those with the duty of locally implementing state policy. In this situation, the inspectors find themselves in a continuously exposed situation in which they are expected to behave professionally while managing emotion-laden circumstances. They are expected to distance themselves in order to present impartial grounds for decisions on indemnity, while empathetically addressing the concerns of the farmers and hunters exposed to property damage. To summarize, members of the public administration become facilitators, interpreters, and mediators expected to balance national regulations, local demands, and public opinion against the background of past and present experiences and anticipations, all of which intuitively inform current policy implementation.

Any research into contentious and politicized situations characterized by polarization between those conceiving of themselves as 'objective' and others not regarded as equally objective requires a methodological approach capable of addressing social, cultural, and political dimensions, as well as knowledge and value structures, as they appear and inform current understandings and opinions. Because of the inherent complexity of controversial issues, establishment of research legitimacy is crucial. We have discussed how local perspectives and the mediating position of the bureaucratic agent tap into one another and how societal conflict over large carnivores apparently taps into the value orientations of public opinion. If we think of these dimensions as explorative entry points through which frames of reference and actors can meet, we can assess how political-institutional structures function as boundary agents (Star and Griesemer, 1989) connecting the micro and macro levels (Rodman, 2003).

Conclusion

Present-day ecological interventions to help endangered species retain their places in the environment are directed by contemporary nature conservation politics. Conservation outcomes, however, are also conditioned by the understanding that predatory animals encroach uninvited on humanized habitats and have consequences for local livelihoods, evoking hostility towards their presence. Since 'it is the local people who are experiencing the costs of living alongside wildlife' (Woodroffe et al., 2005, p. 402), the locally undesired and politically unintended consequences must be acknowledged (Tsing, 2001; Brechin et al., 2003). Even though the decision on the recovery and dispersal of wolves in Sweden was made democratically through parliamentary voting and has been backed by national regulations and international directives (such as the EU Habitat Directive), previous and current research demonstrates that the politics of and underlying reasons for wolf population recovery remain contested.

Drawing on results and findings achieved by exploring local perspectives, the mediating position of authorities and managers, and the value orientations of public opinion, all of which may reflect divergent perceptions of wolves, wolf policy, the local environment, and how to conduct bureaucratic work, lays the groundwork for multidisciplinary research designs capable of addressing both individual and collective dimensions (Table 2.1). The existence of three reference frames – namely, the local, political-institutional, and public opinion

Table 2.1 Reference frames for investigating stakeholder reaction to wolf policy in Sweden

	Local perspectives	*Authorities' intermediary position*	*Matters of public opinion*
Main actors engaged and voiced	Farmers, livestock and pet owners, hunters	County administrative board managers; field staff	Public opinion through surveys
Main challenge	• Wolves do not really belong in the landscape; wolves, wolf advocates and Swedish wolf policy intruding on local lives • The effects of the presence of wolves will further increase the marginalization of rural people and the depopulation of the countryside	• The regulatory framework for protection of wolves embraces an extensive administrative package • However, their work exposes them to contradictory expectations and demands from different parties	• Although there is a clear majority in favour of enacted Swedish wolf policy, urban and rural areas are divided • Public opinion is not unified; several cleavages are present which aggravate representation

(Continued)

Table 2.1 (Continued)

	Local perspectives	Authorities' intermediary position	Matters of public opinion
	• Small-scale farming is considered to be at stake • Wolf presence is understood to endanger biological and cultural diversity	• Local people affected often lose trust in authorities because they feel that they fail in considering local concerns • Expected to harmonize the multiple interactions of scale and level, maintain rules and achieve specific policy targets	
Main quest	The intrusive behaviour of wolves and their residence near villages and farmsteads/pastures must be controlled and restricted; wolf attacks on dogs during hunt must be controlled and restricted	Delicate balance between local interests and sympathy for local actors, on the one side, and responsibility for enforcing rules and regulations, on the other	Achieve democratic legitimacy for wolf policy with regard to strong cleavages

frames – reflects a tangled web of assumptions and values that come into play and shape responses to the political decision that the wolf should be allowed to recover in Swedish forested areas.

The local level, informed by local identity and local community survival, and the institutional level, informed by hierarchical mechanisms of control and decision making, are inextricably nested and chained to the probabilities and attributes of each particular situation. This means that contextual value-driven circumstances, such as local perspectives and public attitudes, as well as sector-specific tools, rules, resources, goals, and norms, are dimensions that research must address.

Note

1 Whereas today we see legal measures to protect the wolf population, in the past, laws and opinions regarded wolves as detrimental to humans and human activities, leading to wolf persecution. Provincial laws from the fifteenth century, for example, established parishioners' obligation to take an active part in wolf battues, with only the women, vicar, and clerk

of the parish being exempt. Such hunting, before the enactment of the wolf preservation act in 1965, was successful, and most wolves were exterminated from Sweden in the nineteenth and twentieth centuries. Using a variety of methods – traps, nets, weapons, and battues – people throughout Sweden did a good job of hunting wolves. In addition, bounties were imposed in 1647 to encourage wolf hunting, and these remained in force until the wolf preservation act was enacted more than two hundred years later.

References

Bernard, H. R. (1994) *Research Methods in Anthropology: Qualitative and Quantitative Approaches*, SAGE Publications, Thousand Oaks, CA.

Brechin, S. R., Wilshusen, P. R., Fortwrangler, C. L., and West, P. C. (eds) (2003) *Contested Nature: Promoting International Biodiversity with Social Justice in the Twenty-first Century*, State University of New York Press, Albany, NY.

Bunge, M. (1998) *Social Science Under Debate: A Philosophical Perspective*, University of Toronto Press, Toronto, ON.

Casey, E. S. (1996) 'How to get from space to place in a fairly short stretch of time: Phenomenological prolegomena', in S. Feld and K. H. Basso (eds) *Senses of Place*, School of American Research Press, Santa Fe, NM.

Cinque, S. (2008) *I vargens spår. Myndigheters handlingsutrymme i förvaltningen av varg* [In the Wolf Track: Administrative Discretion in Wolf Management], PhD thesis, University of Gothenburg, Sweden.

Cinque, S. (2015) 'Collaborative management in wolf licensed hunting: The role of public managers in moving collaboration forward', *Wildlife Biology*, vol 21, no 3, pp157–164.

Cinque, S., Sjölander-Lindqvist, A., and Bendz, A. (2012) 'Värmlänningarna och rovdjursförvaltningen. Uppfattningar om jakt och genetisk förstärkning av vargstammen' [Translation of title], in L. Nilsson, L. Aronsson and P. O. Norell (eds) *Värmländska landskap*, Karlstad University Press, Karlstad, Sweden.

Clark, S. G., and Rutherford, M. B. (eds) (2014) *Large Carnivore Conservation: Integrating Science and Policy in the North American West*, University of Chicago Press, Chicago, IL.

Conley, A., and Moote, M. (2003) 'Evaluating collaborative natural resource management', *Society and Natural Resources*, vol 16, no 5, pp371–386.

Cox, R. (2010) *Environmental Communication and the Public Sphere*, SAGE Publications, Thousand Oaks, CA.

Crumley, C. L. (ed) (2001) *New Directions in Anthropology & Environment*, Altamira Press, Walnut Creek, CA.

Decker, D. J., Riley, S. J., and Siemer, W. F. (eds) (2012) *Human Dimensions of Wildlife Management* (2nd ed.), Johns Hopkins University Press, Baltimore, MD.

Dressel, S., Sandström, C., and Ericsson, G. (2015) 'A meta-analysis of studies on attitudes toward bears and wolves across Europe 1976–2012', *Conservation Biology*, vol 2, no 2, pp565–574.

Duit, A., Galaz, V., and Löf, A. (2009) 'Fragmenterad förvirring eller kreativ arena? Från hierarkisk till förhandlad styrning i svensk naturvårdspolitik', in J. Pierre and G. Sundström (eds) *Samhällsstyrning i förändring*, Liber, Malmö, Sweden.

Ericsson, G., Sandström, C., and Bostedt, G. (2006) 'The problem of spatial scale when studying human dimensions of a natural resource conflict: Human and carnivores as a case study', *International Journal of Biodiversity, Science and Management*, vol 2, no 4, pp334–349.

Eriksson, M. (2016) *Changing Attitudes to Swedish Wolf Policy: Wolf Return, Rural Areas, and Political Alienation*, PhD thesis, Umeå Universitet, Umeå, Sweden.

Feldman, M. S., and Khademian, A. M. (2002) 'To manage is to govern', *Public Administration Review*, vol 62, no 5, pp541–554.

Feldman, M. S., and Khademian, A. M. (2007) 'The role of the public manager in inclusion: Creating communities of participation', *Governance*, vol 20, no 2, pp305–324.

Fineman, S. (1998) 'Street-level bureaucrats and the social construction of environmental control', *Organization Studies*, vol 19, no 6, pp953–974.

Fulton, D. C., Manfredo, M. J., and Lipscomb, J. (1996) 'Wildlife value orientations: A conceptual and measurement approach', *Human Dimensions of Wildlife*, vol 1, no 2, pp24–47.

Geertz, C. (1973) *The Interpretation of Cultures*, Basic Books, New York.

Government Official Report SOU (1999):146 Rovdjursutredningen-Slutbetänkande om en sammanhållen rovdjurspolitik. Miljö-och samhällsbyggnadsdepartementet (Ministry of Environment and Social Structure).

Hallgren, L., and Westberg, L. (2015) 'Adaptive management? Observations of knowledge coordination in the communication practice of Swedish game management', *Wildlife Biology*, vol 21, no 3, pp165–174.

Hooghe, L., Marks, G., and Wilson, C. J. (2002) 'Does left/right structure party positions on European integration', *Comparative Political Studies*, vol 35, no 8, pp965–989.

Huxham, C. (2000) 'What makes partnerships ork', in S. Osborne (ed) *Managing Public – Private Partnerships for Public Services*, Routledge, London.

Ingold, T. (1993) 'The temporality of the landscape', *World Archaeology*, vol 25, no 2, pp152–174.

Ingold, T. (2008) 'Anthropology is not ethnography', *Proceedings of the British Academy*, vol 154, pp69–92.

Klein, G. A. (1998) *Sources of Power: How People Make Decisions*, MIT Press, Cambridge, MA.

Knight, J. (ed) (2000) *Natural Enemies: People – Wildlife Conflicts in Anthropological Perspective*, Routledge, London.

Knight, J. (2003) *Waiting for Wolves in Japan: An Anthropological Study of People – Wildlife Relations*, Oxford University Press, Oxford.

Latour, B., and Woolgar, S. (1986) *Laboratory life: The Construction of Scientific Facts*, Princeton University Press, Princeton, NJ.

Limoges, C. (1993) 'Expert knowledge and decision-making in controversy contexts', *Public Understanding of Science*, vol 2, no 4, pp417–426.

Lipset S. M., and Rokkan, S. (1967) *Party Systems and Voter Alignments: Cross-national Perspectives*, The Free Press, Toronto, ON.

Lipsky, M. (1980) *Street-level Bureaucracy: Dilemmas of the Individual in Public Services*, Russell Sage Foundation, New York.

Lundmark, C., and Matti, S. (2015) 'Exploring the prospects for deliberative practices as a conflict-reducing and legitimacy-enhancing tool: The case of Swedish carnivore management', *Wildlife Biology*, vol 21, no 3, pp147–156.

Lundmark, C., Matti, S., and Sandström, A. (2014) 'Adaptive co-management: How social networks, deliberation and learning affect legitimacy in carnivore management', *European Journal of Wildlife Research*, vol 60, no 4, pp637–644.

Maynard-Moody, S., and Musheno, M. C. (2003) *Cops, Teachers, Counselors: Stories from the Front Lines of Public Service*, University of Michigan Press, Ann Arbor, MI.

McPhee, R., and Zaug, P. (2001) 'Organizational theory, organizational communication, organizational knowledge, and problematic integration', *Journal of Communication*, vol 51, no 3, pp574–591.

Meier, K. J., and O'Toole, L. J. (2006) *Bureaucracy in a Democratic State: A Governance Perspective*, Johns Hopkins University Press, Baltimore, MD.

Page, B. I., and Shapiro, R. Y. (1983) 'Effects of public opinion on policy', *The American Political Science Review*, vol 77, no 1, pp175–190.

Redpath, S., Linnell, J., Festa-Bianchet, M., Boitani, L., Bunnefeld, N., Dickman, A., Gutiérrez, R. J., Irvine, J. I., Johansson, M., Majic, A., McMahon, B., Pooley, S., Sandström, C., Sjölander-Lindqvist, A., Skogen, K., Swenson, J., Trouwborst, A., Young, J., and Milner-Gulland, E. J. (2017) 'Don't forget to look down – Collaborative approaches to predator conservation', *Biological Reviews*, vol 92, no 4, pp2157–2163.

Rodman, M. (2003) 'Empowering place: Multilocality and multivocality', in S. M. Low and D. Lawrence-Zúñiga (eds) *The Anthropology of Space and Place: Locating Culture*, Blackwell, Oxford.

Sandström, C., Pellikka, J., Ratamäki, O., and Sande, A. (2009) 'Management of large carnivores in Fennoscandia: New patterns of regional participation', *Human Dimensions of Wildlife*, vol 14, no 1, pp37–50.

Schama, S. (1995) *Landscape and Memory*, Alfred A. Knopf, New York.

Scott, J. C. (1998) *Seeing Like a State*, Yale University Press, New Haven, CT.

Scott, P., Richards, E., and Martin, B. (1990) 'Captives of controversy: The myth of the neutral social researcher in contemporary scientific controversies', *Science, Technology, and Human Values*, vol 15, no 4, pp474–494.

Simon, H. A. (1947) *Administrative Behavior: A Study of Decision-making Processes in Administrative Organizations*, Free Press, New York.

Simon, H. A. (1987) 'Making management decisions: The role of intuition and emotion', *Academy of Management Executive*, vol 1, no 1, pp57–64.

Sjölander-Lindqvist, A. (2004) 'The effects of environmental uncertainty on farmers' sense of locality and futurity: A Swedish case study', *Journal of Risk Research*, vol 7, no 2, pp185–197.

Sjölander-Lindqvist, A. (2008) 'Local identity, science and politics indivisible: The Swedish wolf controversy deconstructed', *Journal of Environmental Policy and Planning*, vol 10, no 1, pp71–94.

Sjölander-Lindqvist, A. (2009) 'Social – Natural landscape reorganised: Swedish forest-edge farmers and wolf recuperation', *Conservation & Society*, vol 7, no 2, pp130–140.

Sjölander-Lindqvist, A. (2015) 'Targeted removals of wolves: Analysis of the motives for controlled hunting', *Wildlife Biology*, 21, no 3, pp138–146.

Sjölander-Lindqvist, A., and Cinque, S. (2013) 'When wolves harm private property: Decision making on state compensation', *Focaal – Journal of Global and Historical Anthropology*, vol 65, pp114–128.

Sjölander-Lindqvist, A., and Cinque, S. (2014) 'Dynamics of participation: Access, standing and influence in contested natural resource management', *Partecipazione e Conflitto – The Open Journal of Sociopolitical Studies*, vol 7, no 2, pp360–383.

Sjölander-Lindqvist, A., Bendz, A., and Cinque, S. (2008) 'Rör inte mitt får! Västsvenskens uppfattningar om vargförvaltningens utformning', in L. Nilsson and S. Johansson (eds) *Regionen och flernivådemokratin*, SOM-institutet, Göteborgs universitet, Göteborg, Sweden.

Sjölander-Lindqvist, A., Johansson, M., and Sandström, C. (2015) 'Individual and collective responses to large carnivore management: The roles of trust, representation, knowledge spheres, communication and leadership', *Wildlife Biology*, vol 21, no 3, pp175–185.

Sobo, E. J., and de Munck, V. C. (1998) 'The forest of methods', in V. C. de Munck and E. J. Sobo (eds) *Using Methods in the Field: A Practical Introduction and Casebook*, Altamira Press, Walnut Creek, CA.

Soroka, S. N., and Wlezien, C. (2010) *Degrees of Democracy: Politics, Public Opinion and Policy*, Cambridge University Press, New York.

Star, S. L., and Griesemer, J. R. (1989) 'Institutional ecology, "translations" and boundary objects: Amateurs and professionals in Berkeley's museum of vertebrate zoology, 1907–39', *Social Studies of Science*, vol 19, no 3, pp387–420.

Teel, T. L., and Manfredo, M. J. (2009) 'Understanding the diversity of public interests in wildlife conservation', *Conservation Biology*, vol 24, no 1, pp128–139.

Teel, T. L., Manfredo, M. J., Jensen, F. S., Buijs, A. E., Fischer, A., Riepe, C., Arlinghaus, R., and Jacobs, M. H. (2010) 'Understanding the cognitive basis for human – Wildlife relationships as a key to successful protected-area management', *International Journal of Sociology*, vol 40, no 3, pp04–23.

Treves, A., and Karanth, K. U. (2003) 'Human – Carnivore conflicts and perspectives on carnivore management worldwide', *Conservation Biology*, vol 17, no 6, pp1491–1499.

Tsing, A. L. (2001) 'Nature in the making', in C. L. Crumley, A. E. van Deventer and J. J. Fletcher (eds) *New directions in Anthropology and Environment*, Altamira Press, Walnut Creek, CA.

Winter, S. C. (2007) 'Implementation perspectives: Status and reconsideration', in G. Peters and J. Pierre (eds) *The Handbook of Public Administration*, SAGE Publications, London.

Woodroffe, R., Thirgood, S., and Rabinowitz, A. (eds) (2005) *People and Wildlife: Conflict or Coexistence?* Conservation Biology 9, Cambridge University Press, Cambridge.

Vinzant, J. C., and Crothers, L. (1998) *Street-level Leadership: Discretion and Legitimacy in Front-line Public Service*, Georgetown University Press, Washington, DC.

Zinn, J. (2008) 'Heading into the unknown: Everyday strategies for managing risk and uncertainty', *Health, Risk and Society*, vol 10, no 5, pp439–450.

3 Socio-political illegal acts as a challenge for wolf conservation and management

Implications for legitimizing traditional hunting practices

Mari Pohja-Mykrä

Wolf conflict

The re-colonization of wolves in Nordic countries during the past few decades brought along a bitter human-human conflict that previous research has thoroughly examined. Here I will present the core of this present-day wolf conflict by summarizing the development and implications of an attitude conflict among stakeholders and the influence of several barriers that lead to conflict escalation.

Wolf conflict can be defined as an attitude conflict, meaning that the values and emotions about wolves are in conflict in the society. This controversy has long historical roots originating in the transnational environmental ethos that started from Europe and extended around the world during the late 19th and early 20th centuries. Conservation laws were enacted to protect wildlife resources for the recreational purposes of the middle class and to support the growing tourism industry. Previously acceptable local hunting practices for household subsistence were turned into illegal acts (Reiger, 1975; Forsyth and Marckese, 1993). These conservation priorities were imposed in an authoritarian way that promoted urban needs and desires, causing rural discomfort and opposition and coming into conflict with common practices in rural communities (e.g., Jacoby, 2001).

Urbanization caused the polarization of the rural and urban spheres, and can be seen as the main driver of differing attitudes towards wildlife (Breitenmoser, 1998). Lifeworlds differed along with diverging living habits, anthropocentric values collided with ecocentric values and rural people's negative emotions towards harmful species collided with conservationists' emotional-driven need to protect charismatic animals. After the re-colonization of the wolf, rural people had to accommodate their livelihood practices and leisure activities to the presence of wolves in their living territories. Spatial control has led to adaptations such as building fences, introducing guard dogs, and protecting and monitoring hunting dogs (Skogen et al., 2006; Peltola and Heikkilä, 2015). In spite of the coexistence efforts, the species has caused remarkable damage to livestock and dogs, has raised the question of human safety ("biosecurity" after Buller, 2008) in the rural sphere and has made it challenging for humans to share the same space with wolves (Figari and Skogen, 2011; Liberg et al., 2012). Attitudes towards wolves become more negative the longer people coexist with them (Dressel et al., 2014).

In the meantime, animal welfarists and conservationists have integrated wolves inside the *ethical circle* (Rannikko, 2003); wolves are not treated as game species, but their protection is based on their intrinsic value, while conservation success is to be measured on moral grounds. The wild nature of wolves motivates animal welfarists and conservationists to enthusiastically act in favour of the wolf's strict protection. In highly developed areas, wolves have become an "emotional keystone species" for the restoration of indigenous wildlife (Breitenmoser, 1998, p. 279), and, quite paradoxically, wolves have turned into a symbol of urban values and hegemony, instead of representing wilderness (Skogen and Krange, 2003).

Laws and norms in wolf policy are construed to guide human behaviour, but it seems that these have turned into a type of institutional barrier that fails to address and often fuels wolf conflict (Hiedanpää and Bromley, 2011). Conservation arrangements enacted by means of international agreements such as the Habitats Directive and its mandate of strict protection for wolves have aggravated the widespread feeling of injustice (Buller, 2008) and the feeling of weakness and alienation among rural working-class residents and hunters (Mischi, 2013; von Essen et al., 2015). EU conservation policies have been criticized for preventing traditional rural ways of life and hindering the welfare of rural livelihoods (Zabel and Holm-Muller, 2008; Bisi et al., 2010). Conservationists have been empowered by EU conservation policy, whilst rural people have to confront an EU environmentalist agenda ruled by technical-ecological expertise that excludes alternative rationalities (von Essen, 2015). Dissent, resistance and even political mobilization against wolf policies have occurred (von Essen, 2016). The goals set for wolf conservation are undermined not only by a hostile public debate and anti-conservationist movements, but also by the illegal killing of wolves (Liberg et al., 2012). This chapter focuses on bringing to the fore the illegal killing of wolves in Finland, on discussing the implications of research findings for reducing the crime rate and on addressing the need for legitimizing local knowledge and emotions in wolf policies.

Methodological approaches to examining the illegal killing of large carnivores in Finland

There is a growing scientific understanding that motives for illegal hunting of wildlife go beyond gaining direct economic benefits. Instead, a considerable number of drivers for illegal hunting stem from a vast pool of empirical knowledge deriving from historically, socially and politically oppressive administration (Jacoby, 2001; von Essen et al., 2014). It was formerly widely claimed that economic drivers rule the contemporary illegal wildlife trade from Africa and Asia, but more recently, it has been proposed that the main drivers of illegal killing are to be more or less socially and culturally construed, since local communities consider colonial regimes or post-independence states to be domineering rather than acknowledging their needs (Duffy et al., 2016).

Research on the illegal killing of wildlife has mainly focused on instrumental perspectives that rely on an assumption that humans act as economically rational actors: Individuals are driven to illegal acts by seeking economic profits, and they would respond to incentives or penalties associated with the crime (von Essen et al., 2014). Unfortunately, it seems that the main rationale behind this kind of research is a strong protectionist belief anchored in natural science, which is frequently not accompanied by overall comprehensive understanding of human nature and behavior in the economic, social and political spheres (see Dressel et al., 2014; Bennett et al., 2017). Previous research has led to numerous conflict management implications that relied on disciplinary approaches, enhanced surveillance networks and campaigns against crime, as well as sharing of basic ecological information and setting ecocentric values over anthropocentric values (White et al., 2009). However, in the light of recent research, it seems that the rule of this type of "ecological" approach has largely excluded any alternative approaches, and therefore has been failing in terms of addressing effectively human-human conflict management (see Dressel et al., 2014; Bennett et al., 2017; Redpath et al., 2017).

Normative perspectives on illegal hunting brought socialization processes (Green, 1990; Curcione, 1992; Forsyth and Marckese, 1993) and rationalization of deviant behaviour to the fore (Forsyth et al., 1998; Eliason and Dodder, 1999; Eliason, 2004; Bell et al., 2007; Enticott, 2011). Illegal hunting can be defined as a rebellious act and resistance against the ruling elites (Forsyth et al., 1998), a sign of legitimacy deficit in policies (Mischi, 2013) or rural defiance of authorities (Filteau, 2012; Kahler and Gore, 2012). These previous research studies have dealt with illegal fishing and hunting in developed countries. However, no research on the normative perspective on illegal hunting of large carnivores has existed until the very recent years (see, e.g., von Essen, 2016). Precisely how we manage to explore the motivations of illegal hunting and define the core of a crime defines how effectively we can manage the conflict. In the light of previous research findings and aiming at an exploration of illegal killing of large carnivores in the Nordic setting, we have focused our research on illegal killings on the Finnish context.

In Finland, population parameters showed that 25–30% of the total wolf population was missing in 2014 because of assumed illegal hunting, and national management efforts focused on a more punitive regime such as increased penalties (Pohja-Mykrä and Kurki, 2014a). Finland is the most sparsely populated country in the EU region, and in practice, wolf packs always share their living territories with humans, as there are no remote wilderness areas in Finland outside the reindeer-herding area. Finland also has the highest percentage of hunters (a total of 6% of the total population, of which 7% are women) in the EU (Suomen riistakeskus/Finnish Wildlife Agency, 2016). The majority of hunters are members of local hunting clubs that hold hunting rights to most of the private land. Hunters and hunting clubs have tight social networks in rural communities; even when hunters live in a suburban or urban areas, they spend their leisure time hunting in the countryside as a part of a social group. Based on a core knowledge of attitude formation inside social groups and of how information circulates and is exchanged in a social network (Prell et al., 2009), we concluded that hunters have and hold inside information on illegal hunting. Illegal hunting could not have happened at such a scale unless others have known about these actions, and in many cases, unless there has been support to these illegal acts (Pohja-Mykrä and Kurki, 2014b).

In our methodological approach on examining illegal hunting, we have employed innovative data collecting and triangulation of data sources and relied on both empirical-analytical approaches and normative approaches (see Table 3.1 for s synopsis of main aims, data and main outcomes). We were open to local lifeworlds to get an informed understanding of the motives for illegal hunting and, more importantly, to reveal the social context where these illegal acts did emerge.

Profiling hunting violators

To profile hunting violators and to present descriptive statistics on targeted large carnivore species, we requested investigation records of large carnivore illegal hunting from Police Records for a six-year period, 2005–2010. To reveal the motivations behind illegal acts and to categorize these crimes, we requested the decisions in the cases on illegal hunting of large carnivores from the District Courts for the same time period. From these data sources, we could draw an understanding that illegal hunters were male, registered as hunters and members of the Finnish Wildlife Agency, which has a legal and regulatory role in game and hunting management. Illegal acts were more or less the result of organized action carried out mainly in small groups. (Pohja-Mykrä and Kurki, 2014b; Pohja-Mykrä, 2016a.)

The main motive behind illegal hunting was found to be the *disagreement with game policies*, whereas financial benefit, the need for self-protection, accidents and abetting a friend were found as motives less often. At this point, it became clear that illegal hunting in Finland reflects local defiance of wolf management, and therefore, it could be considered to be a socio-political crime. This research refocusing was necessary for differentiating illegal acts against large carnivores from traditional subsistence-driven poaching or poaching driven by economic reasons. The results also reasserted our presumption that illegal hunting would need a much more comprehensive sociological approach and framing, instead of just documenting its effect on wolf population sizes (Pohja-Mykrä, 2016a).

Table 3.1 Methodological approaches to examining the illegal killing of large carnivores in Finland

	Profiling hunting violators (Pohja-Mykrä and Kurki, 2016a)	Non-active role-playing with empathy-based fictitious stories (Pohja-Mykrä and Kurki, 2014b; Pohja-Mykrä, 2016b)	Semi-structured interviews with visual stimuli (Pohja-Mykrä, 2016a)
Main aims	• Present descriptive statistics on targeted large carnivore species • Reveal the motivations behind illegal acts and categorize these crimes	• Examine support of local community members to violators committing illegal acts • Examine neutralization techniques	• Study illegal hunting as the defiance of authorities • Get a deeper understanding of illegal hunting and how conflicts had developed
Data	• Police Records years 2005–2010 (n = 141) • Decisions in the cases on the illegal hunting of large carnivores from the District Courts years 2005–2010 (n = 30)	• Approach the core group of hunting violators, e.g., hunters and women; the core group consisted of people who had inside information about illegal acts and who may have had an effect on the actualization of illegal acts • Respondents were given a brief background story and they were asked to expand on the narratives (N = 236)	• Semi-structured interviews with convicted hunting violators (n =2) and game management officials (n = 2), along with visual stimuli
Main outcomes	• Main motive behind illegal hunting was disagreement with game policies • Illegal hunting reflecting local defiance of wolf management was to be considered to be a socio-political crime • Illegal killing of wolves is a sign of powerful non-communicative rural resistance.	• The core group of hunting violators arose from the rural context, within which there are were signs of a larger cultural protest against top-down conservation regimes at the EU level • Among hunters, it was essential to stick together on group standards, as deviation may pose a threat to their position in the group	• Alienation of hunters from society and authoritative agents, which had occurred alongside the enforcement of conservation programmes • Rationalization of an illegal act as an integral part of the actor's motivation for the act, not just an after-the-fact excuse

Non-active role-playing with empathy-based fictitious stories

The core group of hunting violators

To examine the possible support that local community members offered to violators committing illegal acts, we chose to approach the core group of hunting violators, which consisted of people who had inside information about illegal acts and who may have an effect on the actualization of illegal acts. On the one hand, we considered that other hunters were familiar with illegal acts, and they might have actively decided whether, and on what conditions, they would support illegal hunting. On the other hand, we were especially interested in whether women supported illegal hunting. We assumed that the support should be given by family members: Family cohesiveness is built upon shared attitudes, and women generally are concerned for their family and security (Verchick, 2004). In Finland, parents' associations with active participation of women have been vocal opponents of wolf conservation in response to the perceived risk faced by schoolchildren on the way to school (Hiedanpää et al., 2016). In turn, women have been shown to differ from men in terms of risk assessment and attitudes towards authorities (Verchick, 2004), and there were gender-marked differences in attitudes towards the environment (Gore and Kahler, 2012).

In collecting the data, we utilized a novel approach to take into account the ethical aspects of this sensitive issue: the method of non-active role-playing with empathy-based fictitious stories. Role-playing provides discourses, which include references to actual illegal actions. Since people are usually expected to elaborate on topics based on their own perceptions and knowledge, the method is able to deliver statements that take into account both the rhetorical and social nature of argumentation.

In practice, respondents were given a brief written story and they were asked to respond to it with a written imagined continuation. The story described a person meeting an old acquaintance in possession of an illegally killed large carnivore. It is worth noting that the story gave no information on the species in question, the hunting violator's sex, the setting, the killing method or what had led up to the situation. The story was made as short and as general as possible so as not to restrict the respondents' narratives, and it gave the respondents an opportunity to write about sensitive issues in the third person. There were two versions of the story: one in which the person sees the killed animal and reports the alleged illegal killing to the authorities, and another one in which s/he does not report the hunting violator after having noticed the killed animal. The respondents received one of the two versions of the story randomly. Then, in complete anonymity, they were requested to expand on the narrative by addressing two questions. This was the background story with the two questions that followed:

> A car drove into a yard. An old acquaintance came up, displayed a large carnivore in the trunk of the car, and said that it had been illegally killed. The

person who had been told this *said something/said nothing* to the authorities about the illegal act. Imagine the situation. Describe a) what had led up to the situation, and b) why the person told *said something/said nothing* about the illegal action to the authorities?

We had a twofold theoretical approach to the examination of community support. First, we examined the narratives with the theoretical framework of Billig's rhetorical attitude theory (1996), according to which individuals are set up in a relationship between the social sphere and their personal lifeworld, and argumentation takes place within these relationships. Argumentation, therefore, can be conceptualized as a way of using linguistic power to take a stand in a conflict situation. By analyzing argumentation, written or verbal, it is possible to make interpretations about attitudes. It also makes it possible to interpret socially and culturally accepted aspects related to the environment. From a rhetorical point of view, everyone has an ability to diversify their verbal expressions, and the perceived audience affects the way in which attitudes are expressed, as well as how people want things to be done (Billig, 1996). Rhetorical attitudes were examined using a qualitative attitude analysis method called *argumentation analysis*, which can be seen as a methodical continuation of rhetorical attitude theory based on two assumptions: Classified data can be interpreted as a sample of controversial claims about social reality, and the formation and expression of attitudes can be studied as social phenomena (Vesala and Rantanen, 2007.)

Our results showed that attitudes towards wolves were built on strong primary emotions such as fear of wolves and anger toward wolves, as well as secondary emotions such as frustration towards game management authorities. Both hunters and women gave strong support to illegal killings, and the level of support did not depend on age group or place of residence. Among hunters, it was essential to stick together on group standards, as deviation may pose a threat to their position in the group. This group pressure among hunters was something that women were lacking. Instead, women still had some faith in authorities and their ability to correct wolf policies, if informed correctly about the current local wolf pressure. The fact that support was contingent on the lack of the necessary actions by the game management authorities indicates that this lack necessitates taking matters into one's own hands and thereby granting illegal killing a permissible status (Pohja-Mykrä and Kurki, 2014b).

Neutralization techniques

We built an understanding that illegal hunting of large carnivores was carried out as an activity deliberately executed as part of belonging to a social group. Hunting violators did not see themselves as detached from society or as marginal or outlaw individuals; they are generally committed to the rules and laws of society, but they may rationalize certain exceptions. Rural protests were not expressed by hunting violators alone: Community members who supported hunting violators may also perform acts of neutralization; namely, ways of justifying the illegal acts. In the second phase, our purpose here was to show how the core group of hunting violators negated the perceived shame from the stigma and sanctions associated with violating the law, and to understand how rural communities attempted to sustain alternative ways of regulating their lifeworld, which they saw to be under pressure from wildlife conservation regimes. In accomplishing this objective, our second theoretical approach was to examine neutralization techniques from the narratives produced by the respondents. Neutralization techniques were introduced in sociological literature by Sykes and Matza (1957), and the list with these techniques has been extended since then by Coleman (1994), Klockars (1974) and Minor (1981).

The neutralization techniques employed by respondents allowed us to see rationalizations behind the hunting violations, but they also showed us how the rural identity and way of life were defended and how rural protests against conservation policies were voiced within the context of a pressure put by "Europeanization", which has occurred in recent decades (Enticott, 2011; von Essen et al., 2014). Although the illegal killing of large carnivores would be expected to invite the criminalization of hunters and their stigmatization by the rest of society, those belonging to the core group of hunting violators employed certain strategies of rationalizing and justifying such illegal actions as well as the support given to them. The results implied that the core group of hunting violators arose from the rural context, within which there were signs of a larger cultural protest against top-down conservation regimes applied at the EU level. These regimes were held to be "illegitimate" in terms of local sociocultural legitimacy. Our findings suggested that not only hunting violators but also the core group of hunting violators had non-communicative resistance power, which they had exercised prior to the implementation of the new wolf management plan as manifested in the drastic decrease in the wolf population mentioned previously (Pohja-Mykrä, 2016a, 2016b).

Interviews with convicted hunting violators and game management officials

To get a deeper understanding of illegal hunting, we interviewed convicted hunting violators and game management officials. Interviews were implemented in a semi-structured manner with visual stimuli – pictures related to large carnivores and their management, illegal killing and stakeholders. Interviewing with stimuli gave an opportunity to approach this sensitive issue in a subtle manner. These stimuli provided an atmosphere in which the interviewed person found himself/herself in the position of a third person. Therefore, they did not have to express their own opinions of the species or reveal their own actions concerning illegal killings of large carnivores, but they acted as observers of the consequences of the current policies and expressed their understanding of how conflicts had developed.

The theoretical basis for illustrating illegal hunting from the interview data as rural defiance rested on Sherman's defiance theory (1993). With critical discourse analysis, it was possible to examine the alienation of hunters from society and authoritative agents that had occurred alongside the enforcement of conservation programmes as well as to examine how interviewees found conservation procedures unfair and even stigmatizing and how they neutralized negative behaviour towards wolves and management authorities (Pohja-Mykrä, 2016a; 2016c). Rural space has been differentiated from the rest of society at large, and the rural social sphere was filled with feelings of inequality, injustice, frustration and lack of trust. Hunting violators came to defend a particular rural identity and way of life. As expected, the rationalization of an illegal act was not to be seen as just an after-the-fact excuse that someone invented to justify their own illegal behaviour, but as an integral part of the actor's motivation for the act (Pohja-Mykrä, 2016a; see Coleman, 1994).

Discussion of our findings in relation to reducing the crime rate

Illegal killing of wolves in Finland may be defined as a socio-political crime, as far as it reflects widespread dissent and rural defiance of authorities (see von Essen et al., 2014). Retaliatory killing was found to be a noteworthy rationale in justifying illegal acts: Hunting violators claimed entitlement, for instance, in the case of a loss of hunting dogs, and the core group supported illegal killings because of loss of livestock. However, these justifications primarily reflected the issue of biosecurity, the need of a rural community for a safe living-environment. A woman respondent who supported hunting violations put it this way: "If people have to live in this kind of fear, it's some relief when someone does something to remedy the situation. The authorities won't help!" The wolf-free era that occurred in Finland from the early 1900s up to the 1990s accustomed people to practices that allowed them to exercise hunting with loose dogs, to keep livestock without wolf fences and to let children freely play in yards and go to school on foot. This kind of socio-cultural accommodation that has been developed and actively constructed by humans in the absence of wolves has been challenged alongside the re-colonization of wolves around Europe.

Rural space is constructed on traditional agricultural values, and there is antagonism between these values and conservation policies. In Norway, it has been concluded that, under the pressure of the environmental ethos, rural people living in wolf territories have confronted a feeling of alienation from society at large (Skogen and Krange, 2003). Our results imply that there are signs of rural and cultural resistance against adopted EU conservation regimes. A hunter respondent argued that "Common sense has defeated all those EU bureaucrats." The acts of local hunting violators need to be seen as contingent on national and EU-level conservation policies: In the absence of formal wolf management, informal management takes place out of necessity.

Hunting violators' law-breaking was an outcome of a kind of rational cost-benefit calculation. However, when considering risks and benefits from a normative perspective, it was not economic penalties but social norms and social justice that counted when the legitimacy of law was contested. (von Essen et al., 2014; Duffy et al., 2016.) The demands of the core group, as a smaller unit of the local society reflected the needs and desires of rural society, overall. This was how a woman justified community members' silence considering hunting violations:

> *[T]he hunting violator was doing them all a favour for the sake of community peace, something that everyone else was afraid to do. . . . Besides, he was also thinking of the safety of the locals and of the farm animals.*

What can count when choosing between alternative ways of acting is not only the shared lifeworlds and empirical knowledge of wolves and the shared negative emotions such as frustration towards management authorities, but also a need to support long-held rural traditions and hunting customs and to actively reflect on management policies. It has to be noted, too, that deviating from shared group attitudes might pose a risk of being socially stigmatized by in-group members or excluded from the group, and therefore, the coherence of the local community might be questioned. In a hunter's words: "Those wolves breed like rabbits, but good friends are damn hard to find."

The essence of the rural protest can be captured with the neutralization of illegal acts that are used to qualify hunting violations as acceptable. Hunting violators were considered *good poachers* and they are doing an act in favour of the entire community: Illegal killings of wolves are, indeed, recognized as *acts of justice*. This support was manifested in the form of silent endorsement and approval of the act, through verbal sparring or even encouragement, but also as a common agreement to which the core group is committed. This is how a female respondent presented this disposition: "Because there was silent agreement among the villagers. And the person went along with the agreement. If he had told the authorities about the illegal killing, he would have failed his own community." Hunting violators were affected and bound by moral and social sanctions in their community. Thus, socio-political crime derived its legitimacy not only from the individual who had committed the illegal act, but from the entire local community, which endorsed the act and was ready to support the violator. Social norms aligned the behavior of in-group members with this code of conduct, and therefore, any deviation from the law should not be treated as unjust. This informal code of conduct contested formal regulations and the law (von Essen, 2016). Interestingly, local authorities such as game wardens were also caught between in this social reality, understanding the local setting that may lead to illegal actions, on the one hand, and acknowledging the demands of their position, on the other.

It is interesting to note that research in the context of international illegal wildlife trade has largely rested on enforcement activities based on the assumption that people act economically rationally. More recently, this approach has been contested, and it has been suggested that a more normative approach would reveal the real reasons behind illegal wildlife hunting (Duffy et al., 2016). From the normative point of view, it was also easier to comprehend why crimes committed in rural defiance could not be countered by more severe punishments. Instead, a lack of legitimacy in policies may result in a legitimization of the authorities and possibly an increase in the crime rate (Sherman, 1993; Wellsmith, 2011; Filteau, 2012).Of course, society can direct more efforts to disciplinary approaches, and that has happened also in Finland. The latest wolf population management plan from 2015 states that illegal hunting will be tackled more seriously and supports the idea of increased patrols (MAF, 2015). Certainly, with increased surveillance, there is an opportunity to detect more hunting violations and punish them more harshly. However, these increased punishments do not reduce the support local community members offer to hunting violators. Instead, it is more likely that this support will increase, making local communities cherish hunting violators as *Robin Hoods* in their communities, a development which in turn deepens the divide between the rural sphere and management authorities (see also von Essen et al., 2014).

Dimensions of "poverty"

Since disciplinary approaches are unhelpful and inadequate when formulating effective disincentives to illegal killing and when attempting to increase compliance with conservation regimes, other ways have to be sought to affect the crime rate. To meet this challenge, a start may be to link poverty to wolf management and to show how traditional hunting of wolves in Finland may respond to the poverty challenge. Such a perspective would verify, at the same time, commonly approved management tools.

To decrease the crime rate in international illegal wildlife trade, Duffy et al. (2016) addressed Professor Sen's thoughtful definition of poverty in the context of illegal hunting in the developing countries, which involved "a lack of power, prestige, voice and an inability to define one's future and day-to-day activities" (p. 16). Rural residents may be "poor" in this sense. In the past several decades, rural people have had to accommodate their life to the presence of wolves and to meet the demands set in EU policies, their agricultural values have been disputed, their empirical knowledge of species has been neglected, they have lost power in relation to wolf management, they have been stigmatized as irresponsible in relation to wildlife and nature, they have had to protect their livelihoods, they have been facing constant dog and livestock losses and their sense of a secure living environment has been shaken. Management policies that aim at reversing these dimensions of "poverty" in rural communities may provide positive policy responses and increase compliance with conservation regimes.

Hunters are key actors in executing the illegal killing of wolves. Hunting holds a strong social connotation and supports hunters' stewardship of wildlife. Furthermore, hunting can be defined as an integral part of rural communities' cultural identity. Especially the hunting of large prey species, such as moose and brown bear, includes long cultural traditions and builds hunters' identity as big-game hunters (Kruuk, 2002; Pohja-Mykrä et al., 2015; Watts et al., 2017). Wolf hunting in Finland also has long historical roots, and hunters have been protectors of rural life and tradition, human safety and livestock (Pohja-Mykrä et al., 2005; Mykrä et al., 2017). In regard to historical, social and cultural aspects and legitimate agency in relation to wolves, hunters may be defined as *primordial stakeholders*. Hunters, along with other locals who share their everyday territory with wolves, have spatial proximity to wolves. In addition to proximity, hunters have a legitimate role to play in wolf management, they may easily feel the urgency of management actions and they may enact actual power – either legal or illegal – over wolf populations (Mykrä et al., 2017).

Traditional wolf hunting

The first attempts at supporting this ancient role of hunters have been made by introducing hunting derogations based on wolf population management in Finland in 2015–2016. This initiative acknowledged hunting of wolves, practiced within the local tradition (not culling), as a valid and eligible means of wolf population management as is the case with other game species worldwide. Traditional wolf hunting leans toward responsive and deliberative governance of wildlife and natural resources management and conservation, with the aim of empowering locals in management and policy, especially, by bringing to the fore their sense of wolf management and how it is to be implemented. Dressel et al. (2014) considered that permission of hunting may also have an effect on attitudes, since the ability to hunt brown bears may be one reason bears were regarded more positively than wolves all over Europe.

Conservation of wolves via traditional wolf hunting acknowledges the importance of local knowledge along with ecological-technological expertise (Skogen and Krange, 2003; von Essen, 2015), respects hunters' stewardship and management role (Kaltenborn et al., 2013) and addresses the importance of building trust and legitimacy between stakeholders (Cvetkovich and Winter, 2003). Thus, it may support voluntary compliance with conservation regimes (Stern, 2008). Increased levels of trust may have a positive effect on reducing the fear of wolves (Johansson et al., 2012). In addition, wolf hunting may fulfill a need for interaction with the living environment and "effectance" on objects that belong in it, and therefore, it may also strengthen the hunters' psychological ownership feelings towards wolves. Feelings of ownership may have positive feedback on behaviour, since such feelings might be accompanied by stewardship towards wolves, if wolves are seen as a valuable quarry hunters may promote (see Pierce et al., 2003; Pohja-Mykrä et al., 2015).

Wolf hunting leans towards positive incentives, as traditional hunting involves approximately 50 men per hunt, and hunters are allowed to keep the hunting bag. Strong social connotations, positive emotions and trophies attached to actions may also strengthen hunters' psychological ownership feelings towards wolves (see Pierce et al., 2003; Pohja-Mykrä et al., 2015). The first signs of revived traditional wolf-hunting practices may be already observable, such as hunting with hounds and skiing after prey. Very recent results show that legal hunting seems to decrease illegal killing, at least in the short run (Suutarinen and Kojol, 2017). Keeping in mind the support that hunting violators may receive from their core group, the introduction of traditional wolf hunting makes hunting violators irrelevant to community welfare. This raises the opportunity to utilize voluntary compliance through informal sanctions based on collective moral judgments and the perceived legitimacy of rules (see Gezelius, 2002; Keane et al., 2008). When gaining hunting permits for legal traditional hunting is contingent on a reduced illegal hunting rate, there is a possibility of transforming the core groups' perception of the hunting violator from beneficiary to divergent.

To summarize this section, wolf as a species challenges the well-being of those humans who share their living territories with wolves. Based on the results of earlier attitude research, it seems almost impossible for a strong wolf population to co-exist with humans without violating some basic human needs for a safe living environment. The failure of formal wildlife management initiatives here leads to the inevitable; namely, informal wolf management takes place. By supporting primordial stakeholders' desires and needs, there is a possibility to conduct more socially just and socio-culturally legitimate wolf management, by means of which the tolerance for wolves may improve (see also Bruskotter and Fulton, 2012). Transparent and reflective formal management with legal hunting may act in favor of maintaining sustainable wolf populations, not against it.

Implications for wolf policy and management

Legitimation of traditional hunting practices

Conservation procedures that affect locals' well-being and their accustomed living habits may face legitimacy deficits and should, therefore, take note of local social and cultural traditions. There is some appreciation of cultural self-preservation and expression when allowing the indigenous people to continue with their traditional hunting practices (Nurse, 2016). There is also a growing recognition of the need for carrying out such management implications in the developing countries that support traditional subsistence hunting practices, and research strongly relies on a presumption of the link between poverty and motives for illegal hunting (Duffy et al., 2016). Previous research efforts, however, tend to dismiss the primary reasons for illegal hunting and the complexity of involved networks and cultural and social traditions (Duffy et al., 2016). In European countries and at the EU level, the debate has more or less downplayed long-lasting hunting traditions in rural communities and has been focused on limiting the right for sports hunting. However, sport and trophy hunting practices faced strong critique based on severe ecological, ethical, economic, social and even political concerns. Indeed, the European Court of Human Rights has determined that hunting does not represent a particular lifestyle considered to be essential for one's personal identity (Nurse, 2016). However, the relevant literature involves examples of local hunting traditions being understated with the rise of urbanization and centralized urban-driven conservation procedures. Such a transformation in wildlife management and conservation has turned previously acceptable hunting practices into illegal acts (Okihiro, 1997; Jacoby, 2001; Holmes, 2007).

The Finnish context is especially interesting when discussing hunting as part of cultural heritage. Traditionally, it has been widely possible for all people in Finland to make use of natural resources, regardless of the landowner. Nowadays this is expressed through "Everyman's Rights" (the right to roam), which guarantee free public access to both private and public forests and which include some rights to use them, even for commercial purposes (Matilainen et al., 2017). Wildlife is considered *res nullius*, "no-one's property", and the outtake of game and fish has been based on a fair share for everyone since time immemorial (de Klemm, 1996; see also von Essen et al., 2017). Local hunting clubs lease their hunting grounds basically free of charge from local landowners, who in turn are usually members of the local hunting clubs. There is also a possibility for anyone outside the hunting clubs to lease hunting premises from the state land: The aim is to provide equal hunting opportunities to anyone interested (Watts et al., 2017). These cultural and social norms may reflect the need for cultural continuity with hunting practices and rights regarding game such as large carnivores even today. Examining the adaptability and resistance to wolf policies brings the legitimacy of local people's and stakeholders' empirical and cultural knowledge about wolves to the fore and calls for new platforms for justifiable inclusion of emotions and local empirical knowledge.

Local knowledge in a clash of discourses

For more than three decades, there has been an understanding that traditional, indigenous or local ecological knowledge should be taken into account when implementing natural resource use or conservation, in order to empower locals and respect the social and economic sustainability of natural-resource-based livelihoods (Davis and Wagner, 2003). However, what has not been sufficiently addressed is that it is not only indigenous people who hold strong empirical knowledge of their living environment – so does everyone, and all have a grave respect for their experienced lifeworlds, as well. It can be argued that wolf conservation has suffered so far from an inadequate understanding of legitimacy deficits in social justice (Borgström, 2012; see also Bennett et al., 2017; Redpath et al., 2017).

Local knowledge may not be an unambiguous term in itself, but the positioning of rural communities in a defensive stance against urban interest appears to be a unifying characteristic for the various dimensions behind it (Skogen and Krange, 2003). The first step to respect local knowledge is to comprehend how locals ground their expertise about wolves and their habits by appealing to traditions, personal experience, established social actions and agrarian values. These lifeworld discourses may easily become delegitimized by authorities as emotionally driven, irrational and ill-informed (see Buijs et al., 2014). In contrast, ecological discourse and those appealing to the law, authorities, ecological knowledge or ecocentric values are often seen as respected actors, and their message as neutral and not carrying any political implications (see von Essen, 2016).

Neglecting local knowledge may lead to counteractive behavior. Von Essen (2015) launched the term "barstool biology" to embody the way in which Swedish hunters increasingly adopt a technical-ecological discourse of scientific experts when disputing wolf policies. Hunters believe that they are at a disadvantage when using lifeworld arguments, but at the same time hunters, attempt to narrow and downgrade their own traditional discourse (von Essen, 2015). This kind of adoption of ecological discourse, including an endorsement of *biodiversity conservation* and *favourable conservation status* of species, can be characterized as publicly accepted *flagship argumentation*, and it aims at making a particular opinion acceptable to the target audience. For example, by adopting *hybrid wolves* when discussing about wolf-dog hybrids, people want to express their knowledge as wolf experts in relation to scientists and authorities. In turn, when hunters and local people want to speak to their own people, they contest dominant ecological discourse and may use traditional, emotion-based empirical and cultural argumentation. For example, when debating the genetic purity of free-ranging wolves, a wolf is stigmatized as a "porridge-eating dog" (Pohja-Mykrä, 2016a) or a "tainted immigrant from the east" (von Essen, 2015), instead of being spoken about as a hybrid wolf. Hunters may also resist ecological discourse in arguing that wolves should be seen as ordinary game species instead of ethically untouchable species (von Essen, 2016).

To respect local knowledge, the spatial proximity of individuals to the wolf issue is something that needs to be carefully addressed in planning and implementing wolf policies. The first steps towards this direction have been already taken in Finland since the implementation of the revised wolf management plan, which has introduced wolf-territory-based stakeholder group gatherings that will have an influence on proximate wolf packs (MAF, 2015). Prior to the previously mentioned management plan, a nationwide wolf management forum was set up on the internet. This informal e-participation supported deliberation of wolf policies by bringing local views and opinions into an interactive knowledge sharing platform. The participants of this e-forum provided also practical suggestions on how local-level decision making related to wolves should be conducted (Salo et al., 2017).

Local knowledge is shared in social networks comprised of actors who are tied to one another through socially meaningful relations. Central actors in those social networks not only pass on and circulate information, but may also produce new information (Prell et al., 2009). This new, informal information may be disseminated and dispute other formal types of information and knowledge. Bearing that in mind, one should further cultivate any platforms in which informal information can be shared. In Finland, for example, "local parliaments" that gather traditionally in local gas stations around a cup of coffee may be sources of local knowledge. Such "parliaments" are set by groups of local actors who informally gather together regularly, for instance weekly, to discuss important and locally "hot" matters. The discussion of the wolf issue has been obviously one of the main topics during the past decade. Capturing the informal discourse might shed light on social justice and legal implications for wolf policy and management.

The concept of psychological ownership

Wolf policies and management can easily fail to take into account adequately the emotional responses and dispositions that affect opinions, attitudes and behaviour (see Jacobs, 2012; Buijs and Lawrence, 2013; Hiedanpää et al., 2016). However, emotions motivate people to think explicitly about an issue, and act. Strong emotions towards animals, natural resources, places and place-based processes reflect basic human needs, build human self-identity, and thus engage local actors in a long-term interest in environmental processes (Buijs and Lawrence, 2013).

Humans have emotional bonds with places and place-based processes, and these emotions may have an effect on how people adopt specific large carnivore governance procedures. In previous research studying the emotions and meanings related to natural or wilderness places, concepts applied have included "place attachment" (e.g., Brehm et al., 2013), "place meanings" (e.g., Smith et al., 2011) and "sense of place" (e.g., Jorgensen and Stedman, 2001). The related concept of "psychological ownership" has recently been introduced to natural resource management and wolf management (Pohja-Mykrä et al., 2015; Matilainen et al., 2017). The core feeling of psychological ownership is possession, a state in which individuals perceive the target of ownership as "theirs". This possession can be directed towards objects such as animals, whilst it fulfills generic and socially generated motives and basic human needs of self-identity, efficacy and effectance in relation to the living environment and having a place in which to dwell (Pierce et al., 2003). These motives serve associated functions such as stimulating security, comfort and pleasure (Pierce et al., 2003), all of which are important in ensuring a good quality of life for local people.

Allowing hunters and rural communities to have psychological ownership feelings towards wolves, and supporting these feelings, reduces their "poverty" in terms of lack of power, prestige and voice. The routes leading to the experience of psychological ownership are the power of control over wolves, access to any knowledge related to wolves, and the possibility to invest time and effort in wolf management. Acknowledging, for example, the role of traditional wolf hunting in strengthening these routes may lead to positive behaviors towards wolves, such as a sense of responsibility and a willingness to support the well-being of the species (Pohja-Mykrä et al., 2015). Although local ownership of wildlife may facilitate responsibility towards it (Measham and Lumbasi, 2013), the ownership of *res nullius* property is contested in the post-modern world (von Essen et al., 2017). Psychological ownership, in general, can be seen as a facilitator of sustainable behaviors towards the environment (Suessenbach and Kamleitner, 2017) and has also been noted in biological reviews on more socially just large carnivore policies (Redpath et al., 2017).

It has to be noted that the feeling of psychological ownership towards wolves in itself cannot guarantee any solution to human-human conflicts lying underneath human-wolf conflicts. Instead, it could be that stakeholders holding strong feelings of possession and safeguarding their psychological ownership may execute negative behavior such as resistance to change, disputes over wolf management and all these aspects may impede co-operation (Pohja-Mykrä et al., 2017). Psychological ownership will not resolve wolf conflict in itself, but it can be used to anticipate and manage conflict in a more socially just way (see Matilainen et al., 2017). In addition, the concept of psychological ownership will allow researchers and management authorities to gain a better understanding of emotional responses in wolf issues, as well as put weight on emotional aspects to develop more successful conservation regimes.

References

Bell, S., Hampshire, K., and Topalidou, S. (2007) 'The political culture of poaching: A case study from northern Greece', *Biodiversity and Conservation*, vol 16, pp399–418.

Bennett, N. J., Roth, R., Klain, S. C., Chan, K. M. A., Clark, D. A., Cullman, G., Epstein, G., Nelson, M. P., Stedman, R., Teel, T. L., Thomas, R. E. W., Wyborn, C., Curran, D., Greenberg, A., Sandlos, J., and Veríssimo, D. (2017) 'Mainstreaming the social sciences in conservation', *Conservation Biology*, vol 31, pp56–66.

Billig, M. (1996) *Arguing and Thinking: A Rhetorical Approach to Social Psychology, Revised Edition*, Cambridge University Press, Cambridge.

Bisi, J., Liukkonen, T., Mykrä, S., Pohja-Mykrä, M., and Kurki, S. (2010) 'The good bad wolf – Wolf evaluation reveals the roots of the Finnish wolf conflict', *European Journal of Wildlife Research*, vol 56, no 5, pp771–779.

Borgström, S. (2012) 'Legitimacy issues in Finnish wolf conservation', *Journal of Environmental Law*, pp1–26, doi:10.1093/jel/eqs015.

Brehm, J. M., Eisenhauer, B. W., and Stedman, R. C. (2013) 'Environmental concern: Examining the role of place meaning and place attachment', *Society & Natural Resources*, vol 26, no 5, pp522–538.

Breitenmoser, U. (1998) 'Large predators in the Alps: The fall and rise of man's competitors', *Biological Conservation*, vol 83, pp279–289.

Bruskotter, J. T., and Fulton, D. C. (2012) 'Will hunters steward wolves? A comment on Treves and Martin', *Society & Natural Resources*, vol 25, pp97–102.

Buijs, A., and Lawrence, A. (2013) 'Emotional conflicts in rational forestry: Towards a research agenda for understanding emotions in environmental conflicts', *Forest Policy and Economics*, vol 33, pp104–111.

Buijs, A., Mattijssen, T., and Arts, B. (2014) '"The man, the administration and the counter-discourse": An analysis of the sudden turn in Dutch nature conservation policy', *Land Use Policy*, vol 38, pp676–684.

Buller, H. (2008) 'Safe from the wolf: biosecurity, biodiversity, and competing philosophies of nature', *Environment and Planning*, vol 40, pp1583–1597.

Coleman, J.W. (1994) *The Criminal Elite: The Sociology of White Collar Crime*, St. Martin's Press, New York.

Curcione, N. (1992) 'Deviance as delight: Party-boat poaching in southern California', *Deviant Behavior*, vol 13, pp33–57.

Cvetkovich, G., and Winter, P. L. (2003) 'Trust and social representations of the management of threatened and endangered species', *Environment and Behavior*, vol 35, no 2, pp286–307.

Davis, A., and Wagner, J. R. (2003) 'Who knows? On the importance of identifying "Experts" when researching local ecological knowledge', *Human Ecology*, vol 31, no 3, pp463–489.

de Klemm, C. (1996) 'Compensation for damage caused by wild animals', *Natural Environment*, No 84, Council of Europe, Strasbourg.

Dressel, S., Sandström, C., and Ericsson, G. (2014) 'A meta-analysis of studies on attitudes toward bears and wolves across Europe 1976–2012', *Conservation Biology*, vol 29, no 2, pp565–574.

Duffy, R., St John, F. A. V., Büscher, B., and Brockington, D. (2016) 'Toward a new understanding of the links between poverty and illegal wildlife hunting', *Conservation Biology*, vol 30, no 1, pp14–22.

Eliason, S. L. (2004) 'Accounts of wildlife law violators: Motivations and rationalizations', *Human Dimensions of Wildlife*, vol 9, no 2, pp119–131.

Eliason, S. L., and Dodder, R. (1999) 'Techniques of neutralization used by deer poachers in the western U.S.: A research note', *Journal of Deviant Behavior*, vol 20, pp233–252.

Enticott, G. (2011) 'Techniques of neutralising wildlife crime in rural England and Wales', *Journal of Rural Studies*, vol 27, pp200–208.

Figari, H., and Skogen, K. (2011) 'Social representations of the wolf', *Acta Sociologica*, vol 54, no 4, pp317–332.

Filteau, M. (2012) 'Deterring defiance: "Don't give a poacher a reason to poach"', *Journal of Rural Criminology*, vol 1, pp236–255.

Forsyth, C., and Marckese, T. A. (1993) 'Folk outlaws: Vocabularies of motives', *International Review of Modern Sociology*, vol 23, pp17–31.

Forsyth, C. J., Gramling, R., and Wooddell, G. (1998) 'The game of poaching: Folk crimes in Southwest California', *Society & Natural Resources*, vol 11, no 1, pp25–38.

Gezelius, S. (2002) 'Do norms count? State regulation and compliance in a Norwegian fishing community', *Acta Sociologica*, vol 45, pp305–314.

Gore, M. L., and Kahler, J. S. (2012) 'Gendered risk perceptions associated with human – Wildlife conflict: Implications for participatory conservation', *PloS One*, vol 7, no 3.

Green, G. S. (1990) 'Resurrecting polygraph validation of self-reported crime data: A note on research method and ethics using the deer poacher', *Deviant Behavior*, 11, pp131–137.

Hiedanpää, J., and Bromley, D. W. (2011) 'The harmonization game: Reason and rules in European biodiversity policy', *Environment Policy and Governance*, 21, pp99–111.

Hiedanpää, J., Pellikka, J., and Ojalammi, S. (2016) 'Meet the parents: Emotional regime and the reception of the grey wolf return in Southwestern Finland', *Trace – Finnish Journal for Human-Animal Studies*, vol 2, pp4–27.

Holmes, G. (2007) 'Protection, politics and protest: Understanding resistance to conservation', *Conservation & Society*, vol 5, pp184–201.

Jacobs, M. H. (2012) 'Human emotions toward wildlife', *Human Dimensions of Wildlife*, vol 17, pp1–3.

Jacoby, K. (2001) *Crimes Against Nature Squatters, Poachers, Thieves and the Hidden History of American Conservation*, University of California Press, Berkeley.

Johansson, M., Karlsson, J., Pedersen, E., and Flykt, A. (2012) 'Factors governing human fear of brown bear and wolf', *Human Dimensions of Wildlife*, vol 17, pp68–74.

Jorgensen, B. S., and Stedman, R. C. (2001) 'Sense of place as an attitude: Lakeshore owners attitudes toward their properties', *Journal of Environmental Psychology*, vol 21, no 3, pp233–248.

Kahler, J., and Gore, M. (2012) 'Beyond the cooking pot and pocket book: Factors influencing noncompliance with wildlife poaching rules', *Comparative and Applied Criminal Justice*, vol 35, pp1–18.

Kaltenborn, B. P., Andersen, O., and Linnell, J. D. C. (2013) 'Predators, stewards, or sportsmen – How do Norwegian hunters perceive their role in carnivore management?', *International Journal of Biodiversity Science*, vol 9, no 3, pp239–248.

Keane, A., Jones, J. P. G., Edwards-Jones, G., and Milner-Gulland, E. J. (2008) 'The sleeping policeman: Understanding issues of enforcement and compliance in conservation', *Animal Conservation*, vol 11, no 2, pp75–82.

Klockars, C. B. (1974) *The Professional Fence*, Free Press, New York.

Kruuk, H. (2002) *Hunters and Hunted: Relationships Between Carnivores and People*, Cambridge University Press, Cambridge.

Liberg, O., Chapron, G., Wabakken, P., Pedersen, H. C., Hobbs, N. T., and Sand, H. (2012) 'Shoot, shovel and shut up: Cryptic poaching slows restoration of a large carnivore in Europe', *Proceedings of the Royal Society Series B*, vol 27, pp9910–9915.

MAF (2015) *Management Plan for Wolf Population in Finland*, Ministry of Agriculture and Forestry, Finland.

Matilainen, A., Pohja-Mykrä, M., Lähdesmäki, M., and Kurki, S. (2017) 'I feel it is mine! – Psychological ownership in relation to natural resources', *Journal of Environmental Psychology*, vol 51, pp31–45.

Measham, T. G., and Lumbasi, J. A. (2013) 'Success factors for community-based natural resource management (CBNRM): Lessons from Kenya and Australia', *Environmental Management*, vol 52, no 3, pp649–659.

Minor, W. W. (1981) 'Techniques of neutralization: A reconceptualization and empirical examination', *Journal of Research in Crime and Delinquency*, vol 18, pp295–318.

Mischi, J. (2013) 'Contested rural activities: Class, politics, and shooting in the French countryside', *Ethnography*, vol 14, pp64–84.

Mykrä, S., Pohja-Mykrä, M., and Vuorisalo, T. (2017) 'Hunters' attitudes matter: Diverging bear and wolf population trajectories in Finland in the late 19th century and today', *European Journal of Wildlife Research*, vol 63, p76, doi:10.1007/s10344-017-1134-1.

Nurse, A. (2016) 'Criminalising the right to hunt: European law perspectives on anti-hunting legislation', *Crime Law and Social Change*, vol 67, pp383–399.

Okihiro, N. R. (1997) *Mounties, Moose and Moonshine. The Patterns and Context of Outport Crime*, University of Toronto Press, Toronto, ON.

Peltola, T., and Heikkilä, J. (2015) 'Response-ability in wolf – Dog conflicts', *European Journal of Wildlife Research*, vol 61, no 5, pp711–721.

Pierce, J. L., Kostova, T., and Dirks, K. T. (2003) 'The state of psychological ownership: Integrating and extending a century of research', *Review of General Psychology*, vol 7, no 1, pp84–107.

Pohja-Mykrä, M. (2016a) 'Felony or act of justice? – Illegal killing of large carnivores as defiance of authorities', *Journal of Rural Studies*, vol 44, pp46–54.

Pohja-Mykrä, M. (2016b) 'Community power over conservation regimes: Techniques of neutralizing illegal killing of large carnivores in Finland', *Crime Law and Social Change*, vol 67, pp439–460.

Pohja-Mykrä, M. (2016c) 'Illegal killing as rural defiance', in J Donnermeyer (ed) *The Routledge International Handbook of Rural Criminology*, Routledge, London.

Pohja-Mykrä, M., and Kurki, S. (2014a) 'Evaluation of the Finnish national policy on large carnivores', *Reports 135*. Ruralia Institute, University of Helsinki.

Pohja-Mykrä, M., and Kurki, S. (2014b) 'Strong community support for illegal killings challenges wolf management', *European Journal for Wildlife Research*, vol 60, no 5, pp759–770.

Pohja-Mykrä, M., Kurki, S., and Mykrä, S. (2015) 'Susipolitiikan suunnanmuutos – ohjailusta omistajuuteen', in J. Hiedanpää and O. Ratamäki (eds) *Suden kanssa*, University of Lapland Printing Centre, Rovaniemi.

Pohja-Mykrä, M., Mykrä, S., Matilainen, A., and Lähdesmäki, M. (2017) 'The emergence of collective psychological ownership in stakeholder large carnivore texts and its effect on conservation conflict escalation' [manuscript].

Pohja-Mykrä, M., Vuorisalo, T., and Mykrä, S. (2005) 'Hunting bounties as a key measure for historical wildlife management and game conservation: Finnish bounty schemes in 1647–1975', *Oryx*, vol 39, no 3, pp284–291.

Prell, C., Hubacek, K., and Reed, M. (2009) 'Stakeholder analysis and social network analysis in natural resource management', *Society and Natural Resources*, vol 22, no 6, pp501–518.

Rannikko, P. (2003) 'Oikeudenmukaisuuskysymys suomalaisen ympäristöliikehdinnän aalloissa', in A. Lehtinen and P. Rannikko (eds) *Oikeudenmukaisuus ja ympäristö*, University Press Finland Ltd, Tampere.

Redpath, S., Linnell, J. D. C., Festa-Bianchet, M., Boitani, L., Bunnefeld, N., Dickman, A., Gutiérrez, R. J., Irvine, R. J., Johansson, M., Majić, A., McMahon, B. J., Pooley, S., Sandström, C., Sjölander-Lindqvist, A., Ketil Skogen, K., Swenson, J. E., Trouwborst, A., Young, J., and Milner-Gulland, E. J. (2017) 'Don't forget to look down – Collaborative approaches to predator conservation', *Biological Reviews*, vol 92, pp2157–2163.

Reiger, J. F. (1975) *American Sportsmen and the Origins of Conservation*, Winchester Press, New York.

Salo, M., Hiedanpää, J., Luoma, M., and Pellikka, J. (2017) 'Nudging the Impasse? Lessons from the Nationwide Online Wolf Management Forum in Finland', *Society & Natural Resources*, vol 30, pp1141–1157.

Sherman, L. (1993) 'Defiance, deterrence, and irrelevance: A theory of the criminal sanction', *Journal of Research in Crime and Delinquency*, vol 30, pp445–473.

Skogen, K., and Krange, O. (2003) 'A wolf at the gate: The anti-carnivore alliance and the symbolic construction of community', *Sociological Ruralis*, vol 43, pp309–325.

Skogen, K., Mauz, I., and Krange, O. (2006) 'Wolves and ecopower: A French-Norwegian analysis of the narratives of the return of large carnivores', *Journal of Alpine Research*, vol 94, pp78–87.

Smith, J. W., Davenport, M. A., Anderson, D. H., and Leahy, J. E. (2011) 'Place meanings and desired management outcomes', *Landscape and Urban Planning*, vol 101, no 4, pp359–370.

Stern, M. J. (2008) 'Coercion, voluntary compliance and protest: The role of trust and legitimacy in combating local opposition to protected areas', *Environmental Conservation*, vol 35, pp200–210.

Suessenbach, S., and Kamleitner, B. (2017) 'Psychological ownership as a facilitator of sustainable behaviors', in J. Peck and S. Shu (eds) *Psychological Ownership* (tbd). Springer, Dordrecht.

Suomen riistakeskus/Finnish Wildlife Agency (2016) 'Metsästäjien määrä pysynyt muuttumattomana', https://riista.fi/metsastajien-maara-pysynyt-muuttumattomana/, accessed 4 December 2017.

Suutarinen, J., and Kojol, I. (2017) 'Poaching regulates the legally hunted wolf population in Finland', *Biological Conservation*, vol 215, pp11–18.

Sykes, G. M., and Matza, D. (1957) 'Techniques of neutralisation: A theory of delinquency', *American Sociological Review*, vol 22, no 6, pp664–670.

Verchick, R. M. (2004) 'Feminist theory and environmental justice', in R. Stein (ed) *New Perspectives on Environmental Justice: Gender, Sexuality and Activism*, The State University of New Jersey, Rutgers.

Vesala, K. M., and Rantanen, T. (2007) 'Laadullinen asennetutkimus: lähtökohtia, periaatteita, mahdollisuuksia', in Vesala K and T. Rantanen (eds) *Argumentaatio ja tulkinta: laadullisen asennetutkimuksen lähestymistapa*, Gaudeamus, Helsinki.

von Essen, E. (2015) 'Whose discourse is it anyway? Understanding resistance through the rise of "barstool biology" in nature conservation', *Environmental Communication*, vol 11, pp470–489.

von Essen, E. (2016) *In the Gap between Legality and Legitimacy: Illegal Hunting in Sweden as a Crime of Dissent*, PhD thesis, University of Uppsala, Sweden.

von Essen, E., Allen, M., and Hansen, H. P. (2017) 'Hunters, crown, nobles, and conservation elites: Class antagonism over the ownership of common fauna', *International Journal of Cultural Property*, vol 24, no 2, pp161–186.

von Essen, E., Hansen, H. P., Nordström Källström, H., Peterson, M. N., and Peterson, T. R. (2014) 'Deconstructing the poaching phenomenon – A review of typologies for understanding illegal hunting', *British Journal of Criminology*, vol 54, no 4, pp632–651.

von Essen, E., Hansen, H. P., Nordström Källström, H., Peterson, M. N., and Peterson, T. R. (2015) 'The radicalisation of rural resistance: How hunting counterpublics in the Nordic countries contribute to illegal hunting', *Journal of Rural Studies*, vol 39, pp199–209.

Watts, D., Matilainen, A., Kurki, S., Keskinarkaus, S., and Hunter, C. (2017) 'Hunting cultures and the 'northern periphery': Exploring their relationship in Scotland and Finland', *Journal of Rural Studies*, vol 54, pp255–265.

Wellsmith, M. (2011) 'Wildlife crime: The problems of enforcement', *European Journal on Criminal Policy and Research*, vol17, pp125–148.

White, R. M., Fischer, A., Marshall, K., . . . and Van der Wal, R. (2009) 'Developing an integrated conceptual framework to understand biodiversity conflicts', *Land Use Policy*, vol 26, no 2, pp242–253.

Zabel, A., and Holm-Muller, K. (2008) 'Conservation performance payments for carnivore conservation in Sweden', *Conservation Biology*, vol 22, pp247–251.

4 Situated, reflexive research in practice

Applying feminist methodology to a study of human-bear conflict

Catherine Jampel

Introduction

From November 2009 through November 2011, over 100 incidents of Andean bears (*Tremarctos ornatus*) predating on cattle were recorded in the provinces of Imbabura and Carchi in the northern Ecuadorian Andes (Laguna, 2011). Biologists feared that without some form of intervention, the level of cattle predation would continue or even increase, possibly leading to retribution poaching and resulting in "an unknown number of illegal bear deaths" (Castellanos et al., 2011, p. 17). This study in the rural parishes of Pimampiro, Ecuador began with the cattle, which mediated the most salient contemporary relationship between humans and bears in the region, and posed the question: How has raising cattle become a form of livelihood in the northern Ecuadorian Andes, and how have bears come to represent a threat in this landscape?

The majority of the study findings corresponded with existing literature from elsewhere in the Andes of Ecuador, Bolivia, and Peru (Jampel, 2016, pp. 87–89). That is, varied socio-ecological, socioeconomic, and political factors contributed to a shift away from crop-based livelihoods and to the growth of cattle-based livelihoods. Push factors that decreased residents' commitments to crop-based livelihoods included soil fertility loss, increasingly variable and unpredictable climate, rising agricultural input prices, and out-migration from rural areas, which reduced the available labor force for such labor-intensive work. Pull factors stimulating residents' adoption of cattle-based livelihoods included the relative ease of raising cattle and that they self-reproduce, the increased market demand for dairy products, and small livestock-oriented development programs. As one resident interviewed said, whereas "in planting, you invest 80% of what you will earn, with milk, almost everything is profit." Bears became a threat in the region, specifically in the context of the growth of cattle-based livelihoods on the agricultural frontier and furthermore as government protection improved for Andean bears and their habitat (Castellanos et al., 2010).

The study's chief addition to the existing literature was to highlight the role of attachment to rural place as a key part of the story for the smallholding farmers raising cattle, through which they entered into new relationships with

Andean bears. Whereas many people in the northern Ecuadorian Andes have participated in the Latin America-wide trend of leaving rural areas, some are "stayers," or non-migrants (Kay, 2008; Huijsmans, 2014). In the study area, those remaining in the countryside following five decades of rural outmigration expressed ideals such as autonomy and tranquility as reasons for staying. They turned to raising cattle in order to participate in the steady dairy economy and as a form of wealth generation. In a work of interpretive social science, I used the idea of "taskscape" (Ingold, 1993) to tell the story of cattle-based livelihoods and their consequences for relationships between residents and bears (Jampel, 2016). The present contribution complements that product, "Cattle-based live-lihoods, changes in the taskscape, and human-bear conflict in the Ecuadorian Andes," published in the geography journal *Geoforum*, and elaborates on the research context, design, and methodology – that is, the conditions of knowledge production.[1] Some readers may wish to read the description of the process in this chapter side by side with the product in *Geoforum*. Table 4.1 summarizes the study questions and findings.

The following material primarily focuses on the process from 2011–2015, but also accounts for newer work relevant to qualitative social science research in conservation research (Moon et al., 2016; Pooley et al., 2017) and geography (Brisbois and Polo, 2017; Caretta and Jokinen, 2017). The aim is to relate two key ideas from feminist methodology – situated knowledges and reflexivity – and then demonstrate how situated, reflexive research may contribute to dialogues about the human dimensions of large carnivore conservation. The next section further details the study, explains situated knowledges and reflexivity, and then elaborates on the study process. The following sections provide further detail on how the conditions of knowledge production influenced the interpretation of the data, and the implications of such interpretations for understanding the human dimensions of large carnivore conservation.

Theoretical background and methodology

Study context

The study, initially framed as "Cattle-based livelihoods and the bear 'problem' in northern Ecuador," was a qualitative research project, conducted with two primary goals: to provide insights about the human-bear conflict in Pimampiro, Ecuador, from a critical social science lens, in order to complement conservation biologists' ongoing work there; and in the course of doing so, continue my graduate education in Geography and Women's Studies. The initial conception of this study focused on the bears because my entrée into the region was through one of the Andean Bear Foundation's two conservation biologists. But the second piece, graduate education, is also crucial to understanding the conditions of knowledge production and problem framing. My thesis committee encouraged me to consider how the researcher's focus might influence the importance respondents ascribe to a certain issue. For example, how might the social context

Table 4.1 Summary of study research questions, main findings, and implications

Research questions	Main findings	Implications
How has raising cattle become a form of livelihood in the northern Ecuadorian Andes?	• The dwelling activities and landscape changed significantly over two decades, as many residents turned from crop-based to cattle-based livelihoods for dairy production, transforming the landscape from one dominated by fields to pastures. • Push factors included changes in climatic conditions, soil fertility, prices, and labor availability due to out-migration. • Pull factors included ease of raising cattle, market demand, and small development program activities.	• People in the area turned from tasks with loud and frequent human presence, such as hunting, timber extraction, and large *mingas* (collaborative workgroups), to sowing and harvesting, to leaving bull calves in remote pastures for days. • Out-migration's "flip side" was that "stayers" may have to adapt to different labor availability; "stayers" are "committed" to the countryside. • Patches of forest, where families used to live before migrating, recovered.
How have bears come to represent a threat in this landscape?	• Bears were not as great of a threat to cattle health, well-being, and life as other threats such as quantity and quality of available pasture, falling, climate (e.g., pneumonia risk), and disease (e.g., mastitis). • However, depopulation contributed to the combination of vulnerable cattle and patches of secondary forests where bears may easily access them. • Political and economic drivers such as conservation legislation and production pressures also shaped the agricultural frontier.	• Bears were able to approach cattle in pastures surrounded by forest patches in a now-quiet area without the kinds of human and mechanical sounds that used to reverberate. • Patches of regenerating forest can have mixed rather than uniform results for the extent of "secure" bear habitat. • Effective approaches to large carnivore conservation will include recognizing and addressing cultural and economic pressures beyond the local region.

Note: See also Jampel, 2013, 2016.

(entrée through an organization focused on conservation) and power relations (researchers educated in the country capital and abroad), combined with interviews centered on bears, serve to produce particular stories, whether emphasizing "conflict"[2] or missing other factors contributing to the phenomenon in question? To re-orient the study, I drew from political ecology and livelihoods studies.

The sub-discipline of political ecology in geography includes in its foundations the idea of opening the aperture of analysis to consider local or proximate phenomena in the context of more "global" or distal phenomena. For example, Tom Bassett's (1988) pioneering work elucidated how the livestock development policies of the government of the Ivory Coast in the 1980s contributed to conflict between pastoralists and peasants. Feminist critiques of science and epistemology in the 1980s also influenced some political ecologists to account for the conditions of knowledge production, some of whom adopted the term feminist political ecology (Sundberg, 2017). To open this aperture while also recognizing the underlying purpose of conducting this particular research in this particular place – to understand why and how relationships between residents and Andean bears had recently changed in the region – led to decentering the bears and focusing on the cattle as the non-indigenous species, and to framing the study around the classic question from the livelihoods literature in development studies (Chambers and Conway, 1992; Scoones, 2009): "How do people live in this place?"

Further, because little information existed about this region or transition, especially in the published literature, an open-ended approach was warranted in order to avoid misdirecting the research with inappropriate hypotheses or distorting the lived experiences of people in the study area. Because this study decentered the bears to focus on the livelihood context of the human-bear "conflict," less attention was given to parsing out the social and ecological contributions to the dynamics of the bear-cattle incidents themselves, which other researchers have taken up (Figueroa Pizarro, 2015; Zukowski and Ormsby, 2016).

Most data collection took place over the course of two months in the summer of 2012. I conducted 83 interviews with residents in three communities and a few NGO and government employees, and lived with two host families. I also collected other data that emerged, such as newspaper clippings, satellite images of the area, and data on cattle ownership, and used these data to challenge or corroborate my findings. The following sections describe the theory underpinning various methodological choices and also provide some examples of the theory "in action."

Situated knowledges

Debates within feminist theory and the broader critical social science literature about knowledge production and objectivity informed the study's development and execution. Trenchant critiques of science emerged in the US in the mid-1980s, as scholars of social studies of science and technology (STS) and feminist critics wrote about the social construction of knowledge – meaning that what anyone knows is influenced by the social conditions of its production; that is, by the perspectives, priorities, and privileges of the people generating the knowledge (Harding, 1986; Hacking, 2000; Wylie, 2003). Any researcher or aspiring scientist taking such critiques seriously might find themselves at an impasse: If all knowledge is a product of its social world, does that mean none can be understood as more reliable, and if so, why conduct research? To address the threat of impasse and the dismissal of potentially useful knowledge, feminist scholar Donna Haraway (1988) introduced the concept of *situated knowledges*. She argued that feminists and other researchers have a responsibility *not* to suggest that all knowledge is equally unreliable, since scholarship and knowledge production inform society. Rather, to make "rational knowledge claims," scholars and scientists must account for their situatedness, and the situatedness of the subjects and objects of their research. Doing so will lead to more "faithful accounts of a 'real' world" (Haraway, 1988, p. 579). The social production of knowledge, including the social locations of the researchers, affect question framing, data collection, data analysis, and communication.

For example, a researcher trained as a wildlife manager or in an educational or policy institution (social location) might design a study reflecting the priorities of their training and professional community. As Bruskotter and Shelby (2010, p. 312) write in their case for social science research on large carnivore conservation:

> Managers would not think of making controversial decisions about endangered species without data on such factors as abundance, distribution, and recruitment, yet appear quite comfortable making assumptions about human values, attitudes, and behaviors that are at least as important to the long-term success of conservation efforts.

In the special issue Bruskotter and Shelby edited, "Human Dimensions of Large Carnivore Conservation and Management," Goldman et al. (2010) challenged some of the routine ways of describing attitudes toward wildlife as Western analytical categories, reflecting a sense of knowledges as situated and socially produced. They used ethnographic approaches to study the Maasai and their relationships with lions in Kenya and Tanzania and highlighted negative, positive, and often "complex and sometimes ambivalent ways in which local people think about, and relate with, wildlife" (Goldman et al., 2010, p. 333). Though their paper does not delve into the specifics, the work rests on previous research by geographers and anthropologists and anticipates a collection of articles bridging STS and political ecology (Nadasdy, 2003; Goldman, 2007, Goldman et al., 2011). They contend that a better understanding of people's ambivalent attitudes toward large carnivores can lead to better practices, and that studies of human relations with wildlife benefit from reflecting on the conditions of knowledge production.

Reflexivity and an abductive approach

Accounting for the conditions of knowledge production can be accomplished, in part, through the practice of *reflexivity*. Reflexivity is one of many "situating technologies" that researchers use in order to practice situated science (Rose, 1997, p. 308). Practicing reflexivity means that a researcher considers various aspects of knowledge production, including social and institutional contexts, the assumptions that come with those contexts, and the ways in which social locations influence various research relations. Two important categories are introspective and epistemological reflexivity (Foley, 2002). Introspective reflexivity – also sometimes called confessional or transparent reflexivity – interrogates the researcher's own position and examines the researcher's relationships with her subjects and reality. In contrast, epistemological reflexivity, also sometimes called theoretical reflexivity, focuses not only on the researcher's identity and relationships, but also on the entire knowledge production process.

Introspective reflexivity about a researcher's own position or situatedness means that "we must recognize and take account of our own position, as well as that of our research participants, and write this in our research practice," according to feminist geographer Linda McDowell (1992, p. 409). However, this requires more than most researchers are capable of: a complete knowledge of self and context (Rose, 1997, p. 311). Instead, Gillian Rose (1997, p. 311) proposes the idea of "constitutive negotiation," meaning that research can be understood knowledge that emerges from varied and changing relationships. The researcher is not authoritative but rather "situated, not by what she knows, but by what she uncertainly performs." For example, Juanita Sundberg (2003, 2005), writing about conservation in Latin America, noted that race and gender influenced research relationships for North American and European geographers conducting studies in the region. In one of Sundberg's studies, she found that when working with NGOs, gender seemed less important to her research identity than her status as a US citizen, whereas when working with extension agents in a rural community, gender discrimination was so pervasive that she elected to hire an assistant to complete the interviews. More recently, two researchers from Canada and Ecuador have written about the ways in which their identities influenced their studies in the coastal region of Ecuador (Brisbois and Polo, 2017). Canadian Ben Brisbois reflected on how he came to insights about poverty and poor health outcomes for marginalized Ecuadorians as related not only to racism and neo-colonialism on the part of the "Global North," external to the country, but also to *within*-country racism. He suspected that the in-country elites perpetuating racist ideologies shared their opinions with him because of their assumption that he would be sympathetic, given his identity (white, male, Canadian). Ecuadorian Patricia Polo, university-educated and from urban highland Ecuador, also reported finding that her "social location appeared to create a comfortable space within which [culturally deterministic] narratives could be shared by relatively privileged Costeños [coastal elites]" (Brisbois and Polo, 2017, p. 197).

Epistemological reflexivity requires a researcher to "ground her theoretical constructs in the everyday cultural practices of the subjects," following the work of French sociologist Pierre Bourdieu (Foley, 2002, p. 476). As anthropologist Douglas Foley (2002, p. 476) helpfully summarizes:

> Such a move replaces abstract armchair theorizing about everyday life with an experiential, abductive (deductive and inductive) way of knowing. An abductive ethnographer must tack back and forth mentally between her concrete field experience and her abstract theoretical explanations of that experience. In the end, "theoretical [or epistemological] reflexivity" should produce a reasonably objective, authoritative account of the cultural other.

An *abductive* approach seeks to use a combination of data and theory to provide the best possible explanation to a problem, thus combining both deductive and inductive reasoning, which in many cases are "ideal types" rather than reflective of the realities of research. As an ideal type, deductive reasoning moves from theory to hypothesis development and testing. For example, in their study of the correlation between normative beliefs and support for wolf management options in Italy, Glikman et al. (2010) began with hypotheses based on specific principles, and then conducted 1,611 personal interviews to test their hypotheses. They summarize their findings as "reinforcing the predictive potential of psychological variables when attempting to understand support or opposition for wildlife management issues." In contrast, inductive reasoning moves from observation to theory. When Maan Barua (2014, p. 1463) went to rural northeast India to investigate human-elephant relationships in the context of crop raiding and house damage, he "initially sought to unpack the political drivers of deforestation to investigate how asymmetric social relationships structured and reproduced conflicts." Yet in the course of research, he began to make observations about the role of alcohol in these relationships, as elephants are attracted to *sulaī*, country spirits that many people distill locally. These observations led Barua to develop insights about how alcohol contributes to the phenomenon of human-elephant "conflict," and further use these insights to contribute to theorizing the role of non-human and non-animal materials in political and ecological relationships.

Abduction means not taking a unidirectional approach, either from observation to theory or from theory to testing, but rather an approach that moves back and forth between the two. Rosemary Collard's (2013) dissertation provides an example. Initial observations of the wildlife trade on a trip to China as a teenager led to research on the flows of live, wild-caught animals into the US, and a project in which she specifically sought to shed light on US demand-driven trade. Throughout the work, she discusses beginning with an anti-speciesist position (theory), leading to a commitment to a set of research questions and approaches (design and methodology), leading to data collection, and then to interpretation focused less on "testing" anti-speciesism per se and more on understanding specific phenomena of the wildlife trade, such as how live, wild-caught animals become commodities. Yet relating the process of abduction to readers may be challenging in the publishing unit of a single paper or book chapter. However, many researchers would likely identify the ways in which they practice abduction on a regular basis.

At its core, practicing reflexivity means that researchers consider and communicate how their "position can manifest in the research findings while still yielding useful insights," as Moon et al. (2016, p. 3) put it in their review and assessment of qualitative social science publishing in ecology and conservation journals. Yet attention to reflexivity was reported in only seven of the 146 studies Moon et al. (2016, p. 17) reviewed in their evaluation of the quality of qualitative research in conservation science, making it one of the most weakly reported attributes. In my own published findings, I reported a form of triangulation only in passing (Jampel, 2016, p. 86) but following what I perceived to be conventions, did not report on reflexivity. The next section addresses that gap in detail to address how researchers might practice reflexivity.

Situated and reflexive research in action

The previous three sections described the origins and overview of the study, the social context, and some of the key concepts related to knowledge production. This section provides further detail about execution in terms of both fieldwork and interpretation, drawing on the initial write-up (Jampel, 2013) as well as further reflections based on more recently available work. In qualitative research, the researcher herself is an "instrument of research" (Brodsky, 2008). Further, emotion, vulnerability, and positionality shape how geographers are able to conduct their fieldwork (Caretta and Jokinen, 2017). It is hard for a researcher to truthfully appraise herself, not for want of trying, and my best guess is that my first graduate-level research experience was characterized by some combination of delusion and hubris, on the one hand, and humility and anxiety, on the other, all of which informed the study's execution. Put simply, data collection involved gathering as much information from residents as possible about two questions: Why do you have cattle, and what are the issues with your cattle?

Sites

I chose three focus communities in the *cantón* (political sub-unit) experiencing bear predation of cattle, based on a combination of practical and theoretical considerations. On a practical note, I began my research in the community where someone had helped me find a place to stay, in the former room of a Peace Corps volunteer. Indeed, if I had to credit a single factor for being able to collect data successfully, it would be the enthusiastically favorable views of the Peace Corps volunteer who had recently completed his two years of service in the region. My position as a *gringa* (white person, from the US) often meant that people related to me first by comparing me with the Peace Corps volunteer, greeting me with more warmth than suspicion, and also expressing disappointment that I was not going to be conducting projects akin to those of volunteers. Because the community was so small, I was new to social science research, and there was little else to do, I conducted interviews in every household except one. My learning there inspired me to seek additional perspectives from other towns in the parish in order to check and challenge my initial findings. I chose the other two communities for their practicality (I could walk to one, and was able to find a place to stay in the other), because they all had residents who had lost cattle to bears, and because together, the three communities would represent the breadth of communities in the parish based on number of households, range of wealth and poverty, and extent of livelihood diversification. In addition, I interviewed residents in neighboring communities when the opportunity arose.

Participants

Interview participants were recruited in three focus communities using a combination of snowball sampling and going door-to-door. I sought a sample of maximum variation with a range of people and experiences in terms of gender, age, life stage, livelihood pursuits, and experiences with bears, which I did by keeping a spreadsheet, talking to my host families, and quietly compiling an informal census of each town. The anxiety that I would miss someone important led me to ask others about any household I could not make contact with, asking questions such as whether they had any pastures bordering the forest, or why it was so hard to make contact.

In general, I approached people directly through an introduction or at home, work, or on the road, and following introductions, requested to set up an appointment. In some cases, people suggested that we conduct the interview on the spot, which often had the added benefit of observing or participating in livelihood activities. Participants always chose the location for the interview, and almost all interviews were conducted either in people's homes or yards, with the exception of a few on walks, near playing fields, or in vehicles in transit. Interviews also became opportunities for observation, and I generally paid attention to the specifics of the surroundings (Oberhauser, 1997, Elwood and Martin, 2000). However, I also strived not to make assumptions about wealth or poverty without hearing directly from people about how they perceived their financial status. Interactions in homes also influenced power dynamics in terms of norms of hospitality and reciprocity, and following the first week I brought rolls or cookies to offer, given that many people served me warm milk and sometimes lunch. Keeping in mind the critique of researchers who treat a household as a single, homogenous unit, I also often interviewed more than one member of a household and purposefully did not seek to interview the "head of household."

The dynamics surrounding interviewing, relationship-building, and getting daily work done shaped the data I was able to collect. For example, I was surprised to find that being a woman seemed to be useful and give me greater access to both men and women than I anticipated. In a way, the gender I "uncertainly performed," as a curious woman happy to peel potatoes, gave me access I had not expected. I also was surprised to find that my dependence on the people in the communities where I lived and interviewed people made me appear nonthreatening, since gaining access to transportation in the area, which involved informal routes and walks alone in the dark, put me in a position of vulnerability. My experience echoed that of geographer Elizabeth Chacko (2004, p. 60) in rural India, who explained that "tacit acknowledgment that villagers possessed superior knowledge in areas where I had little was of immense help in breaking the ice and developing relationships with local people in the field." It also shaped what data I was not able to collect in ways that I cannot even presume to understand fully.

Interview data

Semi-structured, open-ended interviews decentered the bear "problem." The interview protocol was developed with assistance from my faculty, peers, the local biologist, and a long-time friend and Spanish teacher. Some elements of language translation continue to create confusion in terms of data collection and interpretation. For example, I had initially conceived of the study in English as about "threats" to livelihoods, bears being among them. Discussion with my Spanish teacher led to choosing the word "riesgos," or "risks," in order to better communicate my questions. However, as the word "risk" started to seep into my writing, it opened up the question of the "risk literature" and "perceptions of risk," which the study design had not incorporated as a dimension to investigate or measure.

With the majority of the people I interviewed, I began by introducing myself as a student researcher "living with host family X" and explained that I was interested in agriculture (crops) and livestock, purposefully leaving the topic open-ended. One woman excitedly exclaimed, "That's great, because that [agriculture and livestock] is my life here!" Often, I also explained that I was working with a biologist seeking to understand changes in the forest, and that my research was complementary. Almost all interviews began with family and household structure, making a living, and how livelihood activities had changed over time. From there, interviews turned to the rewards and challenges (threats, risks) of a household's livelihood pursuits. In the majority of interviews, when asked about the challenges for raising cattle, people did not bring up bears as an issue. Rather, the primary concerns people raised had to do with cattle-specific issues related to their biology (not to the bears' biology or ethology) such as cold, infection, and risk of rolling and falling. Climate was also a concern, as it had been with crops, and people named pneumonia as a persistent issue.

Except in the few cases in which an interviewee brought up the bears, I raised the question of bear presence near the end of the interview. In only a few instances, I shared my specific interest in the bear issue; for example, my conservation-oriented contacts introduced me to my hosts in each of the three communities, and on one occasion I persuaded an important informant to speak with me by highlighting the connection to the bears. Yet once I raised the question of the bears, the topic often stimulated more passionate discussion than concerns about climate and cattle's physical vulnerability, which I suspected was due to the combination of the novelty, the charisma of mega-fauna, and frustration at a lack of government attention, the assumption being that the government should take responsibility for the bears, given their designation as a protected species. Overall, however, the relatively rare mention of bears indicated that the perception of bears as a *primary* threat was concentrated among a small subset of people with direct experience or particular allegiances or personalities.

Although the interview data yielded material for a rich story about the cultural logics of livestock, one of several major limitations of the study included not having systematically collected data on certain phenomena relevant to the wider issue, such as perceptions and knowledge of the Andean bear. My interview data came from an interview protocol that included 14 questions organized by family profile, household livelihoods, challenges to livelihoods, and bears, and interviews usually lasted 30–90 minutes. Contrast this with the data that Zukowski and Ormsby (2016, p. 5) reported in their study in the same region, in which they used a questionnaire with 27 questions "organized by general conservation themes, knowledge of the Andean Bear, the nearby Cayambe-Coca National Park, cattle raising, and the recent bear depredation." Most of their interviews lasted 30–60 minutes. One of their primary aims was to compare "in situ" and "ex situ" residents, meaning those living in the rural "affected" communities and those living in the equivalent of a county seat. The chief contribution of their research was learning that although, as expected, most residents were more in favor of protecting Andean bears in greater Ecuador than in their own community, residents living "ex situ" – in an area without vulnerable cattle – "responded three times more frequently in outright opposition to bear conservation than the rural communities" (Zukowski and Ormsby, 2016, p. 5). The authors speculated that "informal communication," "word-of-mouth," and "gossip" contributed to the strong reactions of people in town, contrary to their hypothesis that those most affected by predation would be more opposed to conservation. They also speculated that the even greater inexperience of people in town with bears might lead to greater generalizations ("all bears" rather than "problem bear"). I might offer another hypothesis, related to sampling. The authors reported: "Participants were randomly selected from various community gatherings and door-to-door canvassing. Interviews took place outdoors or in the interviewees' private residence, away from any observers who might offer bias" (Zukowski and Ormsby, 2016, p. 4). Yet recruitment at community events and door-to-door canvassing in rural communities of hundreds of people is quite different than a town of several thousand, with many people from surrounding communities circulating as well. It is possible that the sample in town could be skewed due to the ease of recruiting participants who are already out and about, *charlando* (chatting), or willing to open their doors, and who have strong personalities and an eagerness to share their opinions.

Interview style

The benefit of hindsight suggests that one of my primary reasons for choosing the semi-structured, open-ended interview style is my own preference for serendipity. Several years later I learned that one term that characterizes my approach is that of "responsive interviewing as an extended conversation" (Rubin and Rubin, 2012). In particular, one feminist research practice related to situated knowledges and reflexivity – providing opportunities for the "gaze to be returned" (Chacko, 2004, Sharp and Dowler, 2011) – led to the primary insight from the study: the role of people's commitment to the countryside. In some cases, people began to ask me questions from the outset, about where I was from, whether I had siblings, and where they were, and we entered into a conversation about family structures and migration. In all cases, I provided an opportunity for people to ask me questions as well, acknowledging that I had been asking them a lot of questions.

Parents in their 60s asked me about my own moving away from home, and wanted to talk about their children's departure for the provincial capital, country capital, and especially Spain, and the changing nature of work in the countryside as the region depopulated. People in their 30s shared their stories of leaving and returning to build a life of their choosing, and frequent questions about my background highlighted how a strong sense of place and identity of making a living on the land influenced their livelihood choices. One man in his early 30s regularly asked me about technology, job opportunities, and potato production and ranching in the US. His curiosity about agricultural practices in the US and his regrets about his limited educational opportunities both threw into relief the implicit power dynamics and inequality (we are about the same age) and also highlighted one of the major themes that emerged with respect to people's commitments the countryside: It was a place where they could pursue a livelihood that did not require an education or having a boss. He had traveled to work in a flower plantation on the outskirts of an urban area and found the work in freezer rooms and with pesticides intolerable. A number of young people who returned to live where they had grown up also had worked for some period in flower industry. Conversations with this person over time gave me a sense of his life story, and how he and his extended family had come to be some of the households most vulnerable to bear predation of their cattle. Not only did they experience predation directly, but also, they had difficulty maintaining a viable livelihood – and raising a few cattle to have milk for the family and as a form of wealth storage was part of their livelihood of last resort.

Reflecting in action

During the course of fieldwork, I noticed what structured the days of the lives of the people around me: milking cows. Unlike the farmers in Barua's (2014) study of human-elephant interactions in India, mentioned previously, who set up crop-guarding shelters and preoccupied themselves with potential elephant raids, the small-scale farmers and ranchers in Pimampiro did not preoccupy themselves daily with the activities of Andean bears. Rather, daily preoccupations included the weather, the long walks of up to two hours each way to tend to cattle, children who were far away, and for children who were close, school expenses. The research design and interview protocol, aimed at decentering the bears and accompanied by purposeful immersion, led to learning about the "everyday cultural practices of the subjects" and the eventual interpretation of the dairy taskscape.

Telling the story of a taskscape

The published findings of this study are organized around anthropologist Tim Ingold's idea of the *taskscape*. The centrality of this concept to the *Geoforum* paper reflects the situatedness of knowledge production. Multiple faculty members in my doctoral program assigned anthropologist Arturo Escobar's paper "Culture Sits in Places" (2001), in which he briefly refers to Ingold's work (p. 152):

> And as Ingold (1993) so perceptively has discussed, places can only have boundaries in relation to the activities of the people (the "taskscapes"), or animals, for whom they are recognized and experienced as such. Even "natural boundaries" such as rivers and mountains follow this logic of construction. That places are also constituted by capital and "the global" should be clear by now.

This reference led me to read Ingold's original paper, well after what I had considered to be the "completion" of the study.

In his paper "The Temporality of the Landscape," Ingold probes how a landscape is a living process: an "enduring record of – and testimony to – the lives and works of past generations who have dwelt within it, and in so doing, have left there something of themselves" (Ingold, 1993, p. 152). Ingold devotes much of his paper to considering how to define landscape, which he then defines as constituted by and through various processes, most importantly "dwelling activities" or tasks that take place over time. Ingold's conceptualization of *task* is important for my purposes. He defines task as "any practical operation, carried out by a skilled agent in an environment, as part of his or her normal business of life." Tasks include all kinds of activities, whether or not they provide income, such as preparing a field to be sown, or walking to visit a friend, or cooking a meal. The task-scape, then, is "the entire ensemble of tasks, in their mutual interlocking." Any place inhabited by "skilled agents" – which we might argue include human animals, non-human animals, and other agents – has a taskscape, or an "array of related activities" (Ingold, 1993, p. 158). This array of related activities contributes to the shaping of the landscape, an array of related features – whether those features are primary forests and logging paths, secondary forests and clearings, or pastures and fences and long horizons. The landscape features are those that we see, the tasks in their "embodied form," and the tasks, for Ingold, are what we hear – the hum of activity. The landscape is the congealed form of historical and contemporary dwelling activities.

In the study area, the dwelling activities and landscape had changed significantly over two decades, as residents turned from crop-based to cattle-based livelihoods, transforming the landscape from one dominated by fields of potatoes, broad beans, barley, corn, and wheat, as well as peas, carrots, beans, parsnips, red onions, and myriad other produce, to pastures for cattle. Residents turned to raising cattle, primarily for dairy production, due to dairy's reliable profits and because it would allow them to remain in the countryside rather than migrate to the city. The new pasture spaces and the cattle inhabiting those spaces mediated relationships between humans and bears. The concept of the taskscape allows for fully considering the history, political economy, culture, and ecology of the region and how they continually shape each other to constitute the landscape. It aids with understanding how both indigenous and *mestizo* (mixed Spanish colonial and indigenous descent) people, many of whom came to the area from large haciendas to make their own independent livings, developed place attachment and cattle attachment.

Findings and implications for large carnivore conservation

The primary contribution of the taskscape as an analytical concept for researchers interested in human–wildlife conflict is that it provides an organizing principle for thinking about livelihood activities – which are often at the crux of such conflicts – as both economic and cultural, and moreover both shaped by and shaping place, place attachment, and interactions in the landscape. For some, these interconnections may appear self-evident. However, it was useful for this study in order to highlight the importance of those connections. Moreover, it provided an opportunity to recognize the potential importance of the auditory landscape as it relates to dwelling and livelihood activities as a contributing factor to human–bear conflict in the study area. People in the area turned from tasks with loud and frequent human presence, such as hunting, timber extraction, and large *mingas* (collaborative workgroups) to sow and harvest, to leaving bull calves in remote pastures for days. The features of the landscape included recently recovering patches of forest where families used to live before migrating. Bears were able to approach cattle in pastures surrounded by these patches in a now-quiet area without the kinds of human and mechanical sounds that used to reverberate. The focus in this research were the dwelling patterns of residents coming into conflict with bears, and that focus on cattle-based livelihoods made the taskscape a more appropriate organizing principle for relating the study's findings than a focus on animal behavior, for example. This has specific implications for Andean bear conservation.

The primary drivers of Andean bears' vulnerability to extinction are poaching, habitat loss and fragmentation, and overall lack of knowledge about their distribution and status (Goldstein et al., 2008). These phenomena are inter-related and part of larger political-economic and cultural systems. Habitat loss and fragmentation are fundamentally landscape and taskscape changes that in this case, have occurred on the agricultural frontier (*la frontera agrícola*), which is both a conceptual term for the space where field or pasture meets forest, as well as a contemporary legal boundary. For thousands of years, biogeophysical char-acteristics and cultural activities, such as fire, burning, and grazing-and-burning, influenced the tree line and agricultural frontier at high altitudes. More recently, production pressures have pushed the agricultural frontier even farther up into forested mountainsides (Sarmiento, 2002). Further, the combination of Ecua-dor's 1981 Forestry Law, the 1999 Environmental Law, and the 2004 Biodiver-sity Law have enforced this boundary between field or pasture and forest. At present, timber extraction is illegal in any place without prior permission and always illegal beyond the agricultural frontier boundary. Hunting is also forbid-den, and so human activity on the "other side" of the agricultural frontier has diminished. These changing arrays of features (landscape), such as the density and type of forest cover on the agricultural frontier, and changing array of activities (taskscape), such as timber extraction and hunting, characterize the everyday characteristics of changes in bear habitat.

Ingold's emphasis on the taskscape being about not only what is seen but also about what is heard highlights the types of insights that can be gained by thinking about the ways in which different livelihood activities correspond with different levels of human activity in a place, not only in a binary sense of whether activity is or is not present. Some people expressed their thinking that with less timber extraction and no hunting, the bears "aren't scared," especially since people are making much less noise with chainsaws. One person said, "Before, people were sawing [wood], but now they [the government] ask that we don't saw, and so the bears can comfortably come down here, because no one is making noise." Further, faraway pastures are where people had left their bull calves, which in some cases they checked on only once or twice a week, creating periods of many days without human noise. The use of the taskscape as a conceptual framework emerged as a result of reading and reflection and part of the knowledge production process in a specific social context. Yet in turn the concept leads to further ways of thinking about deterrence strate-gies for residents committed to reducing bear predation. For example, there may exist reasonable potential for success for regular patrols of people making noise, though of course this presents "hidden costs" for those making the patrols (Barua et al., 2013).

The shift in the array of tasks came with a shift in the array of features on the agricultural frontier, which now included pastures surrounded by regenerating secondary forest. Some interviewees talked about how households had been moving "down the mountain" and the prospect of "everything turning back into forest." During a walking interview with the man in his 30s mentioned in the section on "Theoretical background and methodology," sub-section "Inter-view Style," we visited an abandoned house that was located near a pasture surrounded by secondary forest, and also was the site of a bear attack. The family had moved almost two decades ago after the interviewee's brother was born, and in that time the area had been completely recolonized with trees. This "invisible" forest recovery may provide a comfortable place from which bears can attack cattle and also a place for them to drag the carcasses. Such forest recovery could potentially be an explanation for why bears had become a problem in the past few years and not in the past two decades, as another person puzzled: "We bought our land thirty, twenty years ago. We had cattle. They began to eat about two years ago. So it is not from working on the mountain [in areas we have cleared]." Qualitative data points such as these offer further reason for those studying forest transitions to take seriously how depopulation of an area may increase contact with large carnivores for those remaining, in some cases precipitating conflict (Naughton-Treves and Treves, 2005, p. 253). Patches of regenerating forest can have mixed rather than uniform results for the extent of "secure" bear habitat (Gray and Bilsborrow, 2014; Hecht, 2014; Hecht et al., 2015, Ohrens et al., 2016).

The combination of features (fields of corn, pastures of cattle) and livelihood activities (planting crops, raising cattle) also intersected with government enforcement of bear protection that increased parallel to livelihood shifts. The fact that killing a bear carries a hefty fine affects how people perceive the responsibility of the government. Two people interviewed speculated that Andean bears had ventured into fields in the past and eaten crops, but that the salience and significance of the damage did not provoke the same notice and therefore outrage as cattle predation. Crops were also valuable, but these cases occurred before the government had began protecting Andean bears, which led to people perceiving the government as having assumed responsibility for them and their activities (Castellanos et al., 2010). For example, two women joked, personifying the bears, that "because now it is forbidden to kill the animals, they come down because no one will bother them," and it "seems like they are coming because now they know they are protected, and no one will kill them. Now that they know we won't touch them, they come to eat!"

Drawing connections between livelihoods as dwelling activities or tasks and a landscape where humans and carnivores come into conflict also highlights the importance of thinking about not only the policies but also the scale of governance in terms of carnivore conservation. The people in this study situated their place attachment and preference for rurality in the changing political economic and climatic conditions of their lives: volatile prices for crops in an unpredictable regional and national market; rural outmigration as younger people moved to cities in Ecuador, to extractive areas, or abroad, following the jobs; and changing, more unpredictable rainfall patterns. These trends link to changes at multiple scales. For example, global climate change affects regional rainfall patterns and national economic policy such as Ecuador's "dollarization" in 2000 relate to global economic systems and affect farmers purchasing imported inputs (Sherwood et al., 2005; McDowell and Hess, 2012). The scale of the analysis to address the question of how raising cattle became an important livelihood in the northern Ecuadorian Andes then begins to suggest different scales of intervention. In my view, any educational program or newly introduced policies and practices must account for the astute and accurate perceptions of people raising cattle and angry about bear predation: Many of the forces responsible for the constraints in which they are pursuing viable livelihoods are out of their control. Looking at those perceptions as they relate to an entire world of threats to livelihoods situates the bear threat among others and highlights the importance of considering people's lives in their entirety when designing programs to promote conservation. Again, I offer the example of Zukowski and Ormsby's (2016) paper on the same issue in the same region as a point of contrast for differences in methodology as related to epistemologies (ideas about how we know what we know) and philosophical paradigms (Moon et al., 2016). The lens of analysis for their paper leads to recommending a long-term strategy that "should promote co-existence through education, research, and a strategy to mitigate the financial costs to depredation victims" (2016, p. 13). Even if stakeholders interested in carnivore conservation are not prepared to take a radical stance – etymologically, meaning a stance that addresses the root causes of the issues – recognizing and affirming the experiences of people like the small-scale ranchers and their neighbors in this study would likely contribute to successful interventions.

Conclusion

This chapter has described how research design and methodology informed by feminist, qualitative, and interpretive approaches led to a particular set of insights about the rise of cattle-based livelihoods in northern Ecuador and its consequences for Andean bears and residents' perceptions of them. The insight that attachment to rural place is a key part of the story emerged from a research design and interview protocol that aimed to decenter the bears as a specific problem, open the aperture, and consider the wider social and cultural dynamics contributing to human-bear relationships. Such an approach contributes to Pooley et al.'s (2017) exhortation that those seeking to address "adverse human-predator encounters" engage with cultural dynamics and underlying drivers of "conflicts," including the problem-framing itself.

Researchers seeking to work across disciplines will need to discuss ways of understanding and communicating the conditions of knowledge production. In this chapter, I have attempted to make transparent how I came to tell a particular story and what was gained and lost in such a telling. As Moon et al. (2016) delineate, readers and reviewers can use four criteria to evaluate qualitative research: dependability (how reliable is it?), credibility (is there internal validity?), confirmability (is "objectivity" accounted for?), and transferability (is there external validity, and is it generalizable?). Although researchers often reported on the first two, the authors found (p. 1) that they were "poorly evolved in relation to critical aspects of qualitative social science such as methodology and triangulation including reflexivity." The authors also found that confirmability and transferability were "poorly developed." Moon et al. (2016) have made clear that in terms of sound communication of qualitative research in the field, much work remains, and my own reflection is that their checklists would have been useful to have when designing the project. This chapter sought to provide relatable specifics of how researchers in the field might think about and practice reflexivity and account for situated knowledges rather than idealized objectivity. Perhaps more importantly, this chapter sought to make the case that such an approach can contribute specific and valuable types of knowledges about the human dimensions of large carnivore conservation.

Notes

1 Acknowledgements and thanks again to Andrés Laguna, Carlos Racines, Luz Amores, and Melissa Wright for their essential contributions to this project; to Alex Moulton, Alida Cantor, and Roopa Krithivasan for their assistance with this piece; and to the Conference of Latin American Geographers, the AAG Rural Geography Specialty Group, and the Department of Geography, The Pennsylvania State University, for their funding support.
2 Conservation biology has increasingly adopted an understanding of human-wildlife relations as *relationships* in an intra-connected system, rather than solely as conflict. Not only can conflict language define a situation in an unnecessarily polarizing way, but it is also often inaccurate, since the human-wildlife conflict label can conceal underlying dynamics such as differences in values, priorities, and power among the people involved (Peterson et al., 2010; Hill, 2015; Pooley et al., 2017).

References

Barua, M. (2014) 'Volatile ecologies: Towards a material politics of human-animal relations', *Environment and Planning A*, vol 46, no 6, pp1462–1478.

Barua, M., Bhagwat, S. A., and Jadhav, S. (2013) 'The hidden dimensions of human-wildlife conflict: Health impacts, opportunity and transaction costs', *Biological Conservation*, vol 157, pp309–316.

Bassett, T. J. (1988) 'The political ecology of peasant-herder conflicts in the northern Ivory Coast', *Annals of the Association of American Geographers*, vol 78, no 3, pp453–472.

Brisbois, B. W., and Polo, P. P. (2017) 'Attending to researcher positionality in geographic fieldwork on health in Latin America: Lessons from La Costa Ecuatoriana', *Journal of Latin American Geography*, vol 16, no 1, pp194–201.

Brodsky, A. E. (2008) 'Researcher as instrument', in L. M. Given (ed) *The SAGE Encyclopedia of Qualitative Research Methods*, SAGE Publications, Thousand Oaks, CA.

Bruskotter, J., and Shelby, L. (2010) 'Human dimensions of large carnivore conservation and management: Introduction to the special issue', *Human Dimensions of Wildlife*, vol 15, no 5, pp311–314.

Caretta, M. A., and Jokinen, J. C. (2017) 'Conflating privilege and vulnerability: A reflexive analysis of emotions and positionality in postgraduate fieldwork', *The Professional Geographer*, vol 69, no 2, pp275–283.

Castellanos, A., Cevallos, J., Laguna, A., Achig, L., Viteri, P., and Molina, S. (eds) (2010) *Estrategia Nacional De Conservación Del Oso Andino (National Strategy for the Conservation of the Andean Bear)*, Ministerio del Ambiente del Ecuador (Ministry of Environment), Quito, Ecuador.

Castellanos, A., Laguna, A., and Clifford, S. (2011) 'Suggestions for mitigating cattle depreda-tion and resulting human-bear conflicts in Ecuador', *International Bear News*, vol 20, no 3, pp16–18.

Chacko, E. (2004) 'Positionality and praxis: Fieldwork experiences in rural India', *Singapore Journal of Tropical Geography*, vol 25, no 1, pp51–63.

Chambers, R., and Conway, G. (1992) *Sustainable rural livelihoods: Practical concepts for the 21st century*, Discussion Paper 296, Institute of Development Studies (IDS), Brighton.

Collard, R.-C. (2013) *Animal traffic: Making, remaking, and unmaking commodities in the global live wildlife trade*, PhD dissertation, University of British Columbia, Vancouver.

Elwood, S. A., and Martin, D. G. (2000) '"Placing" interviews: Location and scales of power in qualitative research', *The Professional Geographer*, vol 52, no 4, pp649–657.

Escobar, A. (2001) 'Culture sits in places: Reflections on globalism and subaltern strategies of localization', *Political Geography*, vol 20, no 2, pp139–174.

Figueroa Pizarro, J. (2015) 'Interacciones humano-oso andino Tremarctos ornatus en el Perú: consumo de cultivos y depredación de ganado' (Human-Andean Bear Tremarctos Ornatus Interaction in Peru: Consumption of Crops and Predation on Livestock), *Therya*, vol 6, no 1, pp251–278.

Foley, D. E. (2002) 'Critical ethnography: The reflexive turn', *International Journal of Qualita-tive Studies in Education*, vol 15, no 4, pp469–490.

Glikman, J., Bath, A., and Vaske, J. (2010) 'Segmenting normative beliefs regarding wolf man-agement in central Italy', *Human Dimensions of Wildlife*, vol 15, no 5, pp347–358.

Goldman, M. (2007) 'Tracking wildebeest, locating knowledge: Maasai and conservation biology understandings of wildebeest behavior in Northern Tanzania', *Environment and Planning D: Society and Space*, vol 25, no 2, pp307–331.

Goldman, M., Nadasdy, P., and Turner, M. (eds) (2011) *Knowing Nature: Conversations at the Intersection of Political Ecology and Science Studies*, University of Chicago Press, Chicago.

Goldman, M., Roque De Pinho, J., and Perry, J. (2010) 'Maintaining complex relations with large cats: Maasai and lions in Kenya and Tanzania', *Human Dimensions of Wildlife*, vol 15, no 5, pp332–346.

Goldstein, I. R., Velez-Liendo, X., Paisley, S., and Garshelis, D. L. (2008) *Tremarctos ornatus (Andean Bear, Spectacled Bear)*, IUCN Red List of Threatened Species, www.iucnredlist.org/apps/redlist/details/22066/0, accessed 15 January 2012.

Gray, C. L., and Bilsborrow, R. E. (2014) 'Consequences of out-migration for land use in rural Ecuador', *Land Use Policy*, vol 36, pp182–191.

Hacking, I. (2000) *The Social Construction of What?*, Harvard University Press, Cambridge, MA.

Haraway, D. (1988) 'Situated knowledges: The science question in feminism and the privilege of partial perspective', *Feminist Studies*, vol 14, no 3, pp575–599.

Harding, S. (1986) *The Science Question in Feminism*, Cornell University Press, Ithaca, NY.

Hecht, S., Yang, A. L., Bimbika, S. B., Padoch, C., and Peluso, N. L. (2015) 'People in motion, forests in transition: Trends in migration, urbanization, and remittances and their effects on tropical forests', Occasional Paper 142, Center for International Forestry Research (CIFOR), Bogor, Indonesia.

Hecht, S. B. (2014) 'Forests lost and found in tropical Latin America: The woodland "green revolution"', *The Journal of Peasant Studies*, vol 41, no 5, pp877–909.

Hill, C. M. (2015) 'Perspectives of "conflict" at the wildlife-agriculture boundary: 10 years on', *Human Dimensions of Wildlife*, vol 20, no 4, pp296–301.

Huijsmans, R. (2014) 'Becoming a young migrant or stayer seen through the lens of "house-holding": Households "in flux" and the intersection of relations of gender and seniority', *Geoforum*, vol 51, pp294–304.

Ingold, T. (1993) 'The temporality of the landscape', *World Archaeology*, vol 25, no 2, pp152–174.

Jampel, C. (2013) *Cattle-Based Livelihoods and the Bear "Problem" in Northern Ecuador*, MS thesis, The Pennsylvania State University, State College, PA.

Jampel, C. (2016) 'Cattle-based livelihoods, changes in the taskscape, and human-bear conflict in the Ecuadorian Andes', *Geoforum*, vol 69, pp84–93.

Kay, C. (2008) 'Reflections on Latin American rural studies in the neoliberal globalization period: A new rurality?', *Development and Change*, vol 39, no 6, pp915–943.

Laguna, A. (2011) 'Resultados preliminares del conflicto hombre-oso en el norte de Ecuador (Preliminary results of the human-bear conflict in northern Ecuador)', I Congreso Ecuatoriano de Mastozoologia (First Ecuadorian Mammalogy Conference), Quito, Ecuador.

McDowell, J. Z., and Hess, J. J. (2012) 'Accessing adaptation: Multiple stressors on livelihoods in the Bolivian highlands under a changing climate', *Global Environmental Change*, vol 22, no 2, pp342–352.

McDowell, L. (1992) 'Doing gender: Feminism, feminists and research methods in human geography', *Transactions of the Institute of British Geographers*, vol 17, no 4, pp399–416.

Moon, K., Brewer, T., Januchowski-Hartley, S., Adams, V., and Blackman, D. (2016) 'A guideline to improve qualitative social science publishing in ecology and conservation journals', *Ecology and Society*, vol 21, no 3.

Nadasdy, P. (2003) *Hunters and Bureaucrats: Power, Knowledge, and Aboriginal-State Relations in the Southwest Yukon*, UBC Press, Vancouver.

Naughton-Treves, L., and Treves, A. (2005) 'Socio-ecological factors shaping local support for wildlife: Crop-raiding by elephants and other wildlife in Africa', in R. Woodroffe, S. Thirgood, and A. Rabinowitz (eds) *People and Wildlife: Conflict or Coexistence?*, Cambridge University Press, Cambridge.

Oberhauser, A. M. (1997) 'The home as "field": Households and homework in rural Appalachia', in J. P. Jones, H. J. Nast, and S. M. Roberts (eds) *Thresholds In Feminist Geography: Difference, Methodology, Representation*, Rowman & Littlefield, Lanham, MD.

Ohrens, O., Treves, A., and Bonacic, C. (2016) 'Relationship between rural depopulation and puma-human conflict in the high Andes of Chile', *Environmental Conservation*, vol 43, no 1, pp24–33.

Peterson, M. N., Birckhead, J. L., Leong, K., Peterson, M. J., and Peterson, T. R. (2010) 'Rearticulating the myth of human-wildlife conflict', *Conservation Letters*, vol 3, no 2, pp74–82.

Pooley, S., Barua, M., Beinart, W., Dickman, A., Holmes, G., Lorimer, J., Loveridge, A. J., Macdonald, D. W., Marvin, G., Redpath, S., Sillero-Zubiri, C., Zimmermann, A., and Milner-Gulland, E. J. (2017) 'An interdisciplinary review of current and future approaches to improving human-predator relations: Improving human-predator relations', *Conservation Biology*, vol 31, no 3, pp513–523.

Rose, G. (1997) 'Situating knowledges: Positionality, reflexivities and other tactics', *Progress in Human Geography*, vol 21, no 3, pp305–320.

Rubin, H. J., and Rubin, I. S. (2012) *Qualitative Interviewing: The Art of Hearing Data*, SAGE Publications, Thousand Oaks, CA.

Sarmiento, F. O. (2002) 'Anthropogenic change in the landscapes of highland Ecuador', *Geographical Review*, vol 92, no 2, pp213–234.

Scoones, I. (2009) 'Livelihoods perspectives and rural development', *Journal of Peasant Studies*, vol 36, no 1, pp171–196.

Sharp, J., and Dowler, L. (2011) 'Framing the field', in V. J. D. Casino, Jr., M. E. Thomas, P. Cloke, and R. Panelli (eds) *A Companion to Social Geography*, Wiley-Blackwell, Malden, MA.

Sherwood, S., Cole, D., Crissman, C., and Paredes, M. (2005) 'From pesticides to people: Improving ecosystem health in the northern Andes', in J. N. Pretty (ed) *The Pesticide Detox: Towards a More Sustainable Agriculture*, Earthscan, London.

Sundberg, J. (2003) 'Masculinist epistemologies and the politics of fieldwork in Latin Americanist geography', *Professional Geographer*, vol 55, no 2, pp180–190.

Sundberg, J. (2005) 'Looking for the critical geographer, or why bodies and geographies matter to the emergence of critical geographies of Latin America', *Geoforum*, vol 36, no 1, pp17–28.

Sundberg, J. (2017) 'Feminist political ecology', in D. Richardson, N. Castree, M. F. Goodchild, A. Kobayashi, W. Liu, and R. A. Marston (eds) *International Encyclopedia of Geography: People, the Earth, Environment and Technology*, John Wiley and Sons, Oxford.

Wylie, A. (2003) 'Why standpoint matters', in R. Figueroa and S. Harding (eds) *Science and Other Cultures: Diversity in the Philosophy of Science and Technology*, Routledge, New York.

Zukowski, B., and Ormsby, A. (2016) 'Andean bear livestock depredation and community perceptions in northern Ecuador', *Human Dimensions of Wildlife*, vol 21, no 2, pp111–126.

5 A methodology for stakeholder analysis, consultation and engagement in large carnivore conservation and management

Tasos Hovardas

Introduction

Interaction among stakeholders in large carnivore and conservation unfolds in Europe under three major trends. First, population densities of species are stabilizing or even increasing in some European countries (Chapron et al., 2014). This development presents substantial challenges for rural communities, especially in areas recolonized by large carnivores, where people have not lived with these species for decades. Second, there are several socio-cultural differences that augment stakeholder heterogeneity at the national (Gangaas et al., 2015) and regional levels (Piédallu et al., 2016), which is mediated through inter-group relations and social influence mechanisms (see, for instance, Hovardas, 2005, 2010a; Hovardas and Poirazidis, 2007; Hovardas et al., 2009; see also Hovardas, in this volume). Third, stakeholder heterogeneity is often condensed in two broad coalitions: a pro-carnivore coalition with social actors who would endorse the comeback of large carnivores, and an anti-carnivore coalition who would be much more reluctant to accept any increase in the abundance and range distribution of large carnivore species (e.g., Parker and Feldpausch-Parker, 2013; Lute and Gore, 2014). This reluctance is frequently exemplified by reintroduction narratives voiced by actors in the anti-carnivore coalition, according to which, some pro-carnivore actors are supposed to breed large carnivores and release them in the wild (Hovardas, 2006; Hovardas and Korfiatis, 2012; see also Hovardas, in this volume).

Interaction among stakeholder coalitions frequently eventuates to tension and conflict (Lüchtrath and Schraml, 2015). Indeed, the need to consult and engage stakeholders in wildlife and nature conservation and management appears more urgent where conflict among them is most salient, which presents a major challenge for stakeholder analysis, consultation, and engagement. Therefore, explicitly addressing this conflictual nature of stakeholder interaction has been considered among the main objectives of stakeholder involvement (Young et al., 2013). A series of processes have been developed for taking into account stakeholder input. For instance, stakeholder analysis has been for long promoted as a procedure aiming at conflict resolution (Grimble and Wellard, 1997). However, it has been highlighted that effective stakeholder involvement would necessitate an iterative procedure, which would build on regular stakeholder feedback (Reed et al., 2009), and which would need to foster ongoing and constructive interaction among stakeholders (George and Reed, 2017). This premise denotes that stakeholder involvement needs to extend well beyond the level of stakeholder analysis (Fraser et al., 2006). It has been observed that stakeholder involvement may itself end up in intensifying social conflict, if it is not properly designed and implemented (Young et al., 2013).

Developing a common vision among stakeholders has been highlighted as a prerequisite for large carnivore conservation and management (Burkardt and Ponds, 2006). However, endorsing pluralism in environmental governance models increases substantially the time frame needed for reaching agreement (George and Reed, 2017). Further, stakeholder heterogeneity makes agreement difficult, since common vision needs to be created out of the interaction of a wide array of different stakeholder groups (e.g., Hovardas and Marsden, 2016, 2017). Participatory procedures attempt to resolve a paradox; namely, to facilitate common vision and agreement among stakeholders with apparently divergent values and conflicting interests. Social fragmentation and heterogeneity leads to a seemingly unsolvable problem: how to reach agreement among social actors in disagreement and conflict? If such an agreement did not yet emerge under standard social interaction and social processes that are to be expected in society, overall, then what can stakeholder consultation and engagement add or deliver? And last but not least, how to motivate stakeholders to get involved when non-participation may appear more advantageous?

The need to involve stakeholders already from the early stage of defining management goals has been well documented so that tension and conflict would be dealt with (e.g., Young et al., 2005). However, this has not always been the case (Jolibert and Wesselink, 2012). Previous research highlighted that stakeholder involvement in wildlife and natural resource management was primarily driven by an instrumental aim of providing legitimacy to decisions that had already been taken, instead of building on normative (e.g., aiming at strengthening the democratic mandate) or substantive (e.g., aiming at increasing the knowledge base available to reach better-informed decisions) claims (Young et al., 2013). This predominance of instrumental motives has often discouraged commitment of stakeholders in the process, while it has not allowed for the establishment of trust among them. Alternative approaches to stakeholder engagement lean toward mutual understanding and two-way symmetrical communication among stakeholders (e.g., symmetrical face-to-face stakeholder interaction in workshops), in contrast to an outdated asymmetrical communication model (e.g., one which involved traditional outreach methods, such as pamphlets, brochures, press releases, etc.) that largely aimed at compliance with pre-determined management objectives (see Reed et al., 2009).

The methodology proposed in the present chapter cannot be implemented unless active participation of stakeholders is sought from the start and throughout the process. It builds on human dimension actions, which were implemented within the frame of LIFE projects (http://ec.europa.eu/environment/life/) in Greek protected areas during the last decade, and which have focused on large carnivores. The methodology covers three different stages, which need to be seen as an integrated whole. Namely, stakeholder analysis (first stage) is performed to lead to stakeholder consultation and engagement (second stage), followed by monitoring stakeholder interaction, attitudes and behaviour (third stage). Stakeholder analysis identifies stakeholder groups and delivers existing or possible points of convergence or divergence among stakeholders. Their engagement is sought right afterwards in pronouncing convergence and attempting to downplay divergence. Stakeholders are invited to outline common objectives and develop future scenarios for stakeholder interaction, in order to monitor the allocation of their input in pursuing their shared goals. All these stages of the methodology are detailed in the following sections. For each stage, overall aim and rationale is presented. Next, methods are exhibited. Finally, previous research is reported with examples to illustrate the perspectives to be taken.

Stakeholder analysis: the SWOT template

Overall aim and rationale

Stakeholder analysis focuses on stakeholder perceptions, current behaviour and future behaviour intentions across a number of core topics that relate to large carnivore conservation and management. Social science research methods to be employed in this stage involve semi-structured interviews and focus groups. Stakeholder analysis delivers a Strengths, Weaknesses, Opportunities and Threats Analysis (SWOT Analysis) for each topic that will be addressed. This deliverable will be used later as a starting point for stakeholder consultation and engagement. Stakeholder analysis is necessary before stakeholder involvement so that the potential of adopting good practice will be scanned across a number of crucial issues for large carnivore conservation and management.

Topics to be addressed by stakeholder analysis in large carnivore conservation and management include damage prevention methods, such as use of electric fences or use of livestock guarding dogs. These topics reveal several instances of in-group characteristics and inter-group relations which can prove quite important. For instance, willingness to exchange dogs can be recorded among stock breeders (in-group aspect), while willingness and readiness to offer veterinary care to a stock breeder guarding dog network can be recorded among veterinarians (inter-group aspect). In addition, sharing offspring and training guarding dogs are crucial in preserving and deploying their ability to prevent damage. Many stock breeders fail to keep up with good practice in both reproduction and training of dogs, which leads to fast degeneration of the gene pool of the local population of guarding dogs and to unattended training that degrades the ability of dogs to prevent damage. Competitive behaviour among stock breeders, moreover, often acts as a barrier to share cubs. Stakeholder analysis may determine the potential of attitude change among stock breeders. In the same vein, stakeholder analysis may investigate an array of topics, including compensation schemes for damages caused by large carnivores, use of illegal poisoned baits, waste management systems (e.g., adoption of alternative systems, which would decrease waste exposure and access of bears to waste), willingness to manufacture and pay for carnivore-friendly products, etc.

Methods

The adjusted SWOT template (Table 5.1) outlines in-group aspects such as attitudes, beliefs, knowledge, skills, behaviour intentions, and behaviours, which promote agreement among stakeholders or foster good practice (i.e., "Strengths"); other in-group aspects hinder convergence or discourage stakeholders from adopting good practice (i.e., "Weaknesses"). The template also reveals inter-group aspects that either enable or hinder agreement and diffusion of good practice (i.e., "Opportunities" and "Threats", respectively). Stakeholder analysis scaffolded by means of the SWOT template sheds light on existing or potential convergences and divergences among stakeholder groups, which will prove decisive for stakeholder interaction during the next stages of stakeholder involvement.

The template is completed through semi-structured interviews for in-group aspects and focus group discussions with members of stakeholder groups for inter-group aspects. Through an initial screening procedure, members of targeted stakeholder groups willing to be interviewed are shortlisted. Purposive sampling results in the first semi-structured interviews and then the procedure continues with snowball sampling until information is saturated; namely, until new information added is not providing any novel insight. After interview data have been collected and analyzed for determining in-group aspects, focus groups follow, which will address inter-group aspects.

The SWOT template is an adapted version of the standard template employed to perform a SWOT Analysis (e.g., Baycheva-Merger and Wolfslehner, 2016). Instead of focusing on aspects within an organization and aspects of the environment of an organization, which is the norm in the standard form of SWOT Analysis, the adjusted SWOT template for stakeholder analysis depicts stakeholder groups in columns. Rows present in-group ("Strengths"; "Weaknesses") and inter-group aspects ("Opportunities"; "Threats"). Stakeholder synthesis in Table 5.1 pertains to a case study of bear (*Ursus arctos*) conservation in three Natura 2000 sites in Central Greece. Other case studies have a more or less differentiated stakeholder synthesis. However, the rows of the SWOT template remain the same.

Previous research

In the first column of the SWOT template for bear conservation presented in Table 5.1 (Hovardas, 2010b), stock breeders acknowledged the effectiveness of

Table 5.1 Strengths, Weaknesses, Opportunities and Threats (SWOT) template for bear conservation and management

	Stock Breeders	Hunters	Foresters	Environmental non-governmental organizations (eNGOs)
Strengths (In-group aspects that promote agreement or good practice)	Damage prevention methods are acknowledged as effective	Build on scientific knowledge and data collection protocols	Valuable experience gained through working in the local context	Able to facilitate inter-contextual transfer of good practice in bear conservation
Weaknesses (In-group aspects that hinder agreement or good practice)	There is yet no valid way of certifying livestock guarding dogs	Any damage to hunting dogs is a significant loss for hunters	There are several barriers to be overcome in order to fulfil the mission of the Forest Service adequately	Objectives and demands might not be adequately tailored to the local context
Opportunities (Inter-group aspects that promote agreement or good practice)	A minimum of damage to livestock can be accepted, provided that it is compensated	Wish to be involved in wildlife and nature conservation initiatives	Institution responsible by the law for bear conservation	Strongly support the improvement of damage compensation systems
Threats (Inter-group aspects that hinder agreement or good practice)	Some do not apply for compensation because they believe they would not be fairly compensated	Conflict with stock breeders may escalate and result in illegal use of poisoned baits	There are several difficulties in engaging in wildlife and nature conservation networks	Negative attitudes towards eNGOs still prevail within other stakeholder groups

Source: Hovardas (2010b) reporting on data collected in three Natura 2000 sites in Central Greece (Aspropotamos – GR1440001; Kerketio Oros-Koziakas – GR1440002; Antichasia Ori – Meteora – GR1440003)

damage prevention methods, especially electric fences. This in-group aspect has proven decisive in achieving convergence among stakeholders and adopting good practice, and therefore, it has been incorporated under "Strengths" in the column devoted to stock breeders. Although livestock guarding dogs would also qualify among damage prevention methods, the lack of any valid way of certifying guarding dogs in Greece was underlined as an important weakness (Table 5.1; "Weaknesses" for stock breeders). Stock breeders would accept damage caused by the bear up to a threshold, provided that this damage was compensated (Table 5.1; "Opportunities" for stock breeders). Such an acceptance would demarcate a possible field of convergence between stock breeders and environmental non-governmental organizations. However, there are many stock breeders who refrain from recording damages they suffer, because they believe they will never get any fair compensation (Table 5.1; "Threats" for stock breeders). This aspect has been noted as a point of tension between stock breeders and environmental non-governmental organizations. In an analogous manner, one can follow "Strengths", "Weaknesses", "Opportunities" and "Threats" for all other stakeholder groups in Table 5.1.

Stakeholder consultation and engagement: the mixed-motive template

Overall aim and rationale

The main points of convergence and divergence among stakeholders were reflected in the SWOT template of the previous section. In the next stage of the methodology, stakeholder consultation and interaction builds on these points in order to elaborate on "Strengths" and "Opportunities" (in-group and inter-group aspects, respectively) and address "Weaknesses" and "Threats" (in-group and inter-group aspects, respectively). The main aim at this stage is to foster a structured interaction and negotiation among stakeholders by taking into account both benefits and costs associated with alternative solutions.

Methods

The deliverable of this stage of the methodology is the completed template of the "mixed-motive" perspective, an example of which is presented in Table 5.2. The mixed-motive perspective offers an alternative trajectory to win-win approaches, in that it explicitly addresses both gains and losses related to alternative solutions for stakeholders. Although the mixed-motive approach still envisages gain solutions for all engaged actors, it acknowledges that there will always be a distributive aspect referring to costs that also need to be discussed and settled (Hovardas, 2012a). In Table 5.2, stakeholders have been aggregated in two coalitions with different positioning concerning large carnivore conservation and management. The positioning of each coalition has been processed to include the main common views of stakeholder groups in each side of the

Table 5.2 Template of the mixed-motive perspective for large carnivore conservation and management

	Stakeholders who are reluctant to endorse large carnivore expansion	*Stakeholders who endorse the comeback of large carnivores*
Anticipated gains or benefits	There are ecotourism opportunities related to large carnivores, which may diversify the tourism product	Better and fairer compensation systems for damage caused by large carnivores would promote human-carnivore coexistence and assist in keeping damages below the tolerance level
Anticipated costs or losses	There can be a tolerance threshold, under which damage caused by large carnivores to livestock may be tolerated and provided that this damage is fairly compensated	• "Problem" animals should be removed if they continue causing damage that overrides a tolerance threshold • Carrying capacity of the park for nature tourism and ecotourism has been already approached in some places

Source: Hovardas (2015a) reporting on data collected in the Rodopi Mountain-Range National Park

controversy, and it exemplifies how perceived benefits and costs can be allocated among stakeholders. The template of the mixed-motive perspective can be completed during workshops with input from working groups with members of all stakeholders and it can be further supported by quantitative evidence provided by questionnaires administered to stakeholder members.

Each topic identified previously by stakeholder analysis can be incorporated in the stage of stakeholder consultation and engagement. Stakeholder consultation and interaction can unfold in workshops, which will outline the most crucial points for consideration, and which will result in a road map for stakeholder concerted action. Stakeholder representatives and spokespersons will be invited to workshops. The number of participants would be high enough to address dropouts and secure an adequate representation of all stakeholder groups. Each workshop needs to be coordinated by a facilitator accepted by all stakeholders. During workshops, the focus of stakeholder interaction needs to be on adjusting good practice to fit demands at the local level. The formation of working groups with members of all stakeholders catalyzes stakeholder communication. Working groups will build on the SWOT template, delivered at the previous stage, and concentrate on fostering "Strengths", addressing "Weaknesses", exemplifying "Opportunities", and mitigating "Threats". The suggestions of working groups can be used as input in a round table discussion for determining commonalities. Taking over from these common aspects of working groups, agreement among stakeholders may be formalized. Minutes taken during workshops can be used to develop a questionnaire to be administered to stakeholder populations, in order to reach out to wider stakeholder audiences and offer quantitative evidence for adopting good practice. After having received quantitative feedback from stakeholder groups, advisory groups are established for each topic of concern. These advisory groups should represent the full array of stakeholder interests and they can take over the responsibility of finalizing approaches to be followed later on (see participatory scenario development in the next stage).

Previous research

Sustainability discourse has often built on an uncritical belief in desirable win-win solutions (e.g., Galuppo et al., 2014). However, win-win conceptualizations of stakeholder interaction fall short of addressing the costs that are linked to any alternative. By undermining or even concealing costs, conflict is tempered in the short-term only to re-surface in the mid- and long-term (e.g., McShane et al., 2011). Contrary to this disadvantage of win-win conceptualizations, stakeholder inclusion needs to acknowledge trade-offs and strive toward negotiated comprises (Sunderland et al., 2008). An alternative conceptualization of stakeholder involvement is supported by the notion of distributive justice, which implies the fair allocation of both benefits and burdens among affected actors (Pelletier, 2010). As solutions to large carnivore conservation and management require the balancing of interests among a complex array of participants, and because this can only be achieved through stakeholder interaction and negotiations inevitably associated with both costs and benefits, there is a need or a theoretical and empirical alternative to win-win perspectives, which would address trade-offs as necessary for reaching and respecting agreement among stakeholder groups.

In this direction, stakeholder engagement is promoted by a "mixed-motive" perspective, which diverges from win-win approaches in that it concentrates on both benefits and costs that accompany any given alternative to be taken (Hoffman et al., 1999; Hovardas, 2012a). An example of the mixed-motive perspective is shown in Table 5.2, which reports on a Greek case study in a protected area with wolf and bear populations (Hovardas, 2015a). The mixed-motive perspective showcases the discursive positioning of stakeholders in two main coalitions of actors: those who endorse the increase of large carnivores in number and distribution, and those who are reluctant to do so. Stakeholder groups in each coalition elaborate on contrasting lines of argumentation and they present the trade-offs for current or anticipated developments. It should be underlined that the mixed-motive perspective is markedly mediated by the local context. Although many aspects of Table 5.2 can be found in other case studies, each different location would come along with its peculiar features.

The diversification of the tourism product in the area, through the promotion of ecotourism opportunities related to large carnivores, was a core benefit expected by stakeholders who were reluctant to endorse large carnivore expansion (Table 5.2). Stakeholders in this coalition were willing to discuss and accept a tolerance threshold for damage caused to livestock by large carnivores, provided that this damage was fairly compensated. The coalition of stakeholders who endorsed the comeback of large carnivores argued that better and fairer compensation systems for damage caused by large carnivores would add to human–carnivore coexistence. Such an improvement would assist in keeping damages below the tolerance level demarcated by the other stakeholder coalition. Removal of "problem" animals would be the cost the pro-carnivore coalition should be ready to accept, if damage caused by large carnivores would override the tolerance threshold. In addition, this coalition expressed concern with regard to carrying capacity of the park for nature tourism and ecotourism, which had been already approached in some places.

Table 5.2 showcases how stakeholder interaction under the mixed-motive perspective can promote discursive positioning of engaged actors in a structured negotiation process. A proposal put forward by a coalition of actors may be taken up by the other coalition (e.g., ecotourism development with an emphasis on large carnivores), but this would give rise to new concerns (e.g., carrying capacity levels). Current or future costs and benefits are closely related and negotiated (e.g., removal of "problem" animals, if the tolerance threshold for damage caused by large carnivores would be overridden). Overall, the mixed-motive perspective creates the background conditions for stakeholders to proceed to a collaboratively defined set of objectives for concerted action. Furthermore, it delineates a common frame for building trust among stakeholder groups and a reference base for their commitment to pursue the goals set jointly. After having agreed on such a mixed-motive perspective, stakeholders are expected to be more determined to work collaboratively and attempt co-creation and innovation.

Monitoring stakeholder interaction: participatory scenario development

Overall aim and rationale

In this last stage of the methodology, stakeholders co-design scenarios to monitor their future interactions. Scenarios will guide stakeholders in the allocation of input and resources, and alert them to take corrective action, any time it is needed. This can be supported by the ongoing operation of advisory groups. Additionally, the scenarios co-designed and implemented by stakeholders and monitored by advisory groups will support adopting good practice and anchoring it at the local level. A major aspect that should be highlighted is that participatory scenario development is expected to empower local stakeholders. This empowerment dimension provides valuable insight for the long-term sustainability of relevant implementations, since stakeholders will have the knowledge and expertise to jointly set goals, plan and monitor their interaction.

Methods

The core method to be employed in this stage is participatory scenario development. Scenarios often take the form of storylines of future developments, which are outlined as reasonable projections of current conditions under certain drivers (Haatanen et al., 2014). Since scenarios are frequently used to plan for an uncertain future (Peterson et al., 2003), their primary aim is not to deliver an indisputable forecast, but rather to coordinate present collaboration among stakeholders and assist them in effectively allocating resources to accomplish common objectives (Kok et al., 2007). Different scenarios can be prepared under varying inputs (Varum and Melo, 2010), and this exemplifies the potential trajectories of stakeholder collaborations in time. When created through a participatory procedure, scenarios comprise a collective resource, which may commit stakeholders in joint action (Newig, 2011). Such a commitment will demarcate a willingness of stakeholders to move away from baseline conditions and work together to facilitate this departure.

Scenarios will guide the monitoring of stakeholder concerted action and its effects. Using workshop reports, questionnaire data and the minutes of the first meetings of advisory groups established during stakeholder consultation and engagement, scenarios for adopting good practice can be co-designed by stakeholders. A "business-as-usual" scenario will describe current or anticipated perceptions and behaviour of stakeholders, when the latter would not present any significant departure from current conditions. A "small-effort" scenario will correspond to small-scale adoption of good practice, but it will, nevertheless, demarcate a departure from the "business-as-usual" trajectory. A "best-case" scenario will denote substantial involvement and commitment of stakeholders, as well as substantial change towards adopting good practice. Through the participatory scenario development procedure, scenarios are expected to realistically reflect the full potential to be exploited for adopting good practice.

Surveys on stakeholder perceptions and behaviour can also be conducted as a follow-up of stakeholder consultation and engagement. The partial overlap of instruments used in the second (stakeholder consultation and engagement) and third (monitoring stakeholder interaction) stages of the methodology will allow for an assessment of the impact of stakeholder involvement. Advisory groups will meet regularly and irregularly anytime needed to discuss adoption of good practice and updates in all topical foci of stakeholder interaction. The ongoing operation of advisory groups is expected to shed light on any expected or unforeseen difficulties in adopting good practice or any issues to be tackled at the local level to re-focus stakeholders and adjust good practice to fit local needs and desires. Advisory groups may revise scenarios if this will be necessary and they will use these scenarios to track stakeholder adoption of good practice in time.

Previous research

Table 5.3 presents an example of the template for participatory scenario development, which focuses on promoting livestock guarding dogs as a damage prevention method (Hovardas, 2012b). The template delineates three different scenarios for each topic of concern ("business-as-usual" scenario; "small-effort"

Table 5.3 Participatory scenario development template for promoting livestock guarding dogs

Topics	"Business-as-usual" scenario	"Small-effort" scenario	"Best-case" scenario
Certification of livestock guarding dogs	There has been no formalized certification process	Phenotypic characteristics will be determined to ensure the preservation of guarding dog characteristics	A certification centre will be established by a consortium of stakeholders
Breeding of livestock guarding dogs	There has been no concerted action to promote good practice in breeding of livestock guarding dogs	Good practice in breeding will be promoted by a stock breeder network for exchanging livestock guarding dogs	A breeding centre will be established by a consortium of stakeholders
Development of a network for exchanging livestock guarding dogs	Many stock breeders would be rather reluctant to exchange livestock guarding dogs	A trusted third party (e.g., an environmental non-governmental organization) will initiate the network	Stock breeder associations will take over and maintain the stock breeder network for exchanging livestock guarding dogs

Source: Hovardas (2012b)

scenario; "best-case" scenario; stakeholder input increases from "business-as-usual" towards "best-case"). The topics handled involved certification and breeding of livestock guarding dogs, as well as development of a network for exchanging livestock guarding dogs. The topics were chosen and elaborated upon so that concerted action among stakeholders can be initiated and supported by stock breeder associations, environmental non-governmental organizations, academia, veterinarians and local authorities.

An interesting aspect that featured in the third topic shown in Table 5.3 (third row; development of a network for exchanging livestock guarding dogs) was in-group conflict among stock breeders (Hovardas, 2012b). Namely, it was highlighted that a good guarding dog is taken to be the one that was able to hunt down and kill a dog from another stock breeder. In a series of instances, analogous accounts were not only voiced by stock breeders themselves, but they were cross-referenced by members of other stakeholder groups; for instance, gamekeepers or hunters. Moreover, many stock breeders would not be willing to exchange puppies among each other. The main reason behind this attitude was a latent competition between stock breeders, a kind of rivalry over being the one to have the best dogs in the area. Such a disposition indicates a way of guaranteeing the protection of one's own livestock against any potential damage from large carnivores. However, stock breeders in focus group discussions admitted that this positioning did not allow for the increase of good guarding dogs in the region, and it also proved deleterious for stock breeders as a whole, confining them in a type of "prisoner's dilemma". If only some stock breeders had good guarding dogs, then there would still be considerable damage from large carnivores. But if all stock breeders had good livestock guarding dogs, then overall damage would decrease substantially. This case was exemplary in showcasing how good practice may be hindered in-group rather than inter-group conflict. For this reason, a trusted third party (e.g., environmental non-governmental organization) can initiate the stock breeder network for exchanging livestock guarding dogs until stock breeders themselves take over the network. The case study was also indicative of the deep roots of conservation controversies, which involved complex socio-cultural aspects of the local context.

Implications for large carnivore conservation and management

In-group relations and inter-group dynamics among stakeholder groups shape the ground on which any conflict between humans and large carnivores will emerge. Damages caused by large carnivores are often the triggering effect behind tension, but they are not always the real cause. Recent research revealed how these types of conflicts may question social identities (Lüchtrath and Schraml, 2015). In the same direction, it seems that stakeholder interaction is frequently more crucial than species density or distribution in determining stakeholder attitudes towards large carnivore species (Gangaas et al., 2013). Therefore, there have been several recent calls for a concerted place-based policy, which is necessary at the local level for conflict resolution (e.g., Pohja-Mykrä and Kurki, 2014). This involves multiple vehicles of stakeholder deliberation, such as regional large carnivore committees (Pellikka and Sandström, 2011) or regional wildlife management delegations (Lundmark and Matti, 2015). These structures facilitate trust and informed decision making based on pluralistic perspectives (Sjölander-Lindqvist et al., 2015). The proposed methodology can complement analogous initiatives.

Any initiative to involve stakeholders will unfold in parallel with ongoing stakeholder interaction in society at various scales (e.g., local, regional, national and international). Further, stakeholder consultation and engagement is expected to augment stakeholder interaction and have a marked effect on in-group and inter-group aspects. Due to the dynamic and persistent nature of stakeholder interaction, stakeholder consultation and engagement is anticipated to produce multiple feedback effects, which can be reflected on in-group aspects – such as social representations of stakeholder groups on wildlife and natural resource management (e.g., Hovardas and Stamou, 2006; Buijs et al., 2012) – as well as on inter-group aspects, such as inter-group relations and discursive positioning (e.g., Hovardas and Korfiatis, 2012; Hovardas, 2017; see also Hovardas, in this volume). A crucial question for the effectiveness of stakeholder consultation and engagement methodologies is if stakeholder involvement just fuels pre-existing reasoning and behaviour or if it leads to substantial change of in-group or inter-group elements. Therefore, monitoring how stakeholder involvement will act on in-group and inter-group aspects and how these outcomes will act back on stakeholder interaction should be of primary importance.

Another demanding task added to stakeholder consultation and engagement is transferability of good practice across contexts. Since context-dependency is crucial for any attempt at engaging stakeholders (Durham et al., 2014, p. 10), and if stakeholder involvement is needed to adopt good practice, then good practice cannot be effectively transferred from one context to another unless it is anchored to the local context. This means that any solutions to be implemented by stakeholders need to be first adjusted to the local context – and this presupposes that stakeholders will not just implement good practice gained elsewhere, but that they would need to anchor good practice on local circumstances and develop a sense of ownership of the process (e.g., Durham et al., 2014).

The methodology outlined in this chapter can also facilitate the empowerment of stakeholders to adopt the process itself. In the frame of the proposed methodology, empowerment refers to all engaged actors and there is no divide between "knowers" and "not-knowers". The fact that no stakeholder may claim to know "best", however, does not lead to a relativist regression that "anything goes". First, there are some points where all stakeholder groups can converge; for instance, on banning the illegal use of poisoned baits, and on endorsing the effectiveness of damage prevention methods, such as livestock guarding dogs and electric fences (Hovardas, 2015b). Second, commitment to the mixed-motive perspective would be the result of stakeholder consultation and engagement, which would build on stakeholder analysis, and which would set the stage for monitoring stakeholder interaction. An implication here is that stakeholder interaction may be directed towards a "small-effort" scenario across a series of topics. Indeed, that has been the case in some implementations (Hovardas, 2015c). Even such a development, however, would signify a departure from the "business-as-usual" scenario and it would present a noteworthy improvement over previous norms and practices. Moreover, going through "small-effort" scenarios is the most usual route towards a comprehensive adoption of good practice. In several examples, the first few outcomes of stakeholder interaction stimulate further investment in the process.

The methodology involved a series of templates (e.g., SWOT template, mixed-motive template, template for participatory scenario development), which would be elaborated upon through stakeholder interaction and which would be employed in later stages of stakeholder interaction as a reference base. In this way, engagement reinforces commitment to the process, since stakeholder involvement is inscribed in the outcome of the procedure. The proposed methodology integrates stakeholder interaction within a culture of continuous social experimentation. Any decisions will necessarily have a temporary character, being subject to scrutiny and critical reappraisal. Within this frame, stakeholder deliberation has to be conceptualized as a process that will remain always incomplete and under the need of regular revision (e.g., Buizer and Van Herzele, 2012). The main drivers of the proposed approach include (1) the acknowledgment of the open and experimental character of stakeholder interaction, which supports social sustainability; (2) the acknowledgment of the need to assess and re-schedule planning for large carnivore conservation and management under the light of novel data and developments; and (3) orientation of stakeholder interaction towards concrete outcomes, since assessment and revision would be a crucial premise of the proposed approach.

Stakeholder interaction under the mixed-motive perspective should be conceptualized as a co-creation process among stakeholders, and it is expected to unfold within a culture of social learning (e.g., Durham et al., 2014, p. 10). To facilitate this objective, however, stakeholder consultation and engagement should enable the accomplishment of another demanding and twofold goal: It should be as open as possible to allow for social learning and stakeholder interaction to be reflected on stakeholder concerted action; and at the same time, it should provide a thorough guidance for that same concerted action. These two different needs are often viewed as contradictory, or too difficult to reconcile within one and the same approach. The methodology proposed in this chapter includes three stages in a modular sequence and uses templates to scaffold stakeholder interaction. In that regard, it provides guidance (adds structure to stakeholder interaction) without dictating the content or course of action to be taken, which is left to be decided upon by stakeholders themselves. The templates provide key intermediate steps to be effectively accomplished, and will also allow for monitoring the progress of stakeholder interaction.

The methodology has a constructive character in that stakeholder interaction produces novel outcomes, which are marked by the procedure followed by stakeholders. In that regard, the current contribution wishes to build on a process-oriented approach to social sustainability (Boström, 2012). Moving towards such an approach features as a core assumption for fair and accountable decision-making (e.g., Reed et al., 2009; George and Reed, 2017). Moreover, the constructive element of the methodology reflects a special instance of social learning (e.g., Keen et al., 2005; Newig, 2011), where the transformative effect on stakeholders is among the most characteristic features (Collins, 2014). Indeed, recognition and empowerment of stakeholders have been both presented among the primary aims of the normative approach to stakeholder engagement (Reed et al., 2009). Recognition of multiple perspectives is a requirement for reaching an equitable distribution of social and economic benefits (e.g., Schlosberg, 2007). In addition, empowerment and capacity building through collaborative learning would allow stakeholders to effectively mobilize future assets in order to achieve their shared goals (Diduck et al., 2015). The social outcomes of stakeholder involvement, such as accountability and trust, may be more appreciated than the actually targeted outcomes in large carnivore conservation and management.

References

Baycheva-Merger, T., and Wolfslehner, B. (2016) 'Evaluating the implementation of the Pan-European Criteria and indicators for sustainable forest management – A SWOT analysis', *Ecological Indicators*, vol 60, pp1192–1199.

Boström, M. (2012) 'A missing pillar? Challenges in theorizing and practicing social sustainability: Introduction to the special issue', *Sustainability: Science, Practice, and Policy*, vol 8, no 1, pp3–14.

Buijs, A., Hovardas, T., Figari, H., Castro, P., Devine-Wright, P., Fischer, A., Mouro, C., and Selge, S. (2012) 'Understanding people's ideas on natural resource management: Research on social representations of nature', *Society & Natural Resources*, vol 25, no 11, pp1167–1181.

Buizer, M., and Van Herzele, A. (2012) 'Combining deliberative governance theory and discourse analysis to understand the deliberative incompleteness of centrally formulated plans', *Forest Policy and Economics*, vol 16, pp93–101.

Burkardt, N., and Ponds, P. D. (2006) 'Using role analysis to plan for stakeholder involvement: A Wyoming case study', *Wildlife Society Bulletin*, vol 34, no 5, pp306–1313.

Chapron, G., Kaczensky, P., Linnell, J. D. C., et al. (2014) 'Recovery of large carnivores in Europe's modern human-dominated landscapes', *Science*, vol 346, no 6216, pp1517–1519.

Collins, K. (2014) 'Designing social learning systems for integrating social sciences into policy processes: Some experiences of water managing', in M. J. Manfredo, J. J. Vaske, A. Rechkemmer, and E. A. Duke (eds) *Understanding Society & Natural Resources* (pp. 229–251), Springer, Dordrecht.

Diduck, A., Reed, M. G., and George, C. (2015) 'Participation in environment and resource management', in B. Mitchell (ed) *Resource and Environmental Management in Canada* (5th ed., pp. 142–170), Oxford University Press, Don Mills.

Durham, E., Baker, H., Smith, M., Moore, E., and Morgan, V. (2014) *The BiodivERsA Stakeholder Engagement Handbook*, BiodivERsA, Paris.

Fraser, E. D. G., Dougill, A. J., Mabee, W., Reed, M. S., and McAlpine, P. (2006) 'Bottom up and top down: Analysis of participatory processes for sustainability indicator identification as a pathway to community empowerment and sustainable environmental management', *Journal of Environmental Management*, vol 78, no 2, pp114–127.

Galuppo, L., Gorli, M., Scaratti, G., and Kaneklin, C. (2014) 'Building social sustainability: Multi-stakeholder processes and conflict management', *Social Responsibility Journal*, vol 10, no 4, pp685–701.

Gangaas, K. E., Kaltenborn, B. P., and Andreassen, H. P. (2013) 'Geo-spatial aspects of acceptance of illegal hunting of large carnivores in Scandinavia', *PloS One*, vol 8, no 7, pp1–9.

Gangaas, K. E., Kaltenborn, B. P., and Andreassen, H. P. (2015) 'Environmental attitudes associated with large-scale cultural differences, not local environmental conflicts', *Environmental Conservation*, vol 42, no 1, pp41–50.

George, C., and Reed, M. G. (2017) 'Revealing inadvertent elitism in stakeholder models of environmental governance: Assessing procedural justice in sustainability organizations', *Journal of Environmental Planning and Management*, vol 60, no 1, pp158–177.

Grimble, R., and Wellard, K. (1997) 'Stakeholder methodologies in natural resource management: A review of principles, contexts, experiences and opportunities', *Agricultural Systems*, vol 55, no 2, pp173–193.

Haatanen, A., den Herder, M., Leskinen, P., Lindner, M., Kurttila, M., and Salminen, O. (2014) 'Stakeholder engagement in scenario development process – Bioenergy production and biodiversity conservation in eastern Finland', *Journal of Environmental Management*, vol 135, pp45–53.

Hoffman, A. J., Gillespie, J. J., Moore, D. A., Wade-Benzoni, K. A., Thompson, L. L., and Bazerman, M. H. (1999) 'A mixed-motive perspective on the economics versus environment debate', *American Behavioural Scientist*, vol 42, no 8, pp1254–1276.

Hovardas, T. (2005) *Social Representations on Ecotourism: Scheduling Interventions in Protected Areas*, PhD thesis, Aristotle University of Thessaloniki.

Hovardas, T. (2006) *Baseline study for the production of a school education kit on the wolf*, LIFE COEX – improving coexistence of large carnivores and agriculture in Southern Europe – LIFE04NAT/IT/000144.

Hovardas, T. (2010a) 'The contribution of social science research to the management of the Dadia Forest Reserve: Nature's face in society's mirror', in C. Catsadorakis and Kälander (eds) *The Dadia-Lefkimi-Soufli Forest National Park, Greece: Biodiversity, Management, and Conservation* (pp. 253–263), WWF-International, Athens.

Hovardas, T. (2010b) *Stakeholder analysis*, LIFE EXTRA – Improving the conditions for large carnivore conservation – A transfer of best practices (LIFE07NAT/IT/000502), Report of Action A5.

Hovardas, T. (2012a) 'Can forest management produce new risk situations? A mixed-motive perspective from the Dadia-Soufli-Lefkimi Forest National Park, Greece', in J. Martin-Garcia and J. J. Diez (eds) *Sustainable Forest Management: Case Studies* (pp. 239–258), INTECH, Rijeka.

Hovardas, T. (2012b) *Follow up surveys of stakeholder attitudes*, LIFE EXTRA – Improving the conditions for large carnivore conservation – A transfer of best practices (LIFE07NAT/IT/000502), Report of Action E3.

Hovardas, T. (2015a) *Recommendations to the Management Authority of Rodopi Mountain Range National Park Concerning Stakeholder Consultation and Engagement*, Deliverable 3, Monitoring of knowledge and attitudes of stakeholders in national park management. Management Authority of Rodopi Mountain Range National Park (in Greek).

Hovardas, T. (2015b) *Questionnaire Development and Administration*, Deliverable 2, Monitoring of knowledge and attitudes of stakeholders in national park management, Management Authority of Rodopi Mountain Range National Park (in Greek).

Hovardas, T. (2015c) *Analysis and assessment of the impact of LIFE ARCPIN – Conservation actions for improving conditions of human-bear coexistence in Northern Pindos on stakeholder groups*, LIFE ARCPIN-LIFE12 NAT/GR/000784, Grevena Development Agency (in Greek).

Hovardas, T. (2017) '"Battlefields" of blue flags and sea horses: Acts of fencing and de-fencing place in a gold mining controversy', *Journal of Environmental Psychology*, vol 53, no pp100–111.

Hovardas, T. (in this volume) 'Addressing human dimensions in large carnivore conservation and management: Insights from environmental social science and social psychology', in T. Hovardas (ed) *Large Carnivore Conservation and Management: Human Dimensions*, Earthscan, London.

Hovardas, T., and Korfiatis, K. J. (2012) 'Adolescents' beliefs about the wolf: Investigating the potential of human – Wolf coexistence in the European south', *Society & Natural Resources*, vol 25, no 12, pp1277–1292.

Hovardas, T., Korfiatis, K. J., and Pantis, D. J. (2009) 'Environmental representations of local communities' spokespersons in protected areas', *Journal of Community and Applied Social Psychology*, vol 19, no 6, pp459–472.

Hovardas, T., and Marsden, K. (2016) *Use of rural development funding to support large carnivore-human coexistence measures*, EU Platform on Coexistence between People and Large Carnivores, Platform Secretariat to DG Environment of the European Commission.

Hovardas, T., and Marsden, K. (2017) *Outlining Good Practice in Large Carnivore Conservation and Management*, 23rd International Symposium on Society and Resource Management, Umeå, Sweden.

Hovardas, T., and Poirazidis, K. (2007) 'Environmental policy beliefs of stakeholders in protected area management', *Environmental Management*, vol 39, no 4, pp515–525.

Hovardas, T., and Stamou, G. P. (2006) 'Structural and narrative reconstruction of rural residents' representations of "nature", "wildlife", and "landscape"', *Biodiversity and Conservation*, vol 15, pp1745–1770.

Jolibert, C., and Wesselink, A. (2012) 'Research impacts and impact on research in biodiversity conservation: The influence of stakeholder engagement', *Environmental Science and Policy*, vol 22, pp100–111.

Keen, M., Brown, V. A., and Dybal, R. (2005) *Social Learning: A New Approach to Environmental Management*, Earthscan, London.

Kok, K., Biggs, R., and Zurek, M. (2007) 'Methods for developing multiscale participatory scenarios: Insights from Southern Africa and Europe', *Ecology and Society*, vol 12, no 1, www.ecologyandsociety.org/vol12/iss1/art8/.

Lüchtrath, A., and Schraml, U. (2015) 'The missing lynx – Understanding hunters' opposition to large carnivores', *Wildlife Biology*, vol 21, no 2, pp110–119.

Lundmark, C., and Matti, S. (2015) 'Exploring the prospects for deliberative practices as a conflict-reducing and legitimacy-enhancing tool: The case of Swedish carnivore management', *Wildlife Biology*, vol 21, no 3, pp147–156.

Lute, M. L., and Gore, M. L. (2014) 'Stewardship as a path to cooperation? Exploring the role of identity in intergroup conflict among Michigan wolf stakeholders', *Human Dimensions of Wildlife*, vol 19, no 3, pp267–279.

McShane, T. O., Hirsch, P. D., Trung, T. C., Songorwa, A. N., Kinzig, A., Monteferri, B., Mutekanga, D., Van Thang, H., Dammert, J. L. Pulgar-Vidal, M., Welch-Devine, M., Brosius, J. P., Coppolillo, P., and O'Connor, S. (2011) 'Hard choices: Making trade-offs between biodiversity conservation and human well-being', *Biological Conservation*, vol 144, no 3, pp966–972.

Newig, J. (2011) 'Partizipation und neue Formen der Governance', in M. Gross (Hrsg) *Handbuch Umweltsoziologie* (pp. 485–502), VS Verlag, Wiesbaden.

Parker, I. D., and Feldpausch-Parker, A. M. (2013) 'Yellowstone grizzly delisting rhetoric: An analysis of the online debate', *Wildlife Society Bulletin*, vol 37, no 2, pp248–255.

Pelletier, N. (2010) 'Environmental sustainability as the first principle of distributive justice: Towards an ecological communitarian normative foundation for ecological Economics', *Ecological Economics*, vol 69, no 10, pp1887–1894.

Pellikka, J., and Sandström, C. (2011) 'The role of large carnivore committees in legitimising large carnivore management in Finland and Sweden', *Environmental Management*, vol 48, pp212–228.

Peterson, G. D., Cumming, G. S., and Carpenter, S. R. (2003) 'Scenario planning: A tool for conservation in an uncertain world', *Conservation Biology*, vol 17, no 2, pp358–366.

Piédallu, B., Quenette, P.-Y., Mounet, C., et al. (2016) 'Spatial variation in public attitudes towards brown bears in the French Pyrenees', *Biological Conservation*, vol 197, pp90–97.

Pohja-Mykrä, M., and Kurki, S. (2014) 'Strong community support for illegal killing challenges wolf management', *European Journal of Wildlife Research*, vol 60, no 5, pp759–770.

Reed, M. S., Graves, A., Dandy, N., Posthumus, H., Hubacek, K., Morris, J., Prell, C., Quinn, C. H., and Stringer, L. C. (2009) 'Who's in and why? A typology of stakeholder analysis methods for natural resource management', *Journal of Environmental Management*, vol 90, no 5, pp1933–1949.

Schlosberg, D. (2007) *Defining Environmental Justice: Theories, Movements and Nature*, Oxford University Press, Oxford.

Sjölander-Lindqvist, A., Johansson, M., and Sandström, C. (2015) 'Individual and collective responses to large carnivore management: The roles of trust, representation, knowledge spheres, communication and leadership', *Wildlife Biology*, vol 21, no 3, pp175–185.

Sunderland, T., Ehringhaus, C., and Campbell, B. M. (2008) 'Conservation and development in tropical forest landscapes: A time to face the trade-offs?', *Environmental Conservation*, vol 34, no 4, pp276–279.

Varum, C. A., and Melo, C. (2010) 'Directions in scenario planning literature – A review of the past decades', *Futures*, vol 42, no 4, pp355–369.

Young, J., Watt, A., Nowicki, P., Alard, D., Clitherow, J., Henle, K., Johnson, R., Laczko, E., McCracken, D., Matouch, S., Niemela, J., and Richards, C. (2005) 'Towards sustainable land use: Identifying and managing the conflicts between human activities and biodiversity conservation in Europe', *Biodiversity and Conservation*, vol 14, no 7, pp1641–1661.

Young, J. C., Jordan, A. R., Searle, K., Butler, A. S., Chapman, D., Simmons, P., and Watt, A. D. (2013) 'Does stakeholder involvement really benefit biodiversity conservation?' *Biological Conservation*, vol 158, pp359–370.

Part II

Heterogeneity in perceptions of and behaviour towards large carnivores

6 A community divided

Local perspectives on the reintroduction of Eurasian lynx (*Lynx lynx*) to the UK

Steven Lipscombe, Chris White, Adam Eagle and Erwin van Maanen

Introduction

The recent past has seen the number and range of large mammals, particularly carnivores, increase across Europe (Chapron et al., 2014; Milanesi et al., 2017). Factors including agricultural land abandonment (MacDonald et al., 2000; Sieber et al., 2015) and changes in hunting practices (Redpath et al., 2017) have facilitated this recolonisation. Simultaneously, there is growing appetite for reintroduction projects (Van Heel et al., 2017) and rewilding, defined as 'large-scale conservation committed at restoring and protecting natural processes and states in core wilderness areas, providing effective connectivity between such areas, and protecting or reintroducing apex predators and other keystone species' (Soulé and Terborgh, 1999). The combination of these factors have placed these issues at the forefront of public and legislative awareness (Trouwborst, 2010).

Within the UK, there has been a growing interest in the reintroduction of keystone species, including apex predators, within recent years, encouraged in part by the awareness generated by the reintroductions of extirpated species such as the pine marten (*Martes martes*) and Eurasian beaver (*Castor fiber*) in certain areas, as well as growing interest in the potential for keystone species to encourage ecosystem restoration (Hetherington, 2006). While contemporary public perceptions towards apex predators in the UK remain favourable (Smith et al., 2015a), the reintroduction of the Eurasian wolf (*Canis lupis lupis*) and the Eurasian brown bear (*Ursus arctos arctos*) as historical members of the UK predator assemblage is unlikely to occur in the near future (Wilson, 2004; Nilsen et al., 2007). As cursorial predators that adapt well to human-modified landscapes, wolves are perceived as a threat to humans and livestock amongst rural farming communities (Wilson, 2004; Nilsen et al., 2007; Wagner et al., 2012). Furthermore, given the significant spatial requirements of wolves and bears, it is doubtful if enough suitable connected habitat remains in the UK to establish minimum viable populations and maintain genetic diversity (Wilson, 2004). The Eurasian lynx (hereafter lynx), on the other hand, poses no threat to people and while they are also a potential predator of livestock, this is typically 'low-level' (Angst and Breitenmoser, 2003), although there is potential for lynx to switch prey from deer to sheep depending on the relative abundance of prey species (Odden et al., 2013).

Lynx have returned to areas in Western Europe where they had previously been extirpated, partly due to successful reintroductions but largely due to natural recolonisation by recovering populations (Trouwborst, 2010; Kaczensky et al., 2012); however, the recolonisation of lynx to the UK is prevented due to its isolation from mainland Europe. The reintroduction of lynx has gained considerable attention following recent developments towards an application for a controlled trial reintroduction by the Lynx UK Trust (LUKT). Anthropogenic factors such as severe deforestation, declining deer populations and persecution are likely to have caused the extirpation of the lynx in Britain in the 5th Century AD (Hetherington, 2006). Subsequently, the UK has experienced extensive reforestation (Rackham, 2001), a significant shift in hunting practices and a complete cessation of fur trapping and trading. Additionally, due to the over-abundance of deer species in the UK (Jobin et al., 2000; Odden et al., 2006; Basille et al., 2009), lynx are now likely to thrive in many areas of Scotland and England.

The historical extirpation of large carnivores across many parts of Europe, especially in the UK, has resulted in communities, populations and landscapes with little or no experience of coexistence with large predators (Hetherington, 2006; Heurich et al., 2012; Chapron et al., 2014). While charismatic carnivores have significant cultural symbolism (Hetherington, 2006; Sergio et al., 2006, Van Heel et al., 2017) and are often promoted as flagship species for the wider conservation cause (Simberloff, 1998; Andelman and Fagan, 2000), the reintroduction of charismatic animals presents challenges for conservation practice, not least in terms of managing often vehement opposition (Arts et al., 2012).

Species reintroductions were initially quantified in terms of ecological success (Griffith, 1989), but it became increasingly apparent that public concerns regarding translocations needed to be addressed (Marshall et al., 2007; O'Rourke, 2014) and that successful conservation projects required effective integration of the immediate society (Mascia et al., 2003). Public consultation is now an integral feature of the International Union for Conservation of Nature (IUCN) reintroduction guidelines (IUCN/SSC, 2013). Carnivore reintroductions also require consideration for conservation plans over large spatial scales (Kaczensky et al., 2012), creating a demand for wide-scale consultation.

Wilson (2004) identified that attitudes to reintroductions (and particularly reintroductions of carnivores) tended to be favourable among the general public, but negative among those likely to be adversely affected. For example, the illegal persecution of lynx in Switzerland is the legacy of a reintroduction program in the 1970s that excluded and disenfranchised sheep farmers and hunters (Breitenmoser and Breitenmoser-Würsten, 2004). Switzerland was the first country to authorize the reintroduction of lynx to the Alps following the ratification of a Swiss federal government resolution in 1967 (Breitenmoser and Breitenmoser-Würsten, 2004). However, the subsequent releases were clandestine and initially denied in public. Furthermore, the founder population animals were not radio-collared, and no post-release monitoring was undertaken (Breitenmoser and Breitenmoser-Würsten, 2004). The lack of public consultation and covert nature of the releases created conflict within rural communities and resulted in illegal killing as the most common cause of adult mortality in the early post-reintroduction period (Breitenmoser and Breitenmoser-Würsten, 2004; Schmidt-Posthaus et al., 2002).

As a result of such experiences, it is now accepted that in addition to ecological research, reintroduction outcomes are determined by the attitudes and behaviour of the public and regional stakeholder groups (Marshall et al., 2007; Thirgood and Redpath, 2008). Therefore, a broad-based public consultation is an essential tool to reveal contentious issues and identify perceived or actual threats to the interests of any party. These findings will enable conflict mediators to acknowledge concerns and seek solutions through an inclusive and transparent approach to public engagement.

With this in mind, the authors undertook consultation activities in relation to a trial reintroduction of lynx in the Kielder Forest area of Northumberland, UK. This chapter aims to evaluate public opinion regarding the socioeconomic and environmental impacts of the trial, based on these activities. Here we present background information and outline the methods used and interim results, and examine the lessons learned, with the intention of informing future projects.

Background

A national survey (Smith et al., 2015a) was conducted during 2015–2016 to investigate the public desirability of a lynx reintroduction project, and this work identified broad public support for such a project. In October 2015 the LUKT announced its proposal for a 'controlled, scientific and monitored trial reintroduction of lynx' to England and/or Scotland. Smith et al. (2015b, 2015c) outline the proposed consultation process and details on the feasibility, benefits and opportunities, risks and impacts, and potential mitigation measures. These documents made extensive use of and reference to knowledge and experience gained from mainland European lynx reintroduction projects. It was through these documents and associated consultation activities that an open invitation was extended for all stakeholders to actively participate in a transparent, accessible, unbiased and constructive process of discussion and collaboration.

A subsequent national stakeholder consultation exercise based on the content of these documents (Smith et al., 2016), sought views on pre-project desirability and feasibility; socio-economic and ecological considerations; location of trial sites; planning, preparation and release stages; and post-release activities. The LUKT initially identified five potential trial sites as worthy of further investigation. Of these, three (Cumbria, Thetford Forest and Kintyre Peninsula) were excluded after feedback from national stakeholders identified them as less preferred due to concerns including higher road density, a wider range and higher density of livestock species, and, in the case of the Kintyre Peninsula, added pressure on farmers with existing concerns that reintroduced sea eagles are predating lambs (White et al., 2016a).

After more detailed socio-ecological focused work (including site size and connectivity, woodland cover/density, prey availability, human density, road networks and potential for economic development) on the remaining two sites, the Kielder Forest area, an extensive forest block that straddles the border between England and Scotland, was identified as the most suitable location for further investigation (White et al., 2016a). Consultation activities were conducted in Kielder during August 2016 to April 2017 as a precursor to a project licence application being submitted by LUKT to Natural England and Scottish Natural Heritage. The proposed reintroduction was presented as a time limited, five-year trial of a small number of individuals with the stated aim of establishing the information necessary to support a decision regarding full reintroduction.

The Kielder community consultation process

Convery et al. (2016) set out a detailed plan for local consultation and engagement activities. In brief, open community meetings, informal pub drop-in sessions, door-to-door visits and attendance at local parish meetings were planned and delivered with the aims of introducing the project and giving members of the community a chance to ask questions. For most of these activities a questionnaire was used to collect community members' opinions on the key risks and benefits of the proposed trial. These were used to provide context and statements to develop a Q methods approach. Further focus group meetings were planned with key groups likely to be impacted (business, farming and forestry communities), in order to get input on the development of key areas of project policy and to establish a communications platform for ongoing engagement work. These were only partly delivered at the time of writing. It was also our intention to undertake validation meetings with focus groups representative of the wider communities.

The aim was to complete consultation activities set out in the plan by March 2017 in order to inform the licence application to Natural England. However, due to the approach, which put an emphasis on flexibility and responding to community needs and requests, our timeline inevitably changed and at the time of writing, key areas of the consultation plan (in particular farmer engagement regarding compensation for livestock depredation) had not been addressed, though a licence application had been submitted to Natural England by LUKT. As a previously resident native species, the lynx is a species of a kind which is not ordinarily resident in and is not a regular visitor to Great Britain in a wild state and is therefore a species which may not be released in Great Britain under section 14 of the Wildlife and Countryside Act (1981). However, a licence for such a release may be issued under section 16 of the Wildlife and Countryside Act (WCA, 1981) and the assessment of an application for such a licence will usually be based on the requirements of the IUCN Guidelines for Reintroductions and Other Conservation Translocations. The design and implementation of the Kielder-based lynx reintroduction project was closely aligned to the IUCN requirements for wide ranging biological and socio-economic feasibility studies and risk assessments. The licence was supported by a project plan which positioned the reintroduction within legal and policy frameworks, a statement of the project rationale and a list of reports as appendices including, amongst others, National Consultation Reports (Smith et al., 2015a; Smith et al., 2016), Cost-benefit Analyses (White et al., 2015), a Site Selection Report (White et al., 2016a), and a Disease Risk Assessment (Mayhew et al., 2017, unpublished).

At the time of developing the consultation strategy, the LUKT was considering release sites within the Kielder Forest block (Figure 6.1), which sits largely in the English county of Northumberland, with some segments in the Scottish Borders region (White et al., 2016a). This area had been identified by Hetherington et al. (2008) as a potential lynx habitat network covering 1980 km² (White et al., 2016a). Spatially, a zoned approach to consultation work was adopted, comprising a primary zone and a secondary zone (Figure 6.1). The primary zone comprised communities or individuals identified as most likely to be affected, either directly or indirectly, by the presence of the lynx. The surrounding secondary zone included communities less likely to be affected but which should nevertheless be engaged with and members given the opportunity to respond to the consultation process. In addition, it was stipulated that if a population self-identified as being a relevant stakeholder, then engagement could increase to meet such need.

Direct discussion with the local community was achieved through a number of routes, with the aims of explaining the reintroduction proposal and the consultation process, recording or addressing any initial concerns, collecting data on perceived risks and benefits, and to build local contacts.

Community discussion routes

Parish council meetings

Three Parish Councils were located in the primary zone. Of these, two accepted an invitation to discuss the proposals, resulting in one closed Parish Council meeting and one meeting open to the public.

Open public meetings and drop-in sessions

Open public meetings were held in the main centres of habitation within the primary and secondary zones. Drop-in sessions were held at a centrally located pub or held after public meetings allowed for longer discussions on a one-to-one basis.

Figure 6.1 Location of Kielder forest within the UK and zoned approach to community engagement, indicting primary inner zone (inner cycle) and secondary zone (outer cycle).

© Crown Copyright and Database Right 2018, Ordnance Survey (Digimap Licence)

Door-to-door

Door-to-door consultation focused within the identified primary zone, with some closely neighbouring residential areas falling within the secondary zone also visited. In addition, any further residents within the secondary zone requesting participation were also visited. Approximately 12 volunteers from LUKT, the University of Cumbria Centre for Wildlife Conservation and/or AECOM participated in activities over a period from October 2016 to June 2017. These included daytime and early evening calls on both weekdays and weekends. We aimed to visit all households within the primary zone; however, at the time of writing, these activities had not been completed by the authors due to the cessation of their involvement with the consultation activities. If the resident(s) were at home, we offered a brief introduction to the project as well as an opportunity to ask questions. We also asked to collect information through the risks/benefits questionnaire, which was optional. Where there was no answer, an information leaflet was left, along with contact details to request a visit or further information.

Farming community engagement

A combined approach of individual farm visits and a farm focus group was used to enable detailed discussion and constructive dialogue to inform the licence application in the pre-licencing period and with the intention of establishing a farmer-led forum to contribute to project management in the post-licencing period. Three focus group meetings were held, one including the attendance of the local auctioneer, to focus on the evaluation of practical, site-specific management prescriptions to mitigate lynx predation on livestock and to design a retrospective compensation/conservation payments scheme. Following these initial meetings, the authors' consultation activities ceased, and follow-up meetings did not occur to our knowledge.

Business community engagement

Attendance at three local and regional business forums enabled detailed discussion about the benefits and risks of the project specific to local businesses. Local businesses were also visited to allow for lengthier personal conversations, resulting in a tourism-related business event to discuss opportunities and threats to the local tourism trade. At this meeting, the businesses in attendance showed interest in forming a business forum to enhance the potential benefits of the project. However, the authors' consultation activities ceased, and further communication with this group did not occur to our knowledge.

Friends of the lynx

In response to requests from supportive members of the community, a 'Friends of the lynx' group was established to discuss in more depth the ecology of the lynx and the reintroduction rationale. This gave supporters a forum to discuss how best to voice their support and to commit to specific activities. However, this work did not continue when the authors' consultation activities ceased, and further communication with this group did not occur to our knowledge.

Supplementary activities

Further activities were undertaken when opportunities arose to raise awareness of the consultation and increase participation. These included presentations to local groups such as the Natural History Society and visits to a local school. Efforts were also made to obtain fair coverage in the local press.

Data collection

Data collected throughout the consultation process includes meeting notes and transcripts, interview transcripts and sorts from the Q study (n = 15), and information collected from individuals through an open-ended risk/benefits questionnaire (n = 130) developed in order to provide a snapshot of key community concerns and to feed in to the development of other consultation and project plans.

Q methodological study

Q methodology (QM) is a research tool designed to explore individual values, opinions and beliefs regarding a specific subject area. It is particularly useful in community engagement with smaller groups and has proven particularly useful in identifying 'common ground' in conflict management situations and in capturing interesting, informative and relevant viewpoints relative to the question (Watts and Stenner, 2012). In environmental and conservation research, QM has been used in a wide range of contexts, including wind farm development, public opinion on shale gas, afforestation schemes, wildlife management and landscape restoration. QM typically involves a 1–1.5-hour interview in which the participant ranks a set of statements relevant to the topic depending on how strongly they feel about each. A Q set of statements (Annex 6.1) was extracted from the national and local consultation activities, including an online public survey, notes and recordings from local consultation meetings, and questionnaires. We have completed 15 interviews to date, including participants of various ages and occupations and of varying degrees of support for the project residing in the primary zone. It is recommended that a Q study includes approximately 40–60 participants (Watts and Stenner, 2005). Factor analysis would then be undertaken to interrogate the data set. It is therefore premature to do a full factor analysis on the resultant data and the participant set is too small to make any conclusions. However, an initial examination of the 15 sorts and accompanying interviews does result in some interesting initial findings

Table 6.1 Questionnaire administration and number of questionnaires selected

Questionnaire administration	Number (n = 130)
Door-to-door	86
Open meeting: Tarset	5
Open meeting: Newcastleton	3
Open meeting: Langholm	7
Presentation: Borders Natural History Society	17
Presentation: Bellingham Business Forum	12

and emergent themes which may be of value to future projects, and we there-fore include these in our discussion below.

Open-ended risk/benefits questionnaire

The open-ended risk/benefits questionnaire included basic demographic data and asked participants to list what they believed to be the key risks and benefits of the proposed lynx reintroduction, as well as an open comments field and the option to give contact details in order to be contacted by the project team in the future. In total, 130 people have completed the questionnaire to date, dur-ing either door-to-door activities or after a consultation meeting (Table 6.1).

Demographically, the current sample has accurate gender representation but is under-representative of ages 24 and younger, reflects the 25–64 age group well, and is over-representative of those aged 65 and older, in comparison to demographic data for Bellingham ward (Northumberland County Council, 2011), which includes the Kielder, Tarset and Greystead communities.

Emergent themes from the Kielder community consultation

At the time of writing, the authors were unable to progress with consultation activities relating to the proposed lynx reintroduction, and therefore our data is incomplete. However, we present our interim findings here as the data points to emergent themes which we believe should be addressed in ongoing project activities and because they would be of value to similar projects. These themes have been extracted from the data collected through the risks/benefits ques-tionnaires (n = 130), the Q sorts and interview transcripts (n = 15), and notes and transcripts from meetings.

To create a snapshot of responses from the risks/benefits questionnaire, we grouped comments into themes under risks (15 themes) and benefits (nine themes) (Figures 6.2 and 6.3). Community members were given the opportu-nity to elaborate more fully on these themes during meetings and in Q inter-views. Below we present and discuss the most prevalent of these themes.

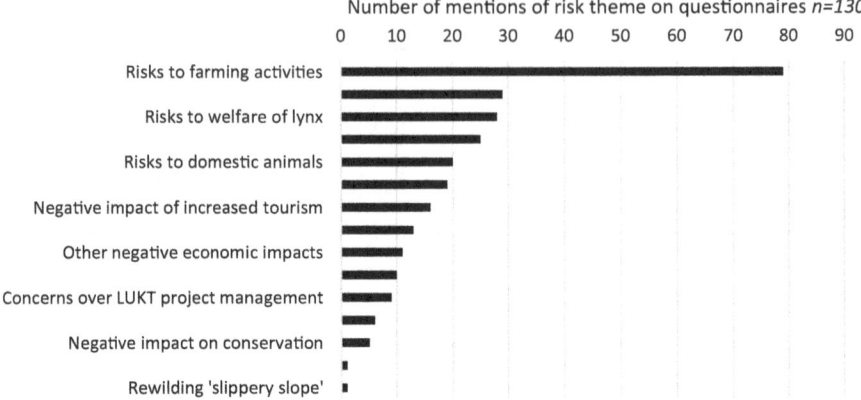

Figure 6.2 Key risks raised on community questionnaires.

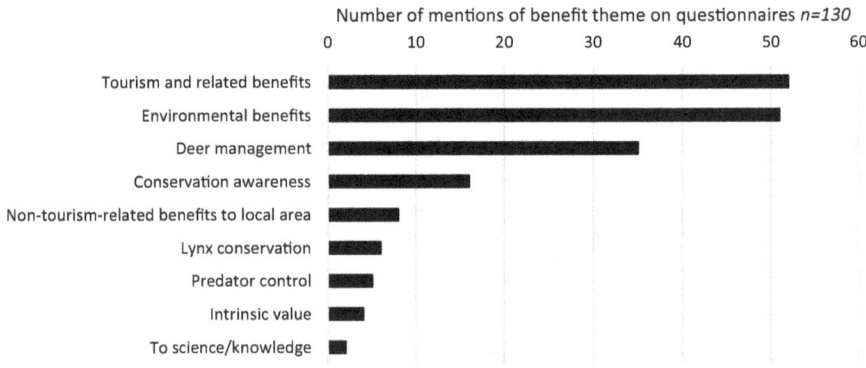

Figure 6.3 Key benefits raised on community questionnaires.

Farming

Our data reflected that hill farming formed an integral part of the culture, economy, landscape and overall sense of place in the Kielder area. Consequently, this theme was prevalent throughout the consultation process and was a talking point among almost all participants, whether for, undecided or against the proposed trial reintroduction.

Given the risk of lynx-related livestock predation, the farming community was considered by project partners as a key stakeholder group to engage. However, during the early stages of the local consultation, members from the farming community expressed their disapproval of the project by refusing to talk to team members at door-to-door visits or by voicing their anger at public meetings. Common themes expressed by farmers at public meetings included a lack of trust and transparency in the LUKT, the potential for lynx to threaten their livelihoods and the need for a compensation scheme, the inability of farmers to control growing lynx populations and a sense of disempowerment that the reintroduction would be imposed on them regardless of their views.

Risks listed on risks/benefits questionnaires were predominately focused around risks to farming (Figure 6.2). This theme includes comments related to *'risks to livestock'* (including predation and worrying; for example, *'no guarantee that they won't attack sheep'*), negative impacts on farmer livelihoods, risks that compensation will not be easily accessible and the impacts on farmer workloads due to compensation or mitigation measures.

During Q sorts, there was strong agreement for the statements *'it is important that adequate compensation agreements are in place, should the lynx cause any destruction'* (Annex 6.1, statement 12), *'I am concerned that lynx will be a threat to livestock during the trial'* (Annex 6.1, statement 8) and *'I am concerned that the lynx will cause economic suffering to farmers and/or countryside managers'* (Annex 6.1, statement 9; one participant strongly disagrees with this statement). This shows that even supporters of the proposal see the need for engaging with the farming community and mitigating possible negative impacts on farming practices. On the risks/benefits questionnaire this is also reflected, as many of those respondents who were *'for it'* could *'still see the point of view of farmers'*.

Accompanying Q interviews, however, indicate a polarity in support for farmers. It is worth noting that none of the 15 participants were directly involved in farming activities. Some individuals were very supportive of farmers and made efforts to justify their concerns:

> There's a feeling that comes across from the Lynx Trust about thinking that farmers don't care about their natural environment . . . or about their sheep but they do passionately.
>
> (Q participant 8, primary zone)

> The big concern is that people, if their livestock is a target, that there will be compensation. People don't want to be seen as money-grabbing farmers that just want the money. It's not about that. It is a livelihood and it is something that could go wrong.
>
> (Q participant 12, primary zone)

> A lot of the farmers in this area are very much on the edge of being able to earn a living, in the first place. Quite a lot of them are tenants, and not landowners themselves. . . . They've got very little possibility of buying somewhere if they retire, if they ever retire. They tend to be quite an elderly population, farming well into their 70s and what they get at market for a lamb just about covers the cost.
>
> (Q participant 6, primary zone)

Others expressed displeasure about how dominant farming is in the area, particularly in decision-making:

> The only thing that really concerns me is the intransigence of the NFU [National Farmers' Union]. . . . They just think 'right we're not talking to you' rather than that we can come to some kind of arrangement where if it does go ahead we can continue the conversation. Farmers I've spoken to and asked whether this is a money issue and they've said 'no because we love our animals so much that we just can't bare them being eaten or even taken'. They're farming because they want money so at the end of the day it is about money. . . . I am concerned there is a no-go area which is why should we reintroduce anything that's going to harm their animals.
>
> (Q participant 1, primary zone)

> The 40 farmers who live in this area are a minority . . . they got £3.6 million in subsidies last year.
>
> (Q participant 9, primary zone)

However, even those who were less supportive of farmers saw the value in compensation as a method for protecting the welfare of the lynx:

> The major industry around here is . . . hill farming, and they're certainly going to need to be persuaded that if they make any losses then they will be compensated, otherwise . . . people are just going to shoot [the lynx].
>
> (Q participant 4, primary zone)

> I don't think it's that important for farmers, but I care about the welfare of the lynx. And without adequate compensation agreements in place the farmers are just going to shoot them aren't they, or poison them. That's the way they think. . . . I don't like to see farmers compensated because they get enough elsewhere. And they will just scam it. . . . I'm surprised that they don't see it that way . . . that it's another cash cow for them.
>
> (Q participant 9, primary zone)

Several other participants raised concern about the possibility of corruption around compensation claims or issues with management and enforcement of a compensation scheme:

> Who's going to calculate whether a lamb that is three weeks old is going to be a prize-winning tup? The farmer will have some idea but whether that will hold any weight . . . we're looking at over £1000 or more for a prize-winning tup against a lamb which is £30/£40. I think it's an absolute nightmare of how on earth it's ever going to be sorted. How do you calculate and compensate for potential?
>
> (Q participant 6, primary zone)

Tourism

The polarised views on the potential increase in tourism in Kielder further supported the theory that urban-to-rural migration, and changing economic activities (Woods et al., 2012) may be the source of much of the conflict among the community. On risk/benefits questionnaires, the potential for tourism and related economic benefits was predominant (Figure 6.3), with respondents listing, for example, '*tourism and recreation income*', '*the project will 'bring in tourism, [it will be] great for the local economy*', with the potential to '*put Kielder on the map as a destination*' and '*raise profile of Kielder for ecotourism*'.

However, others thought that the case for tourism was overstated: '*possible jobs in the area, case not proven*' or that tourists would cause problems locally by '*clogging up the roads*'. The negative impact of tourists was frequently mentioned: '*increase in tourism not necessarily desirable [it could] adversely affect the environment and community*'. Simultaneously, there were concerns that the reintroduction would '*scare tourists off*'. One respondent noted that:

> we do not live in a Transylvanian forest or a theme park. People who choose to live, work and take holidays here do so because it is a quiet, domesticated, typically English landscape.
>
> (questionnaire respondent; primary zone)

This polarisation was reflected in the Q sorts with both strong agreement and disagreement for the statement '*Lynx could beneficially add to the rural economy through eco-tourism*' (Annex 6.1, statement 13). Those that disagreed with the statement could not see the potential for tourism due to the shy nature of lynx:

> Certain people go on and on about millions of pounds coming into the community but we can't see how or why. And certain people don't want thousands of people streaming in. It's supposed to be a national park that's quiet and peaceful and not too busy. There is a balance between people who want tourists and people who don't.
>
> (Q participant 8, primary zone)

> It won't add to the tourism. That was one thing that annoyed me when they [volunteers] came to the door you see, because I'd been saying I was scared about them killing [squirrels] and I was saying this to them and they said 'oh you'll never see them, they never come out of the forest'. So what are tourists going to come and see? It's contradictory. They can't have it both ways.
>
> (Q participant 14, primary zone)

A place for lynx in our landscape and socio-ecological system

Our data reflected that there are conflicting views over whether lynx would fit in the landscape, with concerns raised over habitat suitability, landscape suitability and risks to humans, pets and native wildlife. The potential for landscape and ecosystem restoration was recognised by some. Environmental benefits listed on the risks/benefits questionnaires (Figure 6.3) largely focused on the *'overall benefit to the ecosystem'* and the potential of lynx to act as *'an ecosystem engineer, improving the quality of native woodland'*, with the reintroduction being *'the first step in creating a functioning, diverse ecosystem!'* There was also emphasis on returning a native species – *'it seems right to have them back where they belong'*, *'they have more of a right to be here than us'* – and the potential *'to increase biodiversity, Kielder is a monoculture'*.

The fact that lynx are an *'apex predator [with] long term benefits [to the] predator/prey balance'* and the potential to *'keep deer population under control'*, including *'reducing [the risk of] car accidents'*, was also prevalent. The potential to control numbers of other species, such as fox, was also mentioned. The intrinsic value was recognised, too, such as the *'cultural/spiritual effects of a rewilded landscape'* and *'creating the sense of the wild'*.

However, comments also included concerns about *'interfering with nature'*, such as *'every time man interferes in ecosystems there are unexpected side effects and more times than no they are unwanted'*.

There were conflicting views over whether Kielder would provide suitable habitat for the lynx. The Q sorts indicated there was strong disagreement for the statement *'I do not think this area is suitable for the lynx'* (Annex 6.1, statement 28), indicating that some people felt the habitat in the Kielder area may be suitable. Several comments during interview supported this, such as:

> I imagine once you release the lynx into that [forest] they're just going to retreat into it and you'll never see them again. I don't see any problems with that forest at all.
>
> (Q participant 9, primary zone)

However, this contradicted some concerns raised in the questionnaires about plantation forests being unsuitable habitat for lynx, as well as other comments made during the accompanying interviews, such as:

> This is a man-made forest, not a natural forest, and there are species that weren't here in the first place.
>
> (Q participant 6, primary zone)

> It seems to me that there isn't enough space for them to survive, it's not a very attractive place to be honest, very thick horrible forest and . . . so dense nothing could possibly live in there.
>
> (Q participant 8, primary zone)

At a larger scale, there was direct conflict about the statement *'the British countryside is no longer a suitable place for a sustainable lynx population'* (Annex 6.1, statement 2), with two participants strongly agreeing, and two strongly disagreeing with the statement. The interviews suggested that those in agreement were not necessarily against the proposed reintroduction:

It's no longer suitable for most things. The lynx project should be alongside a much bigger commitment to reforestation and a deep look at the farming practices in this country.

(Q participant 1, primary zone)

The British countryside ... well it needs so much work really. The farmers have really ruined it.

(Q participant 10, primary zone)

There were also inconsistent views over the statement *'Lynx are not compatible with our society today'* (Annex 6.1, statement 16), with three participants strongly disagreeing with and two strongly agreeing with the statement. Comments around this topic in interview reflected the polarisation, but largely focused on landscape issues and people's perception of nature's place within their landscape. Many comments emphasised the need to reform and restore nature within the landscape, with a focus on people's perceptions of nature as part of that process:

It's a landscape issue. People don't understand a wild landscape here, they think it's too messy.

(Q participant 1, primary zone)

If we're talking about putting ecosystems back then we should be talking about a completely different landscape. We wouldn't have this sort of forest in the first place and the weather would be different and everything else would be different. So it's tinkering at the edges of it. It's not really a possible option. Do we go back to woolly mammoths or do we go back beyond that again? We need to look at what we've presently got and whether it's viable and in some cases if it's not viable then we just need to accept it rather than throwing more money and efforts at it.

(Q participant 6, primary zone)

No, they're not compatible with the way we live. And with the wildlife – that must have changed a lot over the years. I just don't think they fit in. I don't think you should go back in history, you can't repeat what's gone. . . . Things evolve, we've evolved, the animals evolve, not always in the right way probably but it is evolution isn't it?

(Q participant 14, primary zone)

However, some did point to acceptance of the current landscape being able to support lynx:

I don't think there's going to be any major problems, lynx have been in this landscape before and we can look at other landscapes and see how the lynx functions in those and they don't seem to have any problems so I don't see there being any problems.

(Q participant 9, primary zone)

Supporters largely expressed excitement at the potential for ecological restoration, and the resultant increase in tourism, for example:

> I'm quite passionate about the introduction of wild species into our landscape. I think it has to be a suitable landscape, and I think here, in and around Kielder, we should really be proud that we have a landscape where they can be reintroduced. If the research has been done and they could flourish here then that would be absolutely amazing. I'm really passionate about the fact that it would create an exciting sense of wilderness . . . I also think it's important in terms of ecosystems and landscape management to actually have wild space and everything that goes with it.
>
> (Q participant 10, primary zone)

On risks/benefits questionnaires, concerns were raised about lynx being *'wild animals'* that are difficult to control (Figure 6.2) and that it would be *'hard to track the young'*, as well as over the expanding lynx population becoming *'out of control'*; *'everything that gets introduced then has to be culled'*.

Another recurring theme was the concern for the impact of lynx on native wildlife, with red squirrels most frequently mentioned, along with the Kielder wild goat population, ground-nesting birds and the recently reintroduced water voles. On risks/benefits questionnaires respondents conveyed fears that lynx could cause the *'downfall of indigenous rarities'*.

In Q sorts, there was strong agreement with the statement *'I am concerned that the lynx will pose a threat to our native wildlife during the trial'* (Annex 6.1, statement 6; only one respondent strongly disagreed with the statement). This was also a common theme in accompanying interviews, with statements such as:

> I'm concerned very much about the wildlife that is already there and that everybody is desperately trying to preserve – squirrels, black grouse, hen harriers. A lot of us are very concerned about that.
>
> (Q participant 8, primary zone)

with some understanding that:

> native wildlife includes roe deer and other species they will prey on, but that's the natural order of things.
>
> (Q participant 1, primary zone)

There were also concerns regarding possible lynx attacks on children; *'the safety of families living on remote farms, especially with babies or young families.'*

In Q sorts, there was strong disagreement across the participant group to the statement *'I am concerned that the lynx will be a threat to people during the trial'* (Annex 6.1, statement 22; nine participants strongly disagreeing). However, two participants strongly agreed with the statement. Throughout the trial, the consultation team communicated that there has been no account of a healthy, wild lynx ever attacking a human being. While the participants, including those for and against the trial, generally accepted this as evidence that lynx were not a threat to humans, there was still some reluctance within the community and therefore further evidence should be sought. One possibility for allaying these concerns further is Gehr et al.'s (2017) 'landscape of coexistence', which suggests that apex predators will avoid human contact, in a similar way to a 'landscape of fear' altering prey behaviour.

The concern that lynx will be *'a danger to dogs'* and they may *'come into the village, eating pets, cats'* was more prevalent. Other people felt reassured by the information provided by LUKT: *'I was scared of walking with dogs at first, but I'm not worried about this anymore.'*

Divisiveness in the local community

As the consultation progressed a risk that began to emerge from the risks/benefits questionnaire was that of divisions in the local community (Figure 6.2), with the project *'causing social divisions in the community'* (*'I've been criticised by locals for being positive'*, *'alienating the farming community'*), or within households (*'I'm for it but my wife is against'*). During Q interviews, where participants were given the opportunity to reflect on this in more detail, it became clear that this division was largely based on polarity of views around farmers' rights, land ownership and land use. These two examples extracted from Q interviews summarise some of these arguments:

> I think there's a lot of the people in this area who are not farmers born and bred and not closely connected with that culture, there's quite a number of them that already don't quite fit in and don't understand, and don't accept that they are coming in to one of the biggest factories there is – a mutton and lamb producing factory. They don't like the mud on the roads, they don't like the fact that the gates are shut at certain times of the year. They came for the pretty countryside but they don't actually like the reality of it. There's people who don't understand the hunting, shooting, fishing aspects and therefore certainly struggle to understand, unless they actually saw a lamb or sheep that had been killed. They've got this sort of fluffy pretty image of nature, and nature isn't like that.
>
> (Q participant 6, primary zone)

in comparison to:

> Your project isn't creating divisions in the community, those divisions were already there and I think your project has highlighted the notion that those that farm the land should be able to dictate what goes on everywhere. It's an outdated sense of ownership that they have over the landscape and it doesn't exist in our modern classless society. But they're so removed from

all the social movements that have created an open and cohesive society that they still think they're living in a feudal landscape and . . . I think most sections of the community are for it, it's just some of the lobbyist movement that represents the farmers and landowners is so vocal and very organised, to the point that they will intimidate everybody else into either not speaking or coming forward, or towing the line with their views. And it's with threats and intimidation.

(Q participant 9, primary zone)

Woods et al. (2012) described urban-to-rural migration and diversification of economic activity in rural areas as potential causes for this type of conflict. Proposing to reintroduce a carnivore within this context may have exacerbated this divide and instigated the extreme views we encountered.

Welfare of the lynx

Our data indicate that concerns for the welfare of the lynx is high (Figure 6.2), with concerns over lynx being harmed in road traffic accidents, an *'increase in illegal poisoning'* (questionnaire respondent) and the risk that farmers might *'club together to shoot lynx'* (questionnaire respondent). During Q interviews, the causes for concern included the threat of conflict with farmers, which links to comments related to the need for a robust compensation scheme to reduce this risk:

> I'd like the project to be introduced, and the welfare and safety of the lynx I think that's the biggest threat to your project. It only takes one or two idiots with a gun or some poison to destroy the whole project. I can see how much money and time and effort's been put into it and the actions of one disgruntled landowner with some poison would in effect destroy hundreds of thousands of pounds worth of work, and all those man hours, and the welfare of that animal. So I think some serious consideration needs to be given to how you're going to ensure those animals aren't trapped or poisoned. It's going to take a lot of security.
>
> (Q respondent 9, primary zone)

Other concerns included placing lynx in an unsuitable habitat and the prospects of the lynx ending up in captivity or being exterminated if the trial were unsuccessful:

> Where do the lynx go and what happens to them? Even if they are successfully rehomed, what impact does that have on them and on wherever they go next?
>
> (Q participant 6, primary zone)

Concerns about project and consultation management

Due to the high-profile nature of the project within the community, the consultation process itself has been under intense scrutiny and has become a talking point among community members. On risks/benefits questionnaires, some concerns were raised regarding consultation activities (Figure 6.2), '*your consultation seems to have been seriously lacking*' and various aspects of the project plan (e.g., '*six* [trial lynx] *is a low number – what if they die soon, will the project be over before it is started?*'). There were also concerns about a lack of funding and '*funding being eaten up by compensation payments*'. Risks to the '*reputation of conservation in general*' and the '*potential to prevent future reintroductions of lynx or other species*' were also raised.

There were also concerns raised regarding the public meetings; for example, following the Tarset public meeting one attendee commented that: '*I object to your patronising and high-handed methods in trying to force your project onto the community. I'm not confident that your consultation or research are impartial.*' Similarly, another attendee stated that '*this proposal seems to be a product of fanatical lynx enthusiasts. It is being imposed on us.*'

Throughout activities, members of the community highlighted the need for scientific rigour and transparency over plans and decision making. In Q sorts, for example, importance was placed on the '*use of biological data and sound science in this trial of introducing lynx*' (Annex 6.1, statement 7), with five participants strongly agreeing and no participants strongly disagreeing. There was also strong agreement for the statement '*all aspects of the trial must be transparent and open for all*' (Annex 6.1, statement 19) from three participants, with no strong disagreement.

Simultaneously, there was a general sense that people were unable to make an informed decision due to a '*general lack of knowledge*'; '*ignorance is the main problem, we don't know anything about lynx*' (both quotes extracted from questionnaires). There was also increasingly a concern about '*misinformation*' causing '*fear without knowledge*'.

The authors themselves encountered misinformation during consultation activities, such as:

> I had heard rumours, or maybe it was in the paper, that if a lamb was found a certain distance from the farm there wouldn't be any compensation.
>
> (Q participant 13, primary zone)

> Having talked to you last time we were talking about the six lynx for five years [the proposed number of released lynx], which I was okay with, that sounds amazing. But when I talked to [my neighbour] who made it to one of your meetings, she said apparently there was actually going to be more than 50.
>
> (Q participant 2, primary zone)

Several probable causes for this were raised by community members during the consultation, including unsuitable communication methods used in consultation leading to misinterpretation of information presented, and misinformation being spread by different parties:

> It's not been communicated to the people around here. If you [don't] people will take their own ideas and their ideas will become defacto truths.
>
> (Q participant 4, primary zone)

> I'm concerned that our politician, Guy Opperman, has become involved and waded in. That he's just representing a very minor section of the community and making claims that 90% of the community are against the lynx and if anything he's done a little bit of anecdotal research.
>
> (Q participant 9, primary zone)

The miscommunication of the level of local support in local (Hexham Courant, 2017) and national press (e.g., Halliday and Parveen, 2017) further exacerbated these concerns and highlighted the difficulty of basing communications on quantifiable and methodically produced evidence when tensions and passions run high.

The aims during early stages of the consultation were to gather opinions, share information on the proposal and build a network of contacts within the community. However, initially at least, discussions at open public meeting were dominated by those who had very strong concerns, which were often expressed in anger, and mutually informative discussions often became impossible. The initial introduction of the project to the community at a public meeting in Kielder in August was particularly heated, with strong representation from National Sheep Association representatives at both a national and local level. Comments in the press (Hexham Courant, 2016) and from members of the public throughout the consultation indicate that this event set the tone for much of the consultation process and it was difficult to overcome the hostility generated by this event, which played a significant part in slowing progress towards consultation aims:

> I don't trust the Lynx Trust, because they put a big report in about how the meeting at Kielder had been invaded by people bussed in, like it was the middle of London which is just ridiculous. But I've asked the people up there about that and they said no, it was just farmers from outlying places that had come in, so that was a complete lie. You can't win an argument with a lie.
>
> (Q participant 14, primary zone)

Smaller meetings were more effective and inclusive, as were discussions during door-to-door visits, but the amount of resource needed to undertake this exercise meant that progress was slow and all homes were not visited. However, data collected through questionnaires during these activities did highlight risks which were useful in planning the ongoing consultation and informing the plan for the trial reintroduction. In hindsight, focused contact with individual stakeholders and small groups would have been more suitable at the start of the process, and this is reiterated by concerns raised by members of the community during the consultation:

> It hasn't been terribly well managed so far, people haven't been listened to and there's a feeling that we know everything and you know nothing. The chief ecologist in the forest hasn't been asked or listened to . . . or the wildlife officer who worked here for years. [He] was at the meeting but was quashed. . . . There are people here who actually know quite a lot and people have been looked down on.
>
> (Q participant 8, primary zone)

> I personally feel it's gone back to front. I would have liked to see it go the other way with questions with the Forestry Commission and the big boys first.
>
> (Q participant 3, primary zone)

The perceived lack of information being communicated about the proposal, and the sense of the project being imposed from outside the community, were most likely related to these unsuitable methods of communication. Berkes (2004) highlights the importance of incorporating local knowledge and perspectives along with scientific information in community-based conservation. Throughout the consultation, evidence from Eurasian lynx present in Europe was used – for example, on sheep predation (White et al., 2015) and increased ecotourism (White et al., 2016b) – to inform the community on potential impacts of the proposal. There is, however, evidence that such a scientific knowledge-based approach can lead to alienation of stakeholders and increased lobbying against reintroductions, resulting in polarisations between a 'science-based technocratic worldview, and its "populist" counterpart that portrays local actors as the victims of external intervention' (Arts et al., 2012). Similarly, Von Essen (2017) highlights the challenges that 'contested knowledge' creates in controversial species reintroductions. Using the example of wolf reintroductions in Sweden, she demonstrates how scientific knowledge can be viewed as hegemonic and patronising from the perspective of rural residents and she argues for a public platform of communication.

A further obstacle in creating avenues for disseminating information that reinforced the purpose of the trial was that many supporters felt unable to express their support. This was a recurrent theme in interviews with more supportive members of the community; for example:

> There is a whole section of the community that is afraid to talk. . . . It's made people go, "wait a minute, why can't I speak out?" And people have

realised they can't because their boss, or landlord, is so against it that they can't say a word. It's frightening really. I haven't been to the village hall since voicing my opinion, because . . . you'd get shunned.

(Q participant 10, primary zone)

This further exacerbated the slow progress with the consultation and made it difficult for those undertaking the consultation work, as well as members of the community, to get an accurate picture of the level of support for the trial. This has eventually resulted in the creation of the "Friends of the Lynx" group.

Discussion and conclusion

The importance of understanding and incorporating social impacts in conservation has long been established (Kaplan-Hallam and Bennet, 2017), and are integral to the IUCN Guidelines for Reintroductions and Other Conservation Translocations (IUCN/SSC, 2013). The main focus of this work has been to examine community attitudes towards the proposed reintroduction. However, although the consultation plan centred on a community-based approach, there were several factors which constrained public participation, information sharing and transparent communications integral to this (Arts et al, 2012, IUCN/SSC, 2013). This chapter presents a reflection on the methods used and data collected from an incomplete community consultation, and provides an opportunity to consider the lessons learnt from the methodological approach.

Reintroductions of charismatic species are often accompanied by vehement controversy (Arts et al., 2012), and the preliminary findings presented in this chapter speak to the polarised nature of the debate around a trial lynx reintroduction. This pattern mirrors responses to elements of both the initial national public survey (Smith et al., 2015a) and the national stakeholders survey (Smith et al., 2016). Strong opinions, both for and against, were held by community members, which demonstrates the strong emotional component to the decision-making process. While this discussion has focused largely on the negative aspects of the consultation process, there was also support and a change in consultation strategies saw the beginnings of constructive, informative dialogue particularly through focused business and farming meetings by invitation and Q interviews. The themes presented here are those most prevalent within the data and should therefore be addressed in ongoing project activities. These themes provide a possible structure around which to approach conversations with more focused stakeholder groups going forward, so that the community can be represented in decisions and solutions to the issues they themselves have highlighted.

IUCN guidelines clearly indicate the need for social support in relation to reintroduction and translocation projects. Annex 6.2, therefore, maps the IUCN guidelines against the consultation activities undertaken to date and provides recommendations on how to address problems by those parties taking the project forward.

Identifying and understanding the structure of a community to work with is challenging and, as Berkes (2004) indicates, community-based conservation 'failure' may be due to the implementation rather than any weakness or impracticality of the concept. Clear devolution of authority and responsibility (Songorwa, 1999; Murphree, 2002) is vital alongside identifying the scale appropriate (Berkes, 2004) in a multi-stakeholder environment. Such an approach takes time, commitment and honesty, and is often messy and complex, but ultimately necessary for conservation success.

Acknowledgements

At the time of developing the consultation plan and collecting data, the authors were working with the LUKT.

Annex 6.1 Q set of statements

No.	Statement
1	The presence of lynx is crucial for the health of forest's ecosystems
2	The British countryside is no longer a suitable place for a sustainable lynx population
3	I feel the resources spent on the trial could be better directed to those species already present in the UK
4	Lynx should be introduced as a natural control of deer
5	Lynx should be introduced to the UK and any uncertainties can and must be dealt with
6	I am concerned that the lynx will pose a threat to our native wildlife during the trial
7	It is critical to use biological data and sound science in this trial of introducing lynx
8	I am concerned that lynx will be a threat to livestock during the trial
9	I am concerned that the lynx will cause economic suffering to farmers and/ or countryside managers
10	I trust the conservationists and scientists behind this trial proposal
11	It is important that the trial has a well-defined exit program in place, which guarantees reversibility
12	It is important that adequate compensation agreements are in place, should the lynx cause any destruction
13	Lynx could beneficially add to the to the rural economy through eco-tourism
14	Having lynx in the countryside would make it more interesting
15	I am happy and excited for the trial to go ahead
16	Lynx are not compatible with our society today
17	A well-designed and regulated trial could help inform decisions on whether lynx should be introduced or not in the future
18	There is too much uncertainty about control measurements during the trial
19	All aspects of the trial must be transparent and open for all
20	The trial's plan of introducing six lynx for five years is too low to be scientifically sustainable
21	Public education and outreach programs for the public should have high priority
22	I am concerned that the lynx will be a threat to people during the trial
23	The trial must engage, consult and involve local communities and stakeholders

No.	Statement
24	We have an obligation to try and restore our natural ecosystem as much as possible. The trial is one step towards that
25	The welfare and safety of the lynx is of highest importance, both during the trial and at the end of the trial
26	There is not enough funding within the trial to implement what needs to be done
27	A clear vision on what the long-term management plans beyond the trial are vital
28	I do not think this area is suitable for the lynx
29	The trial would have negative impacts on my personal situation
30	I believe the trial will have a positive impact on the local community
31	Having lynx in this area would help put Kielder on the map
32	I believe having lynx here, although rarely seen, would create and exciting sense of wilderness
33	I am concerned that the lynx will be a threat to pets during the trial
34	I believe the local community could come to enjoy having lynx here. But it would take a long time
35	I am concerned that the lynx will disperse from the trial release area
36	I believe the lynx will offer more opportunities for employment to the younger generation in particular
37	Lynx would have a negative impact on the venison and/or deer stalking industry
38	I do not know how to voice my opinion to the Lynx UK Trust
39	I have read the trial proposal document
40	I am happy with the way the consultation has been carried out

Annex 6.2 Mapping consultation activity against the 10 IUCN Social Feasibility Guidelines

IUCN Social Feasibility Guideline (IUCN/SSC, 2013)	Lynx consultation	Implications and further initiatives to be undertaken
5.2.1. Any conservation translocation proposal should be developed within national and regional conservation infrastructure, recognising the mandate of existing agencies, legal and policy frameworks, national biodiversity action plans or existing species recovery plans.	Stakeholder consultation has been managed in accordance with national conservation infrastructure. Dialogues were open with Defra, the Scottish government, Scottish Natural Heritage and Natural England as well as other relevant governmental bodies, and there was discussion regarding the approach taken during the consultation.	The approach taken with governmental stakeholders should continue. The full consultation plan, with a focus on national to regional consultation, was logical in conception but with hindsight this approach alienated key regional non-governmental stakeholders. Ongoing plans should seek to involve regional stakeholders in community conversations.
5.2.2. Human communities in or around a release area will have legitimate interests in any translocation. These interests will be varied, and community attitudes can be extreme and internally contradictory. Consequently, translocation planning should accommodate the socio-economic circumstances, community attitudes and values, motivations and expectations, behaviours and behavioural change, and the anticipated costs and benefits of the translocation. Understanding these is the basis for developing public relations activities to orient the public in favour of a translocation.	Data collected reflect the extreme polarity in a series of crucial aspects of lynx reintroduction, which has been often voiced in a polemical fashion.	Complexity in several attitudes and values, especially diverging views on nature, needs to be further investigated. Data collection should continue, using more focused discussion, in order to ensure data represents a wide range of views. The planning and implementation of any project after completion of further data collection must directly address community attitudes and values and this must translate into the execution of any reintroduction rather than being a desk-based or 'on paper' exercise.

5.2.3. Mechanisms for communication, engagement and problem-solving between the public (especially key individuals most likely to be affected by or concerned about the translocation) and translocation managers should be established well in advance of any release.	This is particularly key for the farming community, who are most likely to be adversely affected by the proposal. Methods used initially to attempt to engage the farming community were not widely received; however, more recent focus groups have tentatively opened lines of communication.	It is essential that the farming community is fully involved in the development in key areas of project management such as compensation for livestock predation, mitigation measures to prevent livestock predation, monitoring protocols, measures of success and exit strategy. This work has started, and there were some signs of trust and engagement developing. Perceived lack of information among the wider community also indicates the need for improved forms of communication. Lack of continuity of personnel would tend to undermine this process so a familiar team should be established for the long term to combat this.
5.2.4. No organisms should be removed or released without adequate/conditional measures that address the concerns of relevant interested parties (including local/indigenous communities); this includes any removal as part of an exit strategy.	Data collection during initial consultation activities have highlighted a number of stakeholder concerns in the Kielder area.	Further data collection and engagement activities should be undertaken to ensure that all concerns have been uncovered; to understand how concerns can be adequately met, or whether there are opportunities for negotiation; and that stakeholders have been fully involved in the development of key project areas. Primary areas identified to date include compensation and mitigation, welfare issues, exit measures and measures of success.
5.2.5. If extinction in the proposed destination area occurred long ago, or if conservation introductions are being considered, local communities may have no connection to species unknown to them, and hence oppose their release. In such cases, special effort to counter such attitudes should be made well in advance of any release.	This is a relevant issue with the trial as lynx have been absent from this landscape for over 1300 years. Thus far, the case for reintroduction has largely been based on examples from Europe, the potential economic benefits through tourism and the potential for trophic cascades to control deer numbers and encourage ecosystem restoration (with added benefits to biodiversity).	Based on the data collected, this is a key issue which needs to be addressed in ongoing consultation activities. The project needs to make a much stronger case to the local community, based on extensive ecological feasibility work, for how lynx will fit into the wider Kielder socio-ecological landscape. The theory of the 'landscape of coexistence' should be investigated as a possible mechanism for allaying concerns regarding human–lynx conflict.

(Continued)

IUCN Social Feasibility Guideline (IUCN/SSC, 2013)	Lynx consultation	Implications and further initiatives to be undertaken
5.2.6. Successful translocations may yield economic opportunities, such as through ecotourism, but negative economic impacts may also occur; the design and implementation stages should acknowledge the potential for negative impacts on affected parties or for community opposition; where possible, sustainable economic opportunities should be established for local communities, and especially where communities/regions are challenged economically.	A strong case has been made for the potential net economic benefits of the trial, based on a desktop cost-benefit analysis modelling approach conducted by AECOM.	This work requires further refinement, specifically in relation to the Kielder socio-economic landscape. Linked to 5.2.3., this should include a greater focus on those most likely to be adversely affected. After this, such work must translate into genuine attempts to include targeted aspects in the project plan addressing this point.
5.2.7. Some species are subject to multiple conservation translocations: In these situations, inter-project, inter-regional or international communication and collaboration are encouraged in the interests of making best use of resources and experiences for attaining translocation goals and effective conservation.	Good lines of communication were established with other European countries where lynx are present (e.g., the Harz mountains project and Sweden, the proposed location for the source population), other UK conservation projects (e.g., the pine marten reinforcement project run by Vincent Wildlife Trust) and several academic institutions.	Communication and sharing of experiences should continue, with the network widening to include further knowledge and skills.
5.2.8. Organisational aspects can also be critical for translocation success: Where multiple bodies, such as government agencies, non-government organisations, informal interest groups (some of which may oppose a translocation) all have statutory or legitimate interests in a translocation, it is essential that mechanisms exist for all parties to play suitable and constructive roles. This may require establishment of special teams working outside formal, bureaucratic hierarchies that can guide, oversee and respond swiftly and effectively as management issues arise.	A national forum, targeting national stakeholder organisations, created a mechanism for sharing information and expertise. The intention was to create a similar local stakeholder forum open to any type of body or individual. Given the legal implications of the project, there was a separate dialogue with the government as per 5.2.1.	The national stakeholder forum should continue, with concerted efforts to expand membership, as well as continued communication with governmental agencies. It is also important to ensure that those involved are objective and reasonable when dealing with a broad spectrum of interests and agendas.

5.2.9. The multiple parties involved in most translocations have their own mandates, priorities and agendas; unless these are aligned through effective facilitation and leadership, unproductive conflict may fatally undermine translocation implementation or success.	Agreements were never formalised between project partner organisations, and it was disagreement over project governance that concluded in the cessation of consultation activities by the authors.	Structure and governance of the project and/or the participating organisations should be transparent.
5.2.10. A successful translocation can contribute to a general ethical obligation to conserve species and ecosystems; but the conservation gain from the translocation should be balanced against the obligation to avoid collateral harm to other species, ecosystems or human interests; this is especially important in the case of a conservation introduction.	Data from the consultation highlights that ecological feasibility and impact on native wildlife and human interests are key concerns among the community. Detailed ecological and feasibility work was planned, but remains incomplete to our knowledge.	Detailed ecological feasibility and cost-benefit analyses specific to the proposed area should be completed and published, and should inform project plans and any licence application(s).

References

Andelman, S. J., and Fagan, W. F. (2000) 'Umbrellas and flagships: Efficient conservation surrogates or expensive mistakes?', *Proceedings of the National Academy of Sciences of the United States of America*, vol 97, no 11, pp5954–5959.

Angst, C., and Breitenmoser, U. (2003) 'Eurasian lynx depredation on livestock in Switzerland – A lasting controversy 30 years after the reintroduction', *Environmental Encounters*, vol 58, pp59–60.

Arts, K., Fischer, A., and van der Wal, R. (2012) 'Common stories of reintroduction: A discourse analysis of documents supporting animal reintroductions to Scotland', *Land Use Policy*, vol 29, no 4, pp911–920.

Basille, M., Herfindal, I., Santin-Janin, H., Linnell, J. D. C., Odden, J., Andersen, R., Hogda, K. A., and Gaillard, J. M. (2009) 'What shapes Eurasian lynx distribution in human dominated landscapes: Selecting prey or avoiding people?', *Ecography*, vol 32, no 4, pp683–691.

Berkes, F. (2004) 'Rethinking community-based conservation', *Conservation Biology*, vol 18, no., 3, pp621–630.

Breitenmoser, U., and Breitenmoser-Würsten, C. (2004) 'Switzerland', in M. von Arx, C. Breitenmoser-Würsten, F. Zimmermann and U. Breitenmoser (eds) *Status and conservation of the Eurasian lynx (Lynx lynx) in Europe in 2001*, Kora, Muri.

Chapron, G., Kaczensky, P., Linnell, J. D. C., von Arx, M., Huber, D., Andren, H., Lopez-Bao, J. V., Adamec, M., Alvares, F., Anders, O., Balciauskas, L., Balys, V., Bedo, P., Bego, F., Blanco, J. C., Breitenmoser, U., Broseth, H., Bufka, L., Bunikyte, R., Ciucci, P., Dutsov, A., Engleder, T., Fuxjager, C., Groff, C., Holmala, K., Hoxha, B., Iliopoulos, Y., Ionescu, O., Jeremic, J., Jerina, K., Kluth, G., Knauer, F., Kojola, I., Kos, I., Krofel, M., Kubala, J., Kunovac, S., Kusak, J., Kutal, M., Liberg, O., Majic, A., Mannil, P., Manz, R., Marboutin, E., Marucco, F., Melovski, D., Mersini, K., Mertzanis, Y., Myslajek, R. W., Nowak, S., Odden, J., Ozolins, J., Palomero, G., Paunovic, M., Persson, J., Potocnik, H., Quenette, P.Y., Rauer, G., Reinhardt, I., Rigg, R., Ryser, A., Salvatori, V., Skrbinsek, T., Stojanov, A., Swenson, J. E., Szemethy, L., Trajce, A., Tsingarska-Sedefcheva, E., Vana, M., Veeroja, R., Wabakken, P., Wofl, M., Wolfl, S., Zimmermann, F., Zlatanova, D., and Boitani, L. (2014) 'Recovery of large carnivores in Europe's modern human-dominated landscapes', *Science*, vol 346, no 6216, pp1517–1519.

Convery, I., Smith, D., Brady, D., Hawkins, S., Iversen, S., Mayhew, M., and Eagle, A. (2016) 'Lynx UK trust consultation brief: Kielder', http://insight.cumbria.ac.uk/id/eprint/3453, accessed 4 December 2017.

Gehr, B., Hofer, E. J., Muff, S., Ryser, A., Vimercati, E., Vogt, K., and Keller, L. F. (2017) 'A landscape of coexistence for a large predator in a human dominated landscape', *Oikos*, vol 126, pp1389–1399.

Griffith, B., Scott, J. M., Carpenter, J. W., and Reed, C. (1989) 'Translocation as a species conservation tool: Status and strategy', *Science*, vol 245, no 4917, pp477–480.

Halliday, J., and Parveen, N. (2017) 'Plan to return the lynx splits friends and families in Kielder Forest community', www.theguardian.com/uk-news/2017/feb/03/plan-to-introduce-lynx-to-kielder-forest-angers-farmers, accessed 4 December 2017.

Hetherington, D. (2006) 'The lynx in Britain's past, present and future', *Ecos*, vol 27, no 1, pp66–74.

Hetherington, D. A., Miller, D. R., Macleod, C. D., and Gorman, M. L. (2008) 'A potential habitat network for the Eurasian lynx *Lynx lynx* in Scotland', *Mammal Review*, vol 38, no 4, pp285–303.

Heurich, M., Most, L., Schauberger, G., Reulen, H., Sustr, P., and Hothorn, T. (2012) 'Survival and causes of death of European Roe Deer before and after Eurasian Lynx reintroduction in the Bavarian Forest National Park', *European Journal of Wildlife Research*, vol 58, no 3, pp567–578.

Hexham Courant (2016) 'Chaos erupts at lynx meeting', www.hexham-courant.co.uk/news/bellingham/Chaos-erupts-at-lynx-meeting-45677f1c-956b-414b-8cb0-8dd0e426a85b-ds, accessed 4 December 2017.

Hexham Courant (2017) 'Debate continues over plans to release lynx at Kielder', www.hexham-courant.co.uk/features/Debate-continues-over-plan-to-release-lynx-at-Kielder-26f9ef29-431e-4906-94b6-1d7a15a0de7e-ds, accessed 4 December 2017.

IUCN/SSC (2013) *Guidelines for Reintroductions and Other Conservation Translocations*, IUCN Species Survival Commission, Gland.

Jobin, A., Molinari, P., and Breitenmoser, U. (2000) 'Prey spectrum, prey preference and consumption rates of Eurasian lynx in the Swiss Jura Mountains', *Acta Theriologica*, vol 45, no 2, pp243–252.

Kaczensky, P., Chapron, G., von Arx, M., Huber, D., Andren, H., and Linnell, J. (2012) 'Status, management and distribution of large carnivores – Bears, lynx, wolf & wolverine – In Europe', http://ec.europa.eu/environment/nature/conservation/species/carnivores/conservation_status.htm, accessed 10 November 2017.

Kaplan-Hallam, M., and Bennett, N. J. (2017) 'Adaptive social impact management for conservation and environmental management', *Conservation Biology*, epub ahead of print, doi:10.1111/cobi.12985.

MacDonald, D., Crabtree, J. R., Wiesinger, G., Dax, T., Stamou, N., Fleury, P., Gutierrez Lazpita, J., and Gibon, A. (2000) 'Agricultural abandonment in mountain areas of Europe: Environmental consequences and policy response', *Journal of Environmental Management*, vol 59, no 1, pp47–69.

Marshall, K., White, R., and Fischer, A. (2007) 'Conflicts between humans over wildlife management: On the diversity of stakeholder attitudes and implications for conflict management', *Biodiversity and Conservation*, vol 16, no 11, pp3129–3146.

Mascia, M. B., Brosius, J. P., Dobson, T. A., Forbes, B. C., Horowitz, L., McKean, M. A., and Turner, N. J. (2003) 'Conservation and the social sciences', *Conservation Biology*, vol 17, no 3, pp649–650.

Mayhew, M., Chantrey J., and Morphet, N. (2017) *Disease and Welfare Risk Assessments for the Reintroduction of Eurasian Lynx (Lynx lynx) from Sweden to Kielder Forest*, Northumberland, UK (Unpublished).

Milanesi, P., Breiner, F. T., Puopolo, F., and Holderegger, R. (2017) 'European human-dominated landscapes provide ample space for the recolonization of large carnivore populations under future land change scenarios', *Ecography*, vol 40, no 12, pp1359–1368.

Murphree, M. W. (2002) 'Protected areas and the commons', *Common Property Resource Digest*, vol 60, pp1–3.

Nilsen, E. B., Milner-Gulland, E. J., Schofield, L., Mysterud, A., Stenseth, N. C., and Coulson, T. (2007) 'Wolf reintroduction to Scotland: Public attitudes and consequences for red deer management', *Proceedings of the Royal Society of London B: Biological Sciences*, vol 274, no 1612, pp995–1003.

Northumberland County Council (2011) 'Northumberland knowledge 2011 census factsheet', www.northumberland.gov.uk/NorthumberlandCountyCouncil/media/Northumberland-Knowledge/NK%20place/Parishes%20and%20towns/Parish%20fact%20sheets/FactSheetParishBellingham.pdf, accessed 4 December 2017.

Odden, J., Linnell, J. D. C., and Andersen, R. (2006) 'Diet of Eurasian lynx, *Lynx lynx*, in the boreal forest of southeastern Norway: The relative importance of livestock and hares at low roe deer density', *European Journal of Wildlife Research*, vol 52, no 4, pp237–244.

Odden, J., Nilsen, E. B., and Linnell, J. D. C. (2013) 'Density of wild prey modulates Lynx kill rates on free-ranging domestic sheep', *PLoS ONE*, vol 8, no 11, ppe79261, doi:10.1371/journal.pone.0079261.

O'Rourke, E. (2014) 'The reintroduction of the white-tailed sea eagle to Ireland: People and wildlife', *Land Use Policy*, vol 38, pp129–137.

Rackham, O. (2001) *Trees and Woodlands in the British Landscape*, W&N, London.

Redpath, S. M., Linnell, J. D. C., Festa-Bianchet, M., Boitani, L., Bunnefeld, N., Dickman, A., Gutiérrez, R. J., Irvine, R. J., Johansson, M., Majić, A., McMahon, B. J., Pooley, S., Sandström, C., Sjölander-Lindqvist, A., Skogen, K., Swenson, J. E., Trouwborst, A., Young, J., and Milner-Gulland, E. J. (2017) 'Don't forget to look down – Collaborative approaches to predator conservation', *Biological Reviews*, vol 92, no 4, pp2157–2163.

Schmidt-Posthaus, H., Breitenmoser-Würsten, C., Posthaus, H., Bacciarini, L., and Breitenmoser, U. (2002) 'Causes of mortality in reintroduced Eurasian lynx in Switzerland', *Journal of Wildlife Diseases*, vol 38, no 1, pp84–92.

Sergio, F., Newton, I., Marchesi, L., and Pedrini, P. (2006) 'Ecologically justified charisma: Preservation of top predators delivers biodiversity conservation', *Journal of Applied Ecology*, vol 43, no 6, pp1049–1055.

Sieber, A., Uvarov, N. V., Baskin, L. M., Radeloff, V. C., Bateman, B. L., Pankov, A. B., and Kuemmerle, T. (2015) 'Post-Soviet land-use change effects on large mammals' habitat in European Russia', *Biological Conservation*, vol 191, pp567–576.

Simberloff, D. (1998) 'Flagships, umbrellas, and keystones: Is single-species management passé in the landscape era?', *Biological Conservation*, vol 83, no 3, pp247–257.

Smith, D., O'Donoghue, P., Convery, I., Eagle, A., and Piper, S. (2015a) 'Reintroduction of the Eurasian Lynx to the United Kingdom: Results of a public survey', http://lynxuk.org/publications/lynxinterimsurvey.pdf, accessed 4 December 2017.

Smith, D., O'Donoghue, P., Convery, I., Eagle, A., Piper, S., White, C., and Van Maanen, E. (2015b) 'Application to natural England for the trial reintroduction of Lynx to England', http://lynxuk.org/publications/EngLynxConsult.pdf, accessed 4 December 2017.

Smith, D., O'Donoghue, P., Convery, I., Eagle, A., Piper, S., White, C., and Van Maanen, E. (2015c) 'Application to Scottish Natural Heritage for the trial reintroduction of lynx to Scotland', http://lynxuk.org/publications/ScotLynxConsult.pdf, accessed 4 December 2017.

Smith, D., O'Donoghue, P., Convery, I., Eagle, A., Piper, S., and White, C. (2016) 'Lynx UK Trust – A national stakeholder consultation: An interim consultation document', http://lynxuk.org/publications/lynxinterimdoc.pdf, accessed 4 December 2017.

Songorwa, A. N. (1999) 'Community-based wildlife management (CWM) in Tanzania: Are the communities interested?', *World Development*, vol 27, no 12, pp2061–2079.

Soulé, M., and Terborgh, J. (eds) (1999) *Continental Conservation: Scientific Foundations of Regional Reserve Networks*, Island Press, Washington, DC.

Thirgood, S., and Redpath, S. (2008) 'Hen harriers and red grouse: Science, politics and human-wildlife conflict', *Journal of Applied Ecology*, vol 45, no 5, pp1550–1554.

Trouwborst, A. (2010) 'Managing the carnivore comeback: International and EU species protection law and the return of Lynx, Wolf and bear to Western Europe', *Journal of Environmental Law*, vol 22, no 3, pp347–372.

van Heel, B. F., Boerboom, A. M., Fliervoet, J. M., Lenders, H. J. R., and van den Born, R. J. G. (2017) 'Analysing stakeholders' perceptions of wolf, lynx and fox in a Dutch riverine area', *Biodiversity and Conservation*, vol 26, no 7, pp1723–1743.

von Essen, E. (2017) 'Whose discourse is it anyway? Understanding resistance through the rise of "Barstool Biology" in nature conservation', *Environmental Communication*, vol 11, no.4, pp470–489.

Wagner, C., Holzapfel, M., Kluth, G., Reinhardt, I., and Ansorge, H., (2012) 'Wolf (*Canis lupus*) feeding habits during the first eight years of its occurrence in Germany', *Mammalian Biology-Zeitschrift für Säugetierkunde*, vol 77, no 3, pp196–203.

Watts, S., and Stenner, P. (2005) 'Doing Q methodology: Theory, method and interpretation', *Qualitative Research in Psychology*, vol 2, no 1, pp67–91.

Watts, S., and Stenner, P. (2012) *Doing Q Methodological Research: Theory Method and Interpretation*, SAGE Publications, London.

White, C., Almond, M., Dalton, A., Eves, C., Fessey, M., Heaver, M., Hyatt, E., Rowcroft, P., and Waters, J. (2016b) 'The economic impact of lynx in the Harz mountains', http://lynxuk.org/publications/lynxharz.pdf, accessed 4 December 2017.

White, C., Convery, I., Eagle, A., O'Donoghue, P., Piper, S., Rowcroft, P., Smith, D., and Van Maanen, E. (2015) 'Cost-benefit analysis for the reintroduction of lynx to the UK: Main report', www.aecom.com/uk/wp-content/uploads/2015/09/Cost-benefit-analysis-for-the-reintroduction-of-lynx-to-the-UK-Main-report.pdf, accessed 4 December 2017.

White, C., Waters, J., Eagle, A., O'Donoghue, P., Rowcroft, P., and Wade, M. (2016a) 'Reintroduction of the Eurasian Lynx to the United Kingdom: Trial site selection', http://lynxuk.org/publications/lynxsiteselection.pdf, accessed 4 December 2017.

Wildlife and Countryside Act (1981) 'Wildlife and Countryside Act 1981', www.legislation.gov.uk/ukpga/1981/69/contents, accessed 4 December 2017.

Wilson, C. J. (2004) 'Could we live with reintroduced large carnivores in the UK?', *Mammal Review*, vol 34, no 3, pp211–232.

Woods, M., Heley, J., Richards, C., and Watkins, S. (2012) 'Rural people and the land', in I. Convery, G. Corsane and P. Davis (eds) *Making Sense of Place: Multidisciplinary Perspectives*, Boydell Press, Woodbridge.

7 Heterogeneity in perceptions of large carnivores

Insights from Sanjay Gandhi National Park, Mumbai, and Ladakh

Sunetro Ghosal

Introduction

Large carnivore conservation cuts across disciplinary boundaries and thematic processes. It is contextualised within biological, social, political, economic, and geographical processes. Thus, it requires participation from different groups and sub-groups that have their own interests and unique interactions and relationships with the large carnivore in question, which determines their perception and inter-group interactions. Humans and large carnivores have a complex history in their complex and dynamic relationship that dates back several thousands of years (Cavallo and Blumenschine, 1989; Cavallo, 1990). Heterogeneity in stakeholder perceptions is a reflection of the dynamism and the multiplicity of factors that influence human-carnivore interactions. It is also influenced by different bodies of knowledge accessed by social actors to interpret, adapt, influence, and resist each other's rationale and actions. Knowledge types are neither static nor holistic and often diffuse, compete with, or intersect each other. In addition, there is variability in terms of political economy and managing resources in the landscape, in terms of decision-making and influence over policy. Ecological processes are another important dimension in configuring relationships between people and large carnivores. All the previously mentioned factors are crucial for the behavioural adaption that humans and large carnivores employ to adapt to each other's presence in the landscape.

Given the complexity of all these processes, there are several inherent contradictions, contestations, and challenges in understanding and reconciling this diversity of perceptions. Acknowledging the diversity in stakeholder positions remains a key requirement for biological conservation at large, and is an issue that needs to be integrated in any process for large carnivore conservation. The heterogeneity of stakeholder relationships and perceptions may present both opportunities and threats. These equations have played out in their unique ways in different sites across the world, including lions in Africa, wolves in Europe and North America, and lions, leopards, and tigers in South Asia. Over the last few years, several case studies have emerged that exemplify the diversity of interactions between humans and leopards in India (see Athreya and Venkatesh, 2013; Ghosal et al., 2013; Ghosal and Kjosavik, 2015). Understanding this heterogeneity of perceptions, and their localised dynamics, is the first step in the process that later includes dialogues, negotiations, and actions to enable humans and large carnivores to share space and resources. In this chapter, I examine examples that illustrate the interaction between people and large carnivores in India by taking into account the local context. This includes a setting with people and leopards (*Panthera pardus fusca*) in a metropolitan landscape in suburban Mumbai (Bombay) and Thane, and a second setting in high-altitude steppe and mountain systems of the Indian Trans-Himalayan region of Ladakh, where people share the landscape and its resources with wolves (*Canis lupus chanco*) and snow leopards (*Panthera uncia*).

In the summer of 2014, a leopard entered a school premises in the eastern suburbs of Mumbai, located along the border of Sanjay Gandhi National Park. The school was on holiday, but the leopard's presence caused a large ripple in the city circles. The leopard was soon located in the school hall, as conservation managers tried to devise a way to trap it. Outside, media persons and curious onlookers gathered to witness the proceedings. The media carried spiced up reports of leopards 'straying' out of the park and entering human areas. Various residents in the neighbourhood expressed fear of leopards that were using the area. As an elderly gentleman watched the proceedings of the leopard rescue operation and the crowds that had gathered, from his high-rise apartment that shared a wall with the school, he commented nonchalantly: "I do not see the need for all this fuss. Leopards are nothing new. They have always lived in this area." There were very diverse perceptions and reactions to large carnivore presence evident in the metropolitan city of Mumbai, where an estimated 20,925 people share space with leopards that live in and along the peripheral regions of the 103 km² national park (Deol, 2011).

In the high altitude region of Ladakh in the Indian Trans-Himalayas, wolves (*Canis lupus chanco*) and snow leopards (*Panthera uncia*) share the landscape with livestock herders, farmers, and security agencies. There were many herders and farmers who regarded large carnivores as a menace and repeatedly claimed that they should be eliminated. At the same time, there were numerous others in these communities who took a more benign view of large carnivores and treated them with respect, while acknowledging the reciprocity in the relationship between humans and non-humans. The heterogeneity of stakeholder attitudes toward large carnivores and varying degrees of tolerance and antagonism towards these species transcended rural vs. urban dichotomies. This diversity was illustrative of the complexity and dynamism of human-carnivore interactions, which were influenced by a wide array of factors and are subject to continuous change.

Site descriptions

The Sanjay Gandhi National Park (SGNP) landscape is located in the suburbs of Mumbai, Maharashtra State. It is spread across 103 km², of which 56 km² are within the municipal limits of Mumbai, while the remaining area is located in Thane urban area. Immediately north of the park is the 90 km² Tungareshwar

Wildlife Sanctuary, which is primarily located in rural Thane district. The SGNP landscape is surrounded by suburban Mumbai with an average population density of 20,925/km² (Deol, 2011). The human population is very diverse with variations across ethnicity, class, period of residence, political affiliations, awareness, and physical proximity to the national park. A minimum of 35 individual leopards have been identified in SGNP through a camera-trapping exercise carried out in 2015 (Surve, 2015). This is an exceptionally high number of leopards for the 103 km² protected area. However, the number is not surprising given the high population density of prey animals such as feral dogs, which form a significant part of the diet of Mumbai's leopards (Punjabi, 2013a, 2013b; Surve, 2015). The geography of the interactions between humans and leopards include shrines dedicated to Waghoba (large cat deity), conservation-priority zones, development-priority areas, and forested areas where some developmental activities are allowed (stables for livestock, roads, police training centres, and, more recently, a metro car shed). The region has a long history of human-leopard interactions. In 2003–2004, conflicts between humans and leopards peaked. This intensification has been linked to trapping and translocating leopards, which was practiced earlier (Athreya et al., 2010, 2011), and which is now understood to have been counter-productive as it traumatised individual animals (which tried to return to their home range when relocated), played havoc with the local social structure of leopards, and removed a resident animal familiar with the area to create an opening for a less dominant animal that was unfamiliar with the area (Athreya et al., 2010, 2011). After 2011 and the establishment of guidelines for managing human-leopard interactions (Ministry of Environment and Forests, 2011), conflict has decreased substantially as resident leopards have remained in the area. The Forest Department initiated a project called Mumbaikars for SGNP to conduct research into leopard ecology, understand human-leopard conflict, and engage in dialogue with various stakeholders to facilitate human-leopard coexistence in and around Mumbai.

At the other end of the rural-urban spectrum, we have human-wildlife interactions in Ladakh, especially Chanthang in Jammu and Kashmir State. Ladakh marks the western edge of the Tibetan plateau, and is one of the most sparsely populated areas in India with an estimated two persons/km². It is characterised by its undulating mountainous landscape with high altitude steppe in the east, an arid climatic system (being in the rain shadow of the main Himalayan range) and extreme temperatures that range from −45° C in the winter to 30° C during the summer. Ladakh is located at the altitudinal threshold and above for agriculture at these latitudes. Thus, agro-pastoralism remains the cornerstone of the Ladakhi economy, which brings people into sharp conflict with large carnivore species in the area. These include wolves (*Canis lupus chanco*) and snow leopards (*Panthera uncia*). There are no records of attacks on humans, while human-large carnivore conflicts in Ladakh are over livestock depredation. In retaliation, people have been killing snow leopards and wolves. Some government departments, such as the Animal and Sheep Husbandry Department, have been promoting an increase in production of wool and other products that the herders can sell. At the same time, numerous large protected areas have been established across Ladakh, while conservation practice has been based on landscape-level management that focuses on augmenting human livelihoods to encourage conservation through provision of alternate sources of income, use of renewable energy technology to reduce natural resource extraction, etc. (Takpa, 2017). Unlike other parts of India, people have not been excluded from protected areas and were integrated as the main beneficiaries of conservation measures. This has reduced conflicts between managers and residents, and enabled tolerance towards large carnivores.

The two sites were very different, one being urbanised and at sea-level with a high density of humans and leopards, and the other being rural and at the altitudinal threshold of human habitation (Table 7.1). In addition, there was a history of leopard attacks on humans and domesticated animals in Mumbai, while in Ladakh wolves and snow leopards only attacked domestic animals. However, both sites have several factors in common in that they have a diversity of stakeholder groups sharing the area with large carnivores that are managed by the same formal policy framework for conservation management. Jammu and Kashmir has a special status in India and is governed by a different constitution, which is largely based on the Indian constitution. The main conservation-related policy in India is the Indian Wildlife (Protection) Act (1972) with subsequent amendments, whose equivalent in Jammu and Kashmir is the Jammu and Kashmir Wildlife (Protection) Act, 1978 (amended in 2002).

Table 7.1 Local context and attitudes in Sanjay Gandhi National Park and Changthang, India

	Sanjay Gandhi National Park; suburban Mumbai (Bombay) and Thane, Maharashtra State	*Changthang; eastern part of the Indian Trans-Himalayan region of Ladakh, Jammu and Kashmir State*
Large carnivores	Leopard	Wolf; snow leopard
Local context	• Feral dogs form a significant part of the diet of Mumbai's leopards and sustain a population of a minimum 35 individuals leopards in the 103 km² protected area • Very diverse human population (e.g., ethnicity, class, period of residence, political affiliations, awareness, and physical proximity to the national park) exist around the park; in 2011 the density of human population around the park was 20,925 people/km² • Numerous shrines dedicated to Waghoba (large cat deity)	• A high altitude cold desert region that marks the western edge of the Tibetan plateau • One of the most sparsely populated areas in India with an estimated two persons/km² • Jammu and Kashmir has a special status in India and is governed by a different constitution, which is largely based on the Indian constitution

(*Continued*)

Table 7.1 (Continued)

	Sanjay Gandhi National Park; suburban Mumbai (Bombay) and Thane, Maharashtra State	Changthang; eastern part of the Indian Trans-Himalayan region of Ladakh, Jammu and Kashmir State
Main dimensions of governmental policy	• Human-leopard conflicts peaked mainly due to a counter-productive practice of trapping and translocating leopards, which stopped after 2011 • The Forest Department initiated a research project on leopard ecology, human-leopard conflict, stakeholder consultation, and engagement	• Landscape-level management focused on provision of alternate sources of income and use of renewable energy to reduce natural resource extraction • Unlike other parts of India, people have not been excluded from protected areas and are integrated as the main beneficiaries of conservation measures
Local attitudes towards large carnivores	• Leopards' presence and frequent attacks cause much concern • Residents who have stayed in the area longer believe that leopards belonged there	• Younger herders and farmers regard large carnivores as a menace and claim that they should be eliminated • Elder people acknowledge the reciprocity in the human-carnivore relationship
Main drivers of local attitudes	• History of leopard attacks on humans • National park bordering an urban area • Dualist perspective between humans and nature • Tension catalysed by media discourse	• Resource-poor region with limited alternative livelihoods • Human-large carnivore conflicts in Ladakh are primarily over livestock depredation
Main drivers of attitudes	• In-migration since the 1990s • Insecurity of land tenure • Existing conflicts with the government • Lack of access to information about leopards	• Push to increase production of livestock • Promotion of ecotourism development

Data collection and analysis

Fieldwork has been carried out in the two sites at different periods. In Changthang, data collection started in the summer of 2005 and continued in the winter of 2014 to be concluded in the summer of 2017. In Mumbai, fieldwork started in 2011 and continued until 2014. Data collection included interviews with representatives of different stakeholder groups. The respondents were identified using snowball sampling (i.e., each respondent suggested other interviewees). The interviews were conducted with the use of a semi-structured guide that ensured that specific topics were covered but did not impinge on the order in which they were asked and allowed interviewees to expand on themes, whenever they wished to do so. In total, 150 respondents were interviewed in Mumbai and another 120 respondents were interviewed in Ladakh. For some respondents, follow-up interviews were conducted to clarify points that emerged during data analysis. Interviewees represented a cross-section of various stakeholder groups and sub-groups, taking into account several factors such as livelihood patterns, class, age, gender, education, access to political decision-making processes, and conservation-related roles such as managers, activists, etc. Interviews were recorded if the respondent gave his or her consent by means of an audio-recording device. Where they refused permission to use a recording device, handwritten notes were taken during the interview. During many of the interviews, other people would join the discussion and the interviews would turn into 'natural' focus groups rather than 'formal' focus groups arranged according to predefined categories (Frey and Fontana, 1993). The interviews were conducted in different languages, including English, Urdu, Hindi, Marathi, and Ladakhi. All participants were guaranteed anonymity and therefore, their names have been withheld on purpose. Analysis concentrated on the various 'meanings' attributed by individuals and groups to large carnivores and issues related to sharing resources with large carnivores, which is the primary focus for the present chapter. About 15% of all interviews in each site was selected randomly to be analysed by an independent researcher using the same coding. After inter-coder reliability between the author and the independent coder reached over 85%, mismatches were discussed and used to re-analyse parts of the interview corpus, when this was needed. Insights gained from interviews were correlated with data collected through participant observation and general discussions in both sites.

In addition to interviews, participant observation was used to understand different behavioural patterns of humans and large carnivores. This included observation of human activities and large carnivores in the two landscapes. It also included participation in various activities in the two sites, which provided insights to various social-cultural practices and ecological processes. Data regarding populations of leopards and their prey based in and around SGNP have been collected from projects that are underway. Notes on large carnivores and other wildlife, as well as domesticated livestock and human communities in Ladakh, have been collected from different sources such as the District Administration and other research projects in the area.

Heterogeneity in perceptions of large carnivores

The perceptions of large carnivores among stakeholder groups were heterogeneous as they were influenced by a diversity of factors, which exerted their effect on human-carnivore relationships. For instance, conservation managers were part of an institutionalised value system with a specific conceptualisation of large carnivores, which was quite different from that of indigenous groups that shared the same landscape with large carnivore species (Ghosal and Kjosavik, 2015). This was especially true in both sites in India, where a top-down approach to policy formulation has been followed. Differences in the value system inherent in conservation policy, on the one hand, and value systems that shaped perceptions and actions of local communities, on the other, have often led to clashes and resistance anytime policy implementation was introduced and local people have reacted, with large carnivores often getting dragged into contests between such contrasting worldviews (see Ghosal et al., 2013, for a discussion of the value systems that shape research on, and policy towards, large carnivores in India).

Sanjay Gandhi National Park, Mumbai

Humans and leopards have shared a long history in Mumbai with periodic and intense episodes of conflict, such as the one in 2002–2004, when 84 attacks on humans were reported and at least 96 leopard trappings were recorded (Athreya and Venkatesh, 2013). It is now understood that the trapping and relocation of leopards had contributed to spike in the attacks on humans as has been also reported by previous research (Linnell et al., 2001, 2009; Athreya et al., 2011; Athreya and Venkatesh, 2013).

Mumbai is home to tribal communities such as the Warli and Mahadeo Kolis – but not in other local communities – who perceive leopards as a personification of the deity, Waghoba (Ghosal, 2013; Ghosal and Kjosavik, 2015). This form of socially accepted institutional relationship between leopards and humans in some communities enshrines moral prescriptions and knowledge of leopard ecology and precautions to avoid conflicts. One member of the Warli community explained: "The leopard is our god and we know its habits. We have lived together for very long and we both know how to avoid conflicts." While there are exceptions, most members of these communities were respectful towards leopards and practiced various precautions to reduce the risk of attacks by leopards.

At the other end of the spectrum are individuals from other local communities who regarded the leopard as a dangerous animal and whose presence was perceived to be a threat to humans. During interviews, these respondents justified their fear based on ideas of leopards as man-eaters that dominated certain narratives in popular culture including the writings of Jim Corbett that were popular among urban residents and sensationalised media coverage (Athreya and Venkatesh, 2013). This has been further strengthened by conservation policies that were based on the assumption of a dualism between humans and non-humans (Ghosal et al., 2013). Such a view of leopards logically leads to the conclusion that they were 'out of place' in an urban landscape, where they posed a threat to humans and had to be physically removed.

Many adherents to this dualist perspective have moved into the landscape around SGNP since the 1990s or so, and they have not yet adapted their ideas to the reality of having to live with leopards. Their concern that the largely nocturnal leopards are present in the landscape has increased since 2011 with the installation of closed circuit TVs in these areas. This group included people in the middle and upper classes, who generally reacted with panic and self-righteous outrage at the presence of leopards. This angst was articulated by a resident who exclaimed at a meeting: "The leopard is affecting our way of life. We demand that it be removed from the area immediately!" However, there have been some instances of members of this group changing their perception based on dialogue with researchers and managers, and after an improved access to information about the history of the landscape, human-leopard interactions, and the behaviour of leopards in the area.

Another set of residents have lived in the area for five decades or more, and this has enabled them to become familiar with leopards. However, the relationship between members of this group and leopards can be disrupted by the insecurity of land tenure. This was evident in several interactions with residents of the areas immediately east of SGNP after the state government's revenue department declared 117 acres of built-up area in 2006 as private forest land, which has affected nearly 500,000 residents. The case was only resolved in 2014, after the Supreme Court of India directed the State Government to regularise the buildings after affected residents deposited a penalty. This insecurity over land tenure surfaced during interviews, where discussion on leopards would veer towards the uncertainty that these individuals were experiencing in the area till the issue was finally settled. At a meeting between managers and residents, a resident casually commented: "In any case, we have encroached into the leopard's area and must learn to live with them." Some of the other residents present became agitated with this comment and started shouting: "What do you mean 'encroached? We have all paid for these houses with our hard-earned money! . . . It is not our fault that the government gave permission for these houses initially and is now saying that it is illegal!"

In addition to the inter-group diversity discussed so far, there was diversity of perceptions even within groups owing to individual differences and access to competing forms of knowledge. For instance, conservation managers (Forest Department) are legally responsible for the conservation of leopards in India, both inside and outside protected areas. In general, these officials receive little training in updated knowledge of leopard ecology and behaviour. Their perception of leopards is largely drawn from popular notions, mixed with institutionalised but outdated knowledge of leopards and wildlife, in general. For instance, one official commented: "Well, each leopard requires a territory of 40 [square kilometers] and the main problem we have in the national park is that of over-crowding." This ignores more recent discovery of social structure and territoriality of leopards, wherein territory sizes can vary in relation to the availability of prey. Further, it reflects inadequate or insufficient efforts to update the knowledge base on which conservation managers operate. Another official in 2011 said: "Yes, leopards should not be coming to housing colonies as they pose a threat to humans. We have no option but to trap such leopards." In contrast, there were conservation managers who were also members of local tribal communities and who described leopard ecology and behaviour in great detail, including individuality among leopards. When asked about the source of their knowledge, one official explained: "We have grown up here and our elders knew a lot about leopards and tigers. Now, this knowledge helps me manage human-leopard interactions as I am able to explain different behaviour and help people take necessary precautions." Since 2011, the Ministry of Environment and Forests has adopted formal guidelines to manage human-leopard interactions. In addition, officials at SGNP initiated in 2011 the project Mumbaikars for SGNP, which included workshops for Forest Department personnel on leopard behaviour and ecology. This has increased awareness about the impact of earlier practices of trapping and relocation of leopards.

Changthang

Diversity in attitudes towards large carnivores was also evident in Changthang. For instance, there were individuals who viewed wolves and snow leopards as essential elements of the landscape. They argued that the absence of these large carnivores would ecologically, socially, and economically impoverish the area, its human community included. This was best illustrated by an elderly Changpa nomad who explained that wolves were essential for the existence of human society: "As a result, we herd livestock even if we lose animals occasionally due to depredation. If wolves were absent, we would become lazy and our livestock would soon become feral. This would undermine our economic and social systems." When asked about the source of this idea, the elderly gentleman explained that he learnt this from his parents and went on to explain that humans were part of nature and needed to be responsible in their actions towards other species.

However, other members in the same nomadic community did not share this perception of large carnivores, especially younger people. Many of them demonised wolves – the most common large carnivore in their area – as well as their hunting strategies and their negative impact on their livestock. These members characterised the wolf as a villainous beast, which needs to be eliminated from the landscape. When asked how the wolf harmed them, one middle-aged man explained: "We struggle here in the extreme cold to raise these herds. The wolves just come and take what they want!" This must also been seen in the context of the effort to increase herd sizes and maximise returns from sale of wool and goat hair (*pashm*, which is used to make *pashmina* or *cashmere* products), which has been promoted by government departments such as the Animal and Sheep Husbandry Department – a shift that has occurred since the 1970s and 1980s (Ghosal and Ahmed, 2017). There were anecdotal and unconfirmed reports of younger local people killing wolf pups to control their population (see Herda-Rapp and Goedeke, 2005, for similar trends in Europe and elsewhere). When asked about these reports, respondents expressed ignorance but said this is how they controlled wolf numbers in the past. It was not difficult to justify lethal action to control wolf population based on the perception that wolves were intruders.

In other parts of Ladakh, local residents expressed a degree of tolerance towards snow leopards. In one case that occurred in 2015, a snow leopard had entered a livestock pen, killed 14 sheep and got stuck inside. The following morning, conservation managers anesthetised the snow leopard and removed it from the pen. The whole village, including members of the household in which the pen was located, gathered to watch the operation and when the snow leopard was finally brought out, the villagers were curious but displayed no antagonism. In fact, a mother placed a small child in front of the cage as they both wished to get closer to the anesthetized snow leopard. One resident explained that this has not always been the case:

> Till recently, we lived at the mercy of the snow leopard and never knew when our herd would be attacked. They often kill the whole herd in their blood-thirsty frenzy. Now, we have tourists coming to see snow leopards in the winter, which provides us with an additional income source to offset possible losses to depredation.

The promotion of ecotourism development as a management intervention in the area has been documented by Takpa (2017).

There was a small group of individuals who held relatively romanticised views of snow leopards and wolves in Ladakh. These individuals expressed the idea of large carnivores being noble beasts that embody the spirit of wilderness on a fast-changing planet. They were engaged in semi-institutionalised efforts to conserve wildlife and often described themselves as guardians of wild animals in Ladakh: "We regard wild animals like snow leopards and wolves as our family members and will do everything in our power to ensure that they are conserved," explained one such interviewee. Many of these individuals worked in the tourism sector, while at least one person worked as a conservation manager. They did hold meetings in various places to talk about the importance of conservation and also held clean up drives for waste. However, this group was still in its nascent stage.

Also, similar to the SGNP landscape, conservation managers in Ladakh were a diversified group, since they were drawn from local communities as well as from the state and national forest services, which required qualification through highly competitive public exams. There were some who regarded large carnivore management as part of their job profile and approached conservation as part of their official duty. Others took a more active interest in understanding human-carnivore interactions. One such manager explained:

> We need a new approach to conservation in India. We cannot rely on the protected area model alone for conservation in places like Ladakh. Instead, we need to develop landscape-level models that integrate the improvement of human lives and biodiversity conservation.

These managers took personal interest to expand and update the scope of the current conservation policy framework in India, in their management practices on the ground – a framework which had been centred in the past on protected areas excluding people.

Understanding heterogeneity in perceptions of large carnivores

It is important to note that multiple factors intersect within and between different stakeholder groups increasing stakeholder heterogeneity, overall (see Cronon, 1995 for similar heterogeneity in relation to human-nature relations in the US, and Saberwal and Rangarajan, 2003, for a similar discussion in India). While there is a common denominator that holds each stakeholder group together and influences their perception of large carnivores, there are variations due to individual personality and experience, as well as cross-group membership. A person may belong to a herding community and that same person may also be part of the government machinery or choose to participate in ecotourism-based initiatives for development.

The sources of information and knowledge are another important element that adds to heterogeneity in perceptions of large carnivores. These sources include mythology, popular culture, literature, institutionalised knowledge, traditional ecological knowledge, media, and localised rumours. The impact of these forms of knowledge is contextual and dynamic. For instance, the use of a label such as 'man-eater' has a certain connotation in Mumbai with its history of leopard-related human fatalities that may not have the same implications in Ladakh, where there is no history of attacks by snow leopards and wolves on people. On the other hand, traditional ecological knowledge that identifies leopards as a deity or wolves and snow leopards as neighbours provides specific time-tested moral and behavioural prescriptions to facilitate coexistence between humans and large carnivores.

Another factor which has influenced perceptions of large carnivores was acknowledgment of behavioural reciprocity. Science has traditionally been structured around Cartesian dualisms, which did not recognise agency for nonhumans. Actor network theory has critiqued the reductionism inherent in such dualism and proposed an alternative definition of agency, namely, the ability to influence the actions of other actors (Latour, 2004). This definition allows the idea of agency to be extended to nonhumans, without getting ensnared in controversies surrounding the notion of self-awareness and intentionality. Another approach called ethno-ethology argues for the integration of the behaviour of nonhumans and the manner in which this behaviour is perceived by society as well as the factors that influence these perceptions (e.g., Lescureux, 2006). These ideas have been applied to research on human-nature relations by Callon (1986), Cloke and Jones (2004), and Lescureux (2006), which has opened new avenues of inquiry into human-nature interactions by extending agency to nonhumans. The idea of mutual behavioural reciprocity provides space for negotiation in these interactions, and has given root to the idea of large carnivores as agents responding to human actions and, in turn, influencing human actions. The recognition or denial of reciprocity is an important factor that shaped different value systems. Indeed, many long-term residents around SGNP and members of the previous generation in Ladakh recognised this agency and referred to it in order to negotiate coexistence with large carnivores.

Management practices also influence perceptions of leopards, and in turn, the latter shape management strategies for leopards. Under Indian wildlife laws (including Jammu and Kashmir State policies in the case of Ladakh), all wild animals and biodiversity are regarded as the 'property' of the government. Enforcement of these laws and compensation for losses has been detrimental to traditional ecological knowledge in many areas as the animals are then seen as a property of the government (Ghosal and Kjosavik, 2015). In Mumbai, past management strategies for leopards are known to have affected leopard behaviour and worsened conflicts (Athreya and Venkatesh, 2013). Individual animals trapped in other parts of the state were relocated to SGNP, based on the assumption that a leopard only needs a forest and prey to survive. However, it has been found that many of these animals, already traumatised by the trapping process, attempted to return to their home ranges and re-engaged into conflict with human communities as they attacked people along the way (Athreya et al., 2010). The intense conflict in 2003–2004 strengthened the popular myth of leopards as dangerous man-eaters, especially through sensationalised media reports. The failure to tackle human-leopard conflicts and new research on the impacts of trapping and relocation of leopards prompted the Ministry of Environment and Forests to release management guidelines in 2011, which prohibited random trappings with strict protocols to manage human-leopard interactions (Ministry of Environment and Forests, 2011).

In the past, Forest Department officials were not instructed to consult and engage local communities to generate support for conservation. As a result, there was a divide between management practices and general perceptions of leopards. However, in 2011, the department initiated efforts to engage in dialogue with various stakeholders in and around SGNP to reduce the pressure to trap leopards and mitigate conflicts between people and leopards as well. In Ladakh, the Department of Wildlife Protection and civil society groups have been working with local communities in conservation management practices by promoting eco-tourism and reducing dependence of natural resources through the use of renewable energy technology (Takpa, 2017). These efforts have helped develop a synergy between different stakeholder groups to improve conservation efforts.

While traditional ecological knowledge systems in both sites did recognise behavioural reciprocity, other knowledge systems were also exploring this factor, which has enabled a shift in perception from antagonism to coexistence. This is most evident in Mumbai as housing colonies spring up in the SGNP landscape along the periphery of the national park. The common response of new residents on discovering leopards in their neighbourhood is one of fear, outrage, and demands to trap and relocate leopards. However, dialogues initiated by the Forest Department have helped people take precautions and learn to live with leopards by reducing risks of attacks. As a result of this process, many residents have started to express respect towards leopards. A similar process is also evident in Ladakh. In the past, members of resident communities shared a reciprocal relationship with large carnivores in the landscape, though this would often be tinged with a degree of hostility. However, in many parts of Ladakh, wildlife-centred tourism has become popular and the local government and some civil society groups have supported the development of home-stays, wherein tourists stay in village homes during their travel. In addition, village youth have been trained to act as tourist guides to earn additional income from tourism. At the same time, conservation managers have also provided support to improve livestock pens to ensure that large carnivores are not able to enter them. These measures have not only helped villagers reduce livestock depredation rates; it has also enabled them to earn money from the presence of large carnivores in the region to offset any losses from depredation.

Conclusion

Human-carnivore relations are embedded in socio-ecological processes, which are complex and dynamic. As respondent input in the two sites has illustrated, there are site-specific characteristics in perceptions of large carnivores, largely dependent on the local context, as well as common features, which transcend differences in local history and synthesis of species diversity. Research findings also underline the importance of engaging with this heterogeneity through dialogue and interventions with a primary aim to understand the causes of contestations and then attempting to mitigate conflict. Moreover, the research reported highlights the need for inter-disciplinary research on human-large carnivore relations. Such a perspective would avoid the pitfalls of disciplinary blind spots, while providing a more holistic understanding of the issues in question, provided that a coherent theoretical framework will be used. Future efforts to further develop such a framework can draw valuable insight from actor network theory and ethno-ethology.

Acknowledgements

The research in Mumbai was partly supported by Maharashtra Forest Department. The research in Ladakh was partly funded by Rufford Small Grants and supported by Wildlife Protection Department, Ladakh region.

References

Athreya, V. R., Odden, M., Linnell, J. D. C., and Karanth, K. U. (2010) 'Translocation as a tool for mitigating conflict with Leopards in human-dominated landscapes of India', *Conservation Biology*, vol 25, pp133–141.
Athreya, V. R., Odden, M., Linnell, J. D. C., and Karanth, K. U. (2011) 'Translocation as a tool for mitigating conflict with Leopards in human-dominated landscapes of India', *Conservation Biology*, vol 25, no 1, pp133–141.

Athreya, V. R., and Venkatesh, V. (eds) (2013) *Mumbaikars for SGNP, 2011–2012: Report Submitted to Director and Chief Conservator of Forests*, Sanjay Gandhi National Park, Mumbai.

Callon, M. (1986) 'Some elements of the sociology of translation: Domestication of the scallops and the fishermen of St. Brieuc bay', in J. Law (ed) *Power, Action and Belief: A New Sociology of Knowledge?* Routledge, London.

Cavallo, J. A. (1990) 'Cat in the human cradle', *Natural History*, February 1990, pp53–60.

Cavallo, J. A., and Blumenschine, R. J. (1989) 'Tree-stored leopard kills: Expanding the hominid scavenging niche', *Journal of Human Evolution*, vol 18, pp393–399.

Cloke, P., and Jones, O. (2004) 'Turning in the graveyard: Trees and the hybrid geographies of dwelling, monitoring and resistance in a Bristol cemetery', *Cultural Geographies*, vol 11, pp313–341.

Cronon, W. (ed) (1995) *Uncommon Ground: Towards Reinventing Nature*, W.W. Norton and Co., New York.

Deol, R. S. (2011) *Census of India 2011: Provisional Population Totals (Maharashtra)*, Government of India, New Delhi.

Frey, J. H., and Fontana, A. (1993) 'The group interview in social research', in D. L. Morgan (ed) *Successful Focus Groups: Advancing the State of the Art*. SAGE Publications, Newbury Park, CA.

Ghosal, S. (2013) 'Cats in the city: Narrative analysis of the interactions between people and leopards in the SGNP landscape, Mumbai', in V. R. Athreya and V. Venkatesh (eds) *Mumbaikars for SGNP, 2011–2012: Report Submitted to Director and Chief Conservator of Forests*, Sanjay Gandhi National Park, Mumbai.

Ghosal, S., and Ahmed, M. (2017) 'Pastoralism and wetland resources in Ladakh Changthang plateau', in H. H. T. Prins and T. Namgail (eds) *Bird Migration Across the Himalayas*, Cambridge University Press, Cambridge.

Ghosal, S., Athreya, V. R., Linnell, J. D. C., and Vedeld, P. O. (2013) 'An ontological crisis? A review of large felid conservation in India', *Biodiversity and Conservation*, vol 22, no 11, pp2665–2681.

Ghosal, S., and Kjosavik, D. J. (2015) 'Living with Leopards: Negotiating morality and modernity in Western India', *Society and Natural Resources*, vol 29, pp1092–1107.

Herda-Rapp, A., and Goedeke, T. L. (eds) (2005) *Mad About Wildlife: Looking at Social Conflict Over Wildlife*, Brill Academic Publishers, Leiden.

Latour, B. (2004) *The Politics of Nature: How to Bring Science into Democracy*, Harvard University Press, Cambridge, MA.

Lescureux, N. (2006) 'Towards the necessity of a new interactive approach integrating ethnology, ecology and ethology in the study of the relationship between Kyrgyz stockbreeders and wolves', *Social Science Information*, vol 45, pp463–478.

Linnell, J. D. C., Breitenmoser, U., Breitenmoser-Würsten, C., et al. (2009) 'Recovery of Eurasian lynx in Europe: What part has reintroduction played?', in M. W. Hayward and M. J. Somers (eds) *Reintroduction of Top-Order Predators*, Wiley-Blackwell, Oxford.

Linnell, J. D. C., Swenson, J. E., and Anderson, R. (2001) 'Predators and people: Conservation of large carnivores is possible at high human densities if management policy is favourable', *Animal Conservation*, vol 4, pp345–349.

Ministry of Environment and Forests (2011) *Guidelines for Human-Leopard Conflict Management, April 2011*, Ministry of Environment and Forests, Government of India, http://envfor.nic.in/sites/default/files/moef-guidelines-2011-human-leopard-conflict-management.pdf.

Punjabi, G. (2013a) 'Assessing free-roaming dog (*Canis familiaris*) abundance in a mark-resight framework in Aarey Milk Colony, Mumbai', in V. R. Athreya and V. Venkatesh (eds)

Mumbaikars for SGNP, 2011–2012: Report Submitted to Director and Chief Conservator of Forests, Sanjay Gandhi National Park, Mumbai.

Punjabi, G. (2013b) 'Distribution and abundance of herbivores in SGNP, Mumbai', in V. R. Athreya and V. Venkatesh (eds) *Mumbaikars for SGNP, 2011–2012: Report Submitted to Director and Chief Conservator of Forests*, Sanjay Gandhi National Park, Mumbai.

Saberwal, V., and Rangarajan, M. (eds) (2003) *Battles Over Nature: Science and the Politics of Conservation*, Permanent Black, New Delhi.

Surve, N. (2015) *Ecology of Leopard in Sanjay Gandhi National Park, Maharashtra with Special Reference to Its Abundance, Prey Selection and Food Habits*, Wildlife Institute of India, Dehra Dun and Sanjay Gandhi National Park, Mumbai.

Takpa, J. (2017) 'Ecotourism: A new livelihood alternative and conservation opportunity in Hemis National Park, Western Himalaya, India', in S. Ghosal and T. Dolkar (eds) *Proceedings of Sustainable Mountain Development Summit – V: Water Security and Skills for Development in the Mountains*, Integrated Mountain Initiative and Ladakh Snow Leopard Foundation, Leh.

8 Considering wolves as active agents in understanding stakeholder perceptions and developing management strategies

Nicolas Lescureux, Laurent Garde, and Michel Meuret

Introduction

Humans and grey wolves (*Canis lupus*) have been sharing the same landscape, the same habitats and even some similarities in their hunter's way of life for a long time (Olsen, 1985; Clutton-Brock, 1995). However, it is quite probable that the domestication of ungulate species (cattle, sheep, and goats), which started around 11,500 BC and spread across Eurasia through the neolithisation process (Zeder, 2008; Vigne et al., 2011; Zeder, 2011) drove to keep wolves at distance from human settlements (Clutton-Brock, 1995; Sablin and Khlopachev, 2002; Verginelli et al., 2005).

From this period on, wolves became a potential threat to livestock. They appear as one of the most conflictual species wherever they occur and overlap with herding activities. This conflict has probably been responsible for motivating the past reduction in the number and distribution of large carnivores worldwide (Mech, 1995; Breitenmoser, 1998; Kaczensky, 1999). Nowadays, in many countries, land abandonment, drastic changes in rural land use, and conservation legislation are leading to the recovery of large carnivores in multiple-use landscapes (Linnell et al., 2001; Falcucci et al., 2007; Chapron et al., 2014), and accordingly, many conflicts are currently appearing or increasing in several countries (Skogen et al., 2008; Dressel et al., 2015; Garde, 2015; Mech, 2017).

Some authors, considering large carnivore economic impacts at the national level and risks to human safety being low, have been proposing that conflicts with large carnivores reflect the long-term persistence of misconceptions and negative perceptions from earlier times (Clark et al., 1996; Kellert et al., 1996; Fritts et al., 2003). On the one hand, however, economic impacts vary according to the considered scale (Kaczensky, 1999) and cannot be reduced to killed or injured livestock. Wolf attacks do also influence livestock productivity and impair the breeder's selection process, as well as the herd production and structure. Wolf attacks also imply additional working load, induced stress, and psychological trauma for livestock breeders and shepherds. On the other hand, several studies have shown that risks to human safety used to exist in Europe and still exist in some developing countries (Rajpurohit, 1999; Comincini et al., 2002; Linnell et al., 2003; Löe and Røskaft, 2004; Moriceau, 2007). Some recent cases, even if rare, also implied wolves coming close to people and even mauling or killing people in North America (McNay, 2002; Grooms, 2007; Butler et al., 2011; Mech, 2017). Therefore, it appears actual conflicts probably find their root in negative aspects of human-wolf relationships rather than in misconceptions. It has also been suggested that prolonged sympatry can lead to a form of coexistence where compromises are made by both species and conflicts are not perceived as being so intense (Mech and Boitani, 2003). Such dynamic and interactive vision is supported by the ability of wolves to adapt their behaviour to the environment (Packard, 2003; Peterson and Ciucci, 2003), including human activities and infrastructure (Ciucci et al., 1997; Theuerkauf et al., 2003).

The impact wolves can have on humans has been addressed through studies about wolves in history, mythology, and narratives (see e.g., Lopez, 1978; Bobbé, 1998; Coleman, 2004; Walker, 2005), as well as sociological analysis of wolf impacts on human society (Mauz, 2005; Skogen et al., 2008; Doré, 2010). In most social science studies, animals were generally considered as passive objects. Their ability to influence or be influenced by human practices was rarely taken into account (Brunois, 2005). Yet, the existence of animal agency was already suggested in the 1950s when J. Von Uexküll assessed animals have their own world or *Umwelt* (1956). Progress in ethology, behavioural ecology and cognition drove to reconsider existence of animal personalities (Dall and Griffith, 2014) as well as animal ability to adapt their behaviour, facing fast changing human-modified environments (Griffin et al., 2017). Canine cognition has been the subject of numerous studies asserting their abilities in problem-solving (Miklósi and Kubinyi, 2016). Following anthropological perspectives in the relationship between nature and society (Descola and Pálsson, 1996; Ellen and Katsuyoshi, 1996) and in sociology of science (Latour, 1993), animals have been increasingly recognized as actors and active agents able to influence human social life (Ingold, 2000; Brunois, 2005; Lestel et al., 2005; Descola, 2013). This recognition has permitted the development of new interdisciplinary approaches to human-large carnivore relationships spanning both scientific and local knowledge, thus revealing the influence large carnivore behaviour can have on human perceptions and providing insights into the interactive and dynamic character of human-wolf relationships (Lescureux, 2006; Lescureux and Linnell, 2010, 2013).

If human-wolf relationships are dynamic and may be modified by alterations in either part of the relationship, then social, economic, and political transitions in human society should result in changes to the human-wolf relationship. The rapid social changes like the ones which have occurred in most Eastern European and Central Asian countries at the beginning of the 1990s provided an opportunity to assess the dynamic and interactive nature of human-wolf relationships. Indeed, social changes can have a direct impact on wolf ecology and behaviour through the modification in hunting and livestock husbandry practices, for economic or regulatory reasons. Rapid social changes also create a new ecological and socio-economical context in which the human perception of wolves and their place will be affected, all the more as ecology and behaviour of wolves are adapting to these transformations. In turn, the changes in human perceptions of wolves will influence wolf hunting and livestock husbandry practices. Similarly, the return of the wolves in countries from which they have been away for decades should influence their behaviour since they are confronted with different ecological, social, and legal contexts. In addition, the return of the wolves will provoke changes in several human practices, notably husbandry and hunting practices, which will in turn influence wolf behaviour.

The comparison between three different countries which underwent either rapid social changes through different processes (Kyrgyzstan and Former Yugoslav Republic of Macedonia[1]), or had to deal with the return of the wolves after decades of absence (France) allows us to explore how different contexts have influenced the way human-wolf relationships changed along with these different ecological and social dynamics. Thanks to a combination of historical, geographic, and ethnographic data, as well as data on wolf damages, our analysis shows that changing social and ecological contexts influence wolf behaviour and human-wolf relationships in a dynamic way. Indeed, if the behaviour of one of the agents – human or wolf – is modified, then this modification will directly alter the behaviour of the other agent, and this alteration will soon influence back the first agent. Therefore, it appears that wolves have to be considered as active agents able to adapt their behaviours to human practices. We then propose to take into account wolf ability to adapt when defining and implementing wolf management strategies and conflict management strategies. If wolves are adapting their behaviour to human practices, then it is probably possible to develop practices aiming at changing wolf behaviour. Quite often, conflict management strategies have been focused only on protecting livestock from wolves based on the assumption that human and livestock guarding dog (LGD) presence are sufficient to deter wolf attacks. They have not been directed towards wolves in order to influence their behaviour towards livestock, dogs, and humans, notably associating them with real danger.

From respected enemies to unwanted pests: wolves in Kyrgyzstan[2]

At the time of the Russian revolution, Kyrgyzstan was mainly occupied by nomadic livestock breeders and Russian colonists. After the restitution of colonists' land to the Kyrgyz people in 1921–1922, a process of collectivisation and the creation of kolkhozes was initiated from 1925–1932, which was accompanied by the settlement of the nomads (Jacquesson, 2004). The systems of kolkhozes persisted until the collapse of the USSR. Our investigations highlighted that the institutional and economic crisis following this collapse had a strong impact on livestock breeding and hunting activities, which were mainly dependent on the state (Lescureux, 2006).

A reorganisation of livestock breeding and hunting activities

Sheep breeding was the agricultural activity most affected by this crisis. Before the transition, all of the land was state property and most of the sheep (77%) were owned by the state and collective farms (Van Veen, 1995). At the beginning of the transition process, many flocks were sold or sent to slaughter houses (Jacquesson, 2004) and between 1992 and 2004, the number of sheep was reduced to one-third in Kyrgyzstan, followed by a slight increase after that year (FAO, 2012). Private farms gradually replaced collective and state farms (Giovarelli, 1998). Intensive livestock farming has been abandoned and changes in the distribution of livestock also occurred. Previously, part of the livestock remained for months in high pastures above 3,000 m, called *sïrt*, kept by salaried shepherds, cut off from villages and sometimes supplied by helicopter. Now, villagers' livestock tend to graze around the villages because of transhumance costs and the pastures furthest away from the villages are less densely occupied than before (Jacquesson, 2004).

The impact of post-soviet upheaval on hunting was quite strong. During the Soviet era, wolf hunting was highly organized at the state level and subsidized. Wolves were considered pest animals and trapped, captured in dens, poisoned, and hunted from planes or helicopters in open areas (Bibikov et al., 1983). The economic and logistical means supporting this intensive wolf hunting were no longer available after the collapse of the USSR. In Kyrgyzstan, each kolkhoz used to have its professional hunters, and shepherds were also equipped with rifles. Hunting was partly an economic activity providing meat, fat, and fur. As a consequence, it is highly probable that the hunting pressure on wolves decreased after the independence of Kyrgyzstan, as in other former Soviet countries (e.g., Belarus; cf., Sidorovich et al., 2003). In addition, members of hunting associations dramatically decreased in Kyrgyzstan from 25,900 in 1990 to 8,617 in 2002 as a consequence of Russian emigration. Despite the existence of knowledgeable wolf hunters in the villages and wolf bounty, local people have neither the means nor the time to hunt wolves on a regular basis.

Perceived changes in behaviour of the wolves

During our ethnographic fieldwork in Kyrgyzstan, many informants reported that the wolves have changed their behaviour since the fall of the USSR (Lescureux, 2007). Not only had wolves become more numerous, but they were also bolder and approached the villages. This drawing closer of the predator was explained in two ways. First, the partial abandonment of high pastures and the regrouping of domestic animals around the villages in winter could drive wolves to approach the villages in order to find their prey. Second, hunting pressure on wolves had decreased. Without rifles, shepherds were no longer able to frighten them. Indeed, the overall presence of armed shepherds in the landscape was perceived as ensuring that wolves had to withdraw back into areas with less human activity instead of coming down to villages. As a consequence, villagers were hardly surprised that wolves were less fearful and no longer hesitated to approach villages and even to attack flocks in broad daylight (Lescureux, 2006).

Wolves as one of the main threats to the capital of Kyrgyz villagers

In Kyrgyz villagers' view of the world, there were no strong borders between the human world and the animal world, and wolves were considered as intelligent, conscious, and even useful animals, removing carrion and killing sick ungulates. Kyrgyz villagers had a well-shaped view on the fact wolves should not be eradicated since predators belonged to nature and were regarded as having a sanitary role. The wolf was thus perceived as an *alter ego* engaged in reciprocal relationships with humans (Lescureux, 2006, 2007). However, Kyrgyzstan had to face a hard and dramatic transition process. Many people complained about the economic situation and all the advantages they lost with the collapse of the Soviet Union (cf. Anderson and Pomfret, 2000). With the loss of social security and very low pensions and salaries, livestock became a vital form of capital and most villagers had a few sheep, one or two cows, and sometimes a few horses. Sheep were mainly kept as a form of capital, which could be sold in case of inflexible expenses (school and university fees, hospitalization, etc.). Under these conditions, villagers were always trying to increase the size of their flocks. As a consequence, when wolves attacked livestock, they were threatening the capital of villagers, who had the feeling they could not effectively control wolves anymore. The actual lack of control over wolves was viewed as a loss of reciprocity and a breakdown in the balance of the human–wolf relationship.

Changes in human practices in the context of the Kyrgyz transition to the market economy have had a clear impact on human–wolf relationships. It has made livestock breeders more vulnerable to wolf attacks and more prone to be highly affected by these attacks, since each domestic animal was more valuable than before. The view on the ecological and sanitary role of the wolf was now contested and could be even more contested if this unbalanced relationship persisted. For many villagers, wolves were no longer perceived as a respectable "enemy" they had to regulate, but as an "invader", preventing the increase of a herder's capital and even threatening the future of pastoralism and economy in the country:

> Outside, yaks are eaten by wolves. So we have to keep them in enclosures, like sheep, like horses. So you see neither horses, nor yaks, nor cows outside. So if we don't eliminate the wolves, what do we do with them? The state could give means. If not, where are we going?
>
> (Shepherd from Korgondu-Bulak, 12/2005)

Wolves as an additional threat to a highly weakened activity in the Former Yugoslav Republic of Macedonia[3]

The collapse of Yugoslavia had a strong impact on livestock breeding and hunting activities. However, the socio-economic context was quite different from Kyrgyzstan. Indeed, in Yugoslavia, an economic and cultural policy of centralization as well as an accelerated industrialization starting in the 1960s led to an exodus from the countryside to towns and a concomitant decrease in the agricultural sector (Hadživuković, 1989). Shepherds also went to work in other countries, like in Italy, where their knowledge was appreciated (Pardini and Nori, 2011). The rural population growth rate started to be negative as early as the 1960s in Yugoslavia. The proportion of the population living in rural areas in Macedonia dropped from 76.6% in 1950 to 27.7% in 2010 (FAO, 2012), and Macedonia became mainly urban as early as the 1970s. As a consequence, there has been a strong rural abandonment and a continuous decrease of the rural population (FAO, 2012).

A collapse of sheep breeding and a disorganization of hunting

Sheep breeding in the Former Yugoslav Republic of Macedonia was quite different from the Kyrgyz one. Even before transition, 90% of the sheep were privately owned even if state or collective farms (*Agro-Kombinat*) had large flocks ranging between 1,000 and 25,000 sheep (MAFWE, 2003). Sheep breeding was highly affected by the transition process. The number of sheep was halved between 1992 and 2006, notably because of a EU ban on the import of lamb meat from the country following the outbreak of foot-and-mouth disease in 1996 (Dimitrievski and Ericson, 2010). Moreover, between 1995 and 2007, the domestic consumption of lamb and mutton meat decreased from 10.1 kg to 3.3 kg per household per year and the domestic consumption of ewe milk and cheese decreased from 15,643 to 11,291 tonnes per year. According to livestock breeders, the market for lamb meat was not stable and exposed them to economic risks when they did not manage to sell their lamb in time. Subsequently, the low prices of meat, milk, and wool combined with the high prices of fodder and concentrate feed, as well as the rising of labour cost, pushed the sheep breeders to reduce their flock size (Dimitrievski and Ericson, 2010). Despite the fact that livestock breeding still remains a relatively important economic activity in the country, accounting for 24% of the total agricultural output in the period 1995–2007 (Dimitrievski and Ericson, 2010), livestock breeders are not numerous. Compared to Kyrgyzstan, livestock breeding and especially sheep breeding appears to be a more commercial (rather than subsistence) activity in northwestern Macedonia, focused on cheese and lamb production in a highly seasonal manner.

The impact of the transition process on the organization of hunting was probably less dramatic in former Yugoslavia than in Kyrgyzstan. Macedonia is divided into 249 hunting grounds, and apart from the state hunting grounds, an open competition was held in 2002 to award concessions to the highest bidders (Petkovski et al., 2003). According to Petkovski et al. (2003), hunters were largely unsatisfied since they had to pay for expensive management plans when, at the same time, the legal system did not ensure the punishment of poachers. Therefore, many users were not paying their membership fees and poaching was considerably higher than before independence (Former Yugoslav Republic of Macedonia became independent from Yugoslavia in 1991) as a result of the lack of an organized game warden service. A report from the Ministry of

Environment and Physical Planning (2003) also bemoaned the fact that, despite the existence of hunting management plans and a public enterprise for game wardens and hunting inspections, poaching remained at a high level. The number of wolves reported as being killed has slowly increased since the 1960s and although there are no statistics for the transition period between 1988–1992, it jumped from 200 wolves killed in 1987 to 460 in 1994 (Melovski and Godes, 2002), showing that hunting pressure on wolves did not decrease at that time period. Current harvests (2008–2010) have been between 108 and 188 wolves killed per year (Kaczensky et al., 2012).

Wolves as harmful and uncontrollable animals

Macedonian hunters and livestock breeders did not notice any major changes in wolf behaviour after the collapse of Yugoslavia (Lescureux and Linnell, 2013). However, in a difficult context for Macedonian livestock breeders, wolves were perceived as harmful animals. The way wolves were perceived was strongly influenced by their behaviour and the various encounters between humans and wolves. Notably, interactions with wolves seem to be particularly linked to livestock. Thus, even if wolves were hidden in the deep forest during the day, they often approached humans when attacking sheep in night-time enclosures on summer pastures. They could even come into the villages, notably to attack dogs in winter, boldly crossing physical and symbolic borders that other predators never crossed (cf. Lescureux and Linnell, 2010). The fact that wolves often killed several sheep per attack strengthened the negative perception and many interviewees considered that they enjoyed killing. The reputation of the wolf for excessive killing was widespread and gave rise to some idioms like "the wolf will kill 99 sheep and die at the hundredth" (Elsie, 2001; Lescureux and Linnell, 2010). When the encounter was not linked to livestock, like when hunters encountered wolves, it was often because the later were attacking dogs and thus initiating the interaction. Not only were such hunting dogs highly valuable for hunters, but hunters also noticed the determination of the wolves when attacking dogs. The perceived voracity of wolves did not seem to be limited to domestic animals. Interviewees also blamed them for damage to populations of game animals. Contrary to the bears, to which they were often compared by informants, they were perceived as a rather homogenous population and their harmful characteristics were attributed to the entire species. Because of their perceived damage to wild and domestic animals, and harmfulness to nature in general, wolves were described as "unprofitable monsters" (Lescureux and Linnell, 2010).

Not only were the wolves perceived as harmful animals for both livestock and game animals, but they were also perceived as animals very difficult to control. Even if hunting wolves was authorized and encouraged by the state, it was their behaviour and ecology which made them difficult to hunt. Indeed, even though the informants regarded wolves as being territorial in nature, they also reported that they were always on the move, following their prey or looking for livestock, moving to pastures during the summer and to the vicinity of villages in winter. Hard to localize, wolves were not possible to individualize due to their group living, and they also benefited from a high reproductive rate. Permanent shepherding and LGDs seem to be widely used in the Former Yugoslav Republic of Macedonia (Keçi et al., 2008) and relatively efficient against wolves. However, these mitigation measures were experienced as economic and time constraints by livestock breeders, and they did not function effectively in all cases such as in cases of fog or when wolves managed to slip past LGDs.

As a consequence, human relationships with wolves appeared to be rather unbalanced. On the one hand, these animals often came to take several sheep at a time and their damage could be relatively important at farm or regional levels. In addition, the repeated and fatal intrusions of wolves into domestic space created the impression of an animal that was disrespectful of borders and norms. On the other hand, it was difficult for humans to insure any reciprocity with wolves. The difficulties inherent to wolf hunting in steep, forested habitats hindered the effective implementation of the hunting right.

Redefining the place of the wolf in a changing socio-economic and environmental context

In Macedonia, sheep were not the main capital for most villagers, but rather the resource and occupation of a few professional livestock breeders and their employees. Looking at the results of our investigations, it appears that livestock breeders were mainly complaining about the difficulties inherent to economic activities linked with the transition process and its impact on livestock breeding. Their first complaint was often about the lack of access to markets. There was a widespread view among livestock breeders that it was better before the collapse of Yugoslavia, when the state bought cheese, meat, and wool, paid people to work on pastures, and even provided fertilizers. Extensive herding of sheep had become a marginal activity in the country. Livestock breeders had the feeling they were marginalized and that they were not given proper attention by the state, the politicians, or public opinion in general. As a consequence, there was a complete lack of confidence in the future. Many livestock breeders had the feeling that their activity was coming to an end, and that only the biggest and richest livestock owners having access to subsidies and to the market would survive. As a striking consequence of this downward economic situation, most breeders could not invest in or develop their activity, as this man put it:

> But if we would have more sheep, we would have more expenses! In this situation, it would be worse.
>
> (breeder from Dobri Dol, 10/2007)

The transition process was not just affecting livestock breeders economically. The ongoing rural abandonment process and the crisis in livestock breeding were also affecting the landscape around the villages as well as the social structures in the villages and in the countryside. Therefore, the livestock breeders' perceptions of their social and natural environment were clearly impacted and also their concerns for the future of their occupation. All the villages in the study area from the Shara Mountains were surrounded by fields and orchards which were now mostly abandoned. As a consequence, there was a feeling that the outskirts of the villages were becoming wild. Thus, pastures were suffering shrub encroachment; fields were becoming pastures and for villagers, the former system of order (village, orchard, field, pasture, forest) was being disrupted. The situation was very similar to what Höchtl et al. (2005) had observed in the southwestern Alps.

In this context, any further cost was a weight that could not be tolerated, and the wolf was mainly considered as a burden adding to these costs. Such a cost may be direct, when the wolf was attacking the flock and taking some sheep. However, the cost may also be indirect, through the obligation to keep LGDs, which was also taken into consideration by livestock breeders. Indeed, even if fed with maize flour or old bread mixed with whey, there was a cost to maintain LGDs (around €1 per dog per day), especially when livestock breeders often had to keep five or more dogs per flock.

Wolves were perceived as bloodthirsty and pest animals, destroying livestock, preying upon dogs and game animals. In the harsh economic and social context of rural Macedonia, wolves were not perceived as the main threat to rural life, but were regarded as an additional threat, which symbolized rural people's loss of control over nature, and therefore had to be eliminated, as wolves were absolutely useless for them.

France: the return of a protected carnivore in a pastoral landscape[4]

In Kyrgyzstan and Macedonia, the upheaval in the socio-economic context drove to changes in human-wolf relationships, but the situation in France is quite different (Meuret et al., 2017). Indeed, the main change which occurred in human-wolf relationships was the return of the wolves in the French Southern Alps in the 1990s, after decades of absence. This return was officialized and greeted by the French naturalist's magazine *Terre Sauvage* (Peillon and Carbonne, 1993). Under the Bern Convention and the Habitat's Directive, wolves in France were placed under strict protection. Meanwhile, despite the low number of wolves in the 1990s (fewer than 15), livestock breeders quickly saw the number of killed livestock increase. Wolves were coming back in territories where livestock breeding, and notably sheep breeding, had developed in the absence of predators and following environmental incentives and production choices at the farm level.

Vulnerable breeding systems

In France, following the reform of the Common Agricultural Policy (CAP) in Europe in 1992, environmental public policies started to rely strongly on

pastoral and agro-pastoral systems to deal with landscape dynamics and wildlife habitat management plans (Alphandéry and Billaud, 1996). Restoration
and conservation of summer pastures have been largely entrusted to livestock
breeders through five-year contracts, when the latter have been encouraged
by means of economic incentives to graze on encroached lands or woodlands
prone to wildfire (Léger et al., 1996). In the Natura 2000 management plan
for the Alps, livestock grazing had a central role to conserve the biodiversity of
alpine meadows, exactly where the wolf came back. Livestock breeders were
then confronted with a paradox. On the one side they were encouraged and
even funded to redeploy grazing on encroached lands, scrublands, forest edges,
and undergrowth. On the other side, they were urged to protect themselves
from wolves, for which it was easier to attack flocks in such landscapes.

The situation in the French Mediterranean region and Southern Alps also
results from production choices, where free-range livestock breeding systems
are particularly exposed and vulnerable to predation (Garde, 2015). First, almost
half a million hectares are scrublands or encroached woodlands (Garde et al.,
2014). These areas provide cheap resources for well-adapted livestock and
ensure food security and economic robustness for the breeding system (Hubert
et al., 2014). In addition, the low snow cover allows year-round grazing, contrary to higher or northern places where livestock can graze outside only for
6–8 months. The combination of scrublands and woody pastures with year-
round grazing makes these breeding systems particularly prone to wolf attacks,
notably when considering important topographic constraints.

Another constraint comes from the breeding system itself and production
choices. In southeastern France, sheep production is dominated by sheepfold
lambs production. Only empty or pregnant ewes as well as she-lambs are grazing outside. As their food requirements are limited, shepherds can practice close
herding along daily grazing circuits and take them back for the night in an
enclosure, or sheepfold. However, other breeding systems are producing older
lambs ("tardons"), with both ewes and their lambs grazing throughout the summer period. As suckling ewes' and lambs' requirements are higher, shepherds
cannot practice close herding and have to let them spread on pasture for longer
time periods during the day. Night enclosure becomes quite constraining in
that case. Yet, this breeding system is widespread in Alpes Maritimes, Var and
Alpes de Haute Provence, regions with the most frequent instances of wolf
depredation in France.

A never-ending increase in livestock damage

From 1994 to 2004, the number of killed or deadly injured animals officially
attributed to wolves gradually grew to reach about 3,000 per year. Fewer than
300 livestock protection contracts were signed during this first period, but the
number quadrupled between 2004 and 2009. Livestock breeders have been able
to employ shepherd assistants and to buy LGDs and electric fences for night
enclosures. Although the number of animal losses reached 4,000 in 2005, it

decreased and stabilized around 3,000 per year from 2006–2009.[5] This trend let people assume that protection means were sufficient enough at least to stabilize the situation, even in the case of well-established wolf packs. However, the following seven years (2009–2016) came to contradict this assumption. Even though the number of protected flocks grew from 796 to 2,238, the number of killed or deadly injured animals has tripled, up to 10,000 per year, while the estimated number of wolves was only increased by a factor of 1.5. How can one explain these numbers? A possible explanation is wolf colonization of new territories. Indeed, from 1993–2016, wolves clearly expanded their distribution (Duchamp et al., 2017). They now occupy the entire French Alps and have progressed towards northeastern France, south Massif Central and the eastern Pyrenees. The mainland French Departments with temporal permanent wolf presence currently total 39 out of 95. The hypothesis is that wolves colonize new territories where livestock breeders are not prepared to protect their flocks, which are somehow naïve prey for wolves. However, the number of animal losses continues to grow even in the Alpes Maritimes department, where wolves had arrived already in the 1990s, and the first seven departments colonized by wolves still cumulate more than 90% of the losses in 2015. In addition, about 90% of killed or deadly injured animals belong to protected flocks. These results strongly question the efficiency of livestock protection measures.

Livestock protection measures showing their limits

For livestock breeders and shepherds, the situation is perceived as a loss of control despite the money spent in various measures accompanying the return of the wolf and livestock protection (more than €20 million in 2015). Protection measures have been quickly promoted and financed by the French State. Conceived and tested in US, Italy, and Sweden (Mech, 1995), these measures have been based on three postulates:

First, additional human presence – even if not armed – is sufficient to keep wolves at distance: The shepherd assistant. The shepherd assistant is highly valuable. His work consists of helping the main shepherd in supplementary tasks linked with flock protection. The working load has been evaluated at around 200 hours per month in the case of a collective sheep flock in a summer pasture (Silhol et al., 2007) and 100 hours in the case of an individual sheep or goat flock outside a summer pasture (Garde et al., 2007).

Second, an additional obstacle, more attentive and vigilant than humans, deters bold wolves: The livestock guarding dog (LGD). After a century of absence in the central and eastern parts of France, LGDs have been introduced urgently in the French Alps and Provence starting from the end of the 1990s. LGDs' efficiency is highly variable according to numerous parameters (landscape, breeding system, number of dogs, dog skills, size of the wolf pack, wolf experience and skills, etc.). In addition, considering that the French Alps are among the most prominent tourist destinations in Europe, these conditions require that the dogs be at the same time highly reactive towards wolves and relatively indifferent towards humans, who can be quite numerous (hikers, bikers, hunters, etc.) in and around summer pastures. Therefore, LGDs are becoming a major concern in multi-use landscapes. The number of complaints of people frightened or injured by LGDs is growing (Vincent, 2011), while some livestock breeders have been brought before the court due to such conflicting encounters. Mayors are more and more concerned about dealing with what is becoming a problem of public safety. In highly touristic places, some mayors have even forbidden the use of LGDs on their territory.

Third, a flock regrouped and kept in a night enclosure under human and dog supervision is attacked no more: The night pen. With large night pens with double enclosure close to LGDs and attentive humans, night penning can be efficient, if it is feasible in the first place. However, it does not come without heavy constraints. In hilly or mountainous areas, regular night penning is not compatible with sustainable grazing resources management (Savini et al., 2014), prevention of soil erosion (Vincent, 2011), and sheep health. Flock movement back and forth the night pen can decrease grazing time 15–25%, notably during the summer heat. In addition, night penning is actually put into question as it apparently drove wolves to concentrate their attacks within the day.

Wolves adapting to human practices and presence

Considering the continuous and dramatic increase in killed livestock despite the large scale adoption of protection measures, and also considering that these measures have proven to be relatively efficient for a few years (2006–2009), we wonder whether wolves have learnt to deal with these measures or not, and adapt their behaviour accordingly. Several elements tend to support the ability of wolves to adapt their behaviour to damage prevention methods. One striking fact is that wolves are actually transferring their attacks on sheep flocks during the daytime, after having faced the difficulty of attacking them in night pens. In 2010, only 29% of livestock were killed during daytime, whereas in 2015, this proportion has reached over 50%.[6] In addition, wolf attacks are increasing in plains and valleys, and some wolves do not hesitate to attack flocks very close to human settlements.

Apart from their damages on livestock, a recent survey in a village from Alpes de Hautes Provence has investigated a case study focusing on a pack of wolves that showed aggressive behaviour towards a teenager (Garde and Meuret, 2017). During this survey, numerous testimonies from livestock breeders or shepherds reported encounters with wolves showing no fear of humans, and it appeared that several people had already met this pack at distance but on the same place several times. In other words, under a strict protection status, it seems that wolves tend to adapt their behaviour to different protection means. Even under a strict protection status as in the Italian case, a high level of poaching (Galaverni et al., 2016; Hindrikson et al., 2016) might act as if wolves were hunted. Once wolves have crossed the border to France, the human pressure has strongly decreased on them. However, one of the core constituents of livestock protection measures has always been that human presence would deter wolf attacks on livestock. The crucial question, then, is how long can wolves maintain their fear of humans under a strict protection status?

Conclusions and management implications

Through these different examples, we showed that human–wolf relationships are 1) highly dependent on multiple contexts (e.g., socio-economic, cultural, political, ecological) and 2) dynamic and prone to change in case of shifting contexts. In order to understand human–wolf relationships, it is therefore necessary to grasp the broader context as well as its dynamic. In the three countries, the observed developments have influenced human–wolf relationships in two ways (Table 8.1). In Kyrgyzstan and Macedonia, the transition process has generated multiple socio-economic and environmental trends, by which the place of the wolf has been modified when compared to the previous periods under the USSR and Yugoslavia regimes. Such changes seem to have been beyond the scope and reach of most wolf management strategies attempted so far, including communication or awareness and outreach campaigns. Indeed, we do not believe in communication campaigns as an effective means to resolve human–wolf conflicts. Based on our experience, these conflicts cannot be explained by so-called lack of knowledge or persistence of misconceptions about wolves among local people. Therefore, trying to act on knowledge and perceptions without changing the terms of human–wolf relationships appears to be ineffective. Instead, it often seems to be instrumental to directly address the human–wolf relationship itself. As Forbes (2009: 234) put it:

> [R]elationships are as fundamental as places and things. Conservationists made a strategic error in assuming that our work is more a legal act than a cultural act, in assuming that we can protect land from people through laws, as opposed to with people through relationships.

The transition processes in Kyrgyzstan clearly affected both livestock breeding and hunting practices, then they influenced the human–wolf relationship, and consequently, they have altered wolf behaviour, since wolves seem to have adapted to new practices. The situation is quite different in France, where the return of the wolf has been the major change affecting hunting and livestock breeding activities. While several damage prevention methods were employed to protect flocks from wolves, the damages are still increasing, and it seems that wolves are adapting one way or another to the different barriers put between them and the livestock. In addition, some wolves are showing bolder behaviour under human presence.

In the context of strict protection of the wolf in France or strong decrease of wolf hunting in Kyrgyzstan, one could address the first postulate, on which protection measures have been based, i.e., that additional human presence would be sufficient to keep wolves at distance. Building on the concept of "landscape

Table 8.1 A synopsis of major socio-economic trends, main drivers of the human-wolf conflict, and core challenges and concerns across the three countries studied

	Kyrgyzstan	*Former Yugoslav Republic of Macedonia, Shara Mountains*	*France, Southern Alps*
Major socio-economic trends	After the transition period (1992), livestock, and sheep, especially, comprises a vital form of capital for most villagers, which may be used to cover of inflexible expenses (school and university fees, hospitalization, etc.)	Livestock breeders feeling marginalized and not given proper attention by the state, the politicians, or public opinion in general; lack of confidence in the future and lack of investment in one's own activity	Livestock grazing was promoted (1992) as a central tool for conserving the biodiversity of mountainous and Mediterranean pastures (encroached lands or woodlands prone to wildfire); economic incentives provided to breeders through five-year contracts
Main drivers of the human-wolf conflict	The partial abandonment of high pastures and the regrouping of domestic animals around the villages in winter could drive wolves to approach the villages	Wolves often approached humans when attacking sheep in night-time enclosures on summer pastures; they could even come into the villages to attack dogs in winter	High and increasing level of damages on livestock despite the generalized use of protective methods; landscape and farming systems making livestock highly vulnerable to wolf depredation
Core challenges and concerns	• Although former perceptions of the wolf reflected a reciprocal human-wolf relationship and tolerance, lack of control over wolves viewed as a loss of reciprocity and a break-down in the balance of the human-wolf relationship • Wolf depredation endangering capital deposited in the form of livestock; the wolf portrayed as an "invader"	• Although not a main threat, wolves adding to rural problems directly (i.e., through depredation) or indirectly (e.g., through a necessary investment in livestock guarding dogs); symbolizing loss of control over nature • Wolves perceived as "voracious" and thought to cross physical and symbolic borders that other predators never did	• Damage prevention methods questioned: Night enclosures highly constraining or even non-feasible for breeders using several fenced pastures at a time; livestock guarding dogs in conflict with other land uses, e.g., tourism; human presence not preventing wolves' attacks, even at daytime and close to villages • Considerable indications that wolves are rapidly adapting to human practices and presence

of fear" (cf. Lescureux and Linnell, 2010), which refers to predation risk and any related anti-predator response, Oriol-Cotterill et al. (2015) have conceptualized a "landscape of coexistence" for humans and large carnivores, maintained by means of human-caused mortality risk for large carnivores, through which the latter would act to avoid encounters with humans, for instance, by spatiotemporal partitioning of their activities. However, the risk of human-caused mortality is close to zero under strict protection (poaching excepted). So, do wolves fear human presence in any case? Or do wolves fear humans because there have been centuries of hunting regulations, or even persecution? What can be the influence of long-term strict protection of the species on wolf fear of humans?

In the French Alps, there are numerous people sharing the landscape with wolves. Some are hiking, some are biking, and others are mostly looking for birds or flowers. Most humans are not a threat to wolves and it could well be that humans are less and less perceived as such by wolves. For the wolf, the reward for attacking a sheep flock is quite high: easy and fat meat, and no need to spend much time in searching, stalking, and chasing prey. Comparatively, the risks are quite low. Of course, from time to time, wolves are detected and pursued by a few big dogs they can generally outrun, but they are neither injured nor killed in most of these encounters.

Considering wolves as active agents would entail recognizing their ability to understand part of the context and the other agents they are confronted with, and their ability to adjust their behaviour to the context and to the agents. Livestock protection means are notably based on the assumption that wolves are afraid of humans, but wolf fear of humans may not be an intrinsic or everlasting feature of wolf behavioural patterns. It is rather something that has to be maintained through associating human presence with real risks of human-caused injuries to wolves or human-caused mortality. We believe that the efficiency of most protection means is, indeed, dependent on maintaining wolf fear of humans and that the combination of lethal and non-lethal measures is probably the most efficient solution (Bangs et al., 2006). Therefore, we suggest that wolf control through harassment, chasing, and killing should be part of conflict management strategies and would reinforce the efficiency of livestock protection through maintaining risks associated with approaching human settlements, humans themselves, and livestock.

Some recent studies or reviews have questioned the positive effect of hunting on the decrease of predation on livestock (Wielgus and Peebles, 2014; Treves et al., 2016), even though there is rather a lack of correctly designed studies on that topic (Allen et al., 2017) and the effect of predator control will vary according to the wolf population numbers and distribution (Mech, 2017). Yet, hunting could disrupt wolf social structures and potentially drive wolves to an increase of predation on livestock (Brainerd et al., 2008). Thus, Wielgus and Peebles (2014) reported a positive correlation between the number of sheep depredated and the number of wolves killed the previous year. However, their analysis has been contested in a replication (Poudyal et al., 2016), according to whom the increase in sheep depredation would be a short-term effect confined within the year when culling takes place, followed by a decrease of expected cattle and sheep depredation in the next year, for each individual wolf removed within the previous year. Other studies have shown the effective decrease of livestock depredation following a full or partial wolf pack removal, notably if done quickly after depredation (Bradley et al., 2015). As Treves et al. (2016) put it, predator control should not be "a shot in the dark" and in order to be effective, it has to be correctly designed, implemented, and evaluated according to the expected goal, i.e., maintaining distance with wolves, regulating the population, applying zoning strategies, etc.

Notably, the context has to be taken seriously into account: The husbandry system, the landscape, and of course the wolf control means and the way they will be implemented. To be efficient, wolf control has to be targeted on individuals or packs known to attack livestock in order to (1) eliminate "problem" wolves or packs and (2) to counter select bolder animals. Such control has to be clearly associated with humans so that humans will be associated with injury/mortality risk by wolves. The creation of a wolf squad (*brigade loup*) in 2015 by the French Ministry of Environment is going into that direction.[7] This squad can be quickly operational on the field right after an attack on livestock, and thus it can specifically target wolves responsible for an attack within hours or days following the attack.

Direct action on wolves responsible for attacks on livestock may somehow appear as an act of *reciprocity*, which has been defined as a condition for coexistence between humans and large carnivores (Lescureux, 2006; Lescureux and Linnell, 2010). Reciprocity means that management of conflicts between humans and wolves shall imply not only adaptation of humans to wolf presence through livestock protection means, but also adaptation of wolves to human presence and activities through maintaining pressure on wolves in order to keep a distant relationship, when and where it would be necessary.

Notes

1 Editor's note: Registered as "The former Yugoslav Republic of Macedonia" among United Nations Member States (www.un.org/en/member-states/index.html) and among EU Candidate countries (https://ec.europa.eu/neighbourhood-enlargement/countries/detailed-country-information/fyrom_en).

2 This section is based on results from a 10-month ethnographic survey undertaken between 2003 and 2007 (Lescureux, 2007). Several villages from Naryn and Issyk-Köl regions were investigated through participant observations and semi-directive interviews (n = 91) using snowball sampling, and focus was put on local ecological knowledge on animals in general and wolves in particular.

3 This section is based on the results from seven months of fieldwork in the Former Yugoslav Republic of Macedonia, undertaken by the first author between 2007 and 2011, and focusing on local knowledge on bears, wolves, and lynxes as well as the use of livestock guarding dogs. The ethnographic survey took place mainly in Shara Mountains in the northwestern part of the Former Yugoslav Republic of Macedonia. Snowball sampling was used, and 63 hunters and livestock breeders were interviewed in 33 villages.

4 This section is based on regular field and technical work undertaken by L. Garde and M. Meuret as well as analysis of official data on wolf damages on livestock (Direction régionale de l'Environnement de l'Aménagement et du Logement Auvergne Rhône Alpes (DREAL-AURA) and GéoLoup database, a database with restricted access directly depending on the French Ministry for Environment).
5 Data on wolf damages on livestock come from Direction régionale de l'Environnement de l'Aménagement et du Logement Auvergne Rhône Alpes (DREAL-AURA) and from GéoLoup database, a database with restricted access directly depending on the French Ministry for Environment. Raw data from GéoLoup database were accessed by Centre d'Etude et de réalisation Pastorales Alpes Méditerranées (CERPAM). Analysis and comparison (protected vs. non-protected; daylight vs. night, etc.) have been undertaken by the authors.
6 Analysis undertaken by CERPAM and based on the national database of Géoloup. Cf. note 5.
7 Squad of support to livestock breeders against wolf attacks. Their mission is to participate to livestock protection with defensive shots, to support the agents from the National Office for Hunting and Wildlife management (ONCFS) in their actions to scare or remove wolves, and to participate in documenting and evaluating wolf predation on livestock.

References

Allen, B. L., Allen, L. R., Andrén, H., Ballard, G., Boitani, L., et al. (2017) 'Can we save large carnivores without losing large carnivore science?', *Food Webs,* vol 12, pp64–75.

Alphandéry P., and Billaud, J. P. (1996) 'Cultiver la nature'. *Etudes Rurales,* vol 141–142, p238.

Anderson, K., and Pomfret, R. (2000) 'Living standards during transition to a market economy: The Kyrgyz republic in 1993 and 1996', *Journal of Comparative Economics,* vol 28, pp502–523.

Bangs, E. E., Jimenez, M. D., Niemeyer, C. C., and Fontaine, J. A. (2006) 'Non-lethal and lethal tools to manage wolf-livestock conflict in the Northwestern United States', in R. M. Timm and M. O'Brien (eds) *Proceedings of the Vertebrate Pest Conference,* University of California, Davis.

Bibikov, D. I., Ovsyannikov, N. G., and Filimonov, A. N. (1983) 'The status and management of the wolf population in the USSR', *Acta Zoologica Fennica,* vol 174, pp269–271.

Bobbé, S. (1998) 'Du folklore à la science: Analyse anthropologique des figures de l'ours et du loup dans l'imaginaire occidental', *Ruralia,* vol 1998.

Bradley, E. H., Robinson, H. S., Bangs, E. E., Kunkel, K., Jimenez, M. D., et al. (2015) 'Effects of wolf removal on livestock depredation recurrence and wolf recovery in Montana, Idaho, and Wyoming', *The Journal of Wildlife Management,* vol 79, pp1337–1346.

Brainerd, S. M., Andrén, H., Bangs, E. E., Bradley, E. H., Fontaine, J. A., et al. (2008) 'The effects of breeder loss on wolves', *Journal of Wildlife Management,* vol 72, pp89–98.

Breitenmoser, U. (1998) 'Large predators in the Alps: The fall and rise of man's competitors', *Biological Conservation,* vol 83, pp279–289.

Brunois, F. (2005) 'Pour une approche interactive des savoirs locaux: l'ethno-éthologie', *Journal de la société des océanistes,* vol 120–121, pp31–40.

Butler, L., Dale, B., Beckmen, K., and Farley, S. (2011) *Findings Related to the March 2010 Fatal Wolf Attack Near Chignik Lake,* Alaska, ADF&G, Division of Wildlife Conservation, Palmer, Alaska.

Chapron, G., Kaczensky, P., Linnell, J. D. C., von Arx, M., Huber, D., et al. (2014) 'Recovery of large carnivores in Europe's modern human-dominated landscapes', *Science,* vol 346, pp1517–1519.

Ciucci, P., Boitani, L., Francisci, F., and Andreoli, G. (1997) 'Home range, activity and movements of a wolf pack in Central Italy', *Journal of Zoology, London*, vol 243, pp803–819.

Clark, T. W., Curlee, A. P., and Reading, R. P. (1996) 'Crafting effective solution to the large carnivore conservation problem', *Conservation Biology*, vol 10, pp940–948.

Clutton-Brock, J. (1995) 'Origins of the dog: Domestication and early history', in J. Serpell (ed) *The Domestic Dog: Its Evolution, Behaviour and Interactions with People*, University Press, Cambridge.

Coleman, J. T. (2004) *Vicious: Wolves and Men in America*, Yale University Press, New Haven, CT.

Comincini, M., Oriani, A., Morbioli, C., Castiglioni, C., and Martinoli, A. (2002) *L'uomo e la "bestia antropofaga". Storia del lupo nell'Italia settentrionale dal XV al XIX secolo*, Unicopli, Milan, Italy.

Dall, S., and Griffith, S. (2014) 'An empiricist guide to animal personality variation in ecology and evolution', *Frontiers in Ecology and Evolution*, vol 2.

Descola, P. (2013) *Beyond Nature and Culture*, University of Chicago Press, Chicago.

Descola, P., and Pálsson, G. (1996) *Nature and Society: Anthropological Perspectives*, Routledge, London.

Dimitrievski, D., and Ericson, T. (2010) *Sector Study – Macedonian Agriculture in the Period 1995–2007*, University Ss Cyril and Methodius & Swedish University of Agricultural Sciences, Skopje.

Doré, A. (2010) 'L'histoire dans les méandres des publics: quand les "méchants loups" ressurgissent du passé', in J.-M. Moriceau and P. Madeline (eds) *Repenser le sauvage grâce au retour du loup. Les sciences humaines interpellées*, Presses Universitaires de Caen, Caen.

Dressel, S., Sandström, C., and Ericsson, G. (2015) 'A meta-analysis of studies on attitudes toward bears and wolves across Europe 1976–2012', *Conservation Biology*, vol 29, pp565–574.

Duchamp, C., Chapron, G, Gimenez, O., Robert, A., Sarrazin, F., et al. (2017) *Expertise collective scientifique sur la viabilité et le devenir de la population de loups en France à long terme*, ONCFS & MNHN, Paris.

Ellen, R., and Katsuyoshi, F. (1996) *Redefining Nature: Ecology, Culture and Domestication*, Berg, Oxford.

Elsie, R. (ed) (2001) *A Dictionary of Albanian Religion, Mythology, and Folk Culture*, New York University Press, New York.

Falcucci, A., Maiorano, L., and Boitani, L. (2007) 'Changes in land-use/land-cover patterns in Italy and their implication for biodiversity conservation', *Landscape Ecology*, vol 22, pp617–631.

FAO (2012) *FAOSTAT. Food and Agriculture Organization of the United Nations*, Roma, Italy.

Forbes, P. (2009) 'Reciprocity: Toward a new relationship', in R. L. Knight and C. White (eds) *Conservation for a New Generation. Redefining Natural Resources Management*, Island Press, Washington, DC.

Fritts, S. H., Stephenson, R. O., Hayes, R. D., and Boitani, L. (2003) 'Wolves and Humans', in L. D. Mech and L. Boitani (eds) *Wolves: Behavior, ecology, and conservation*, The University of Chicago Press, Chicago.

Galaverni, M., Caniglia, R., Fabbri, E., Milanesi, P., and Randi, E. (2016) 'One, no one, or one hundred thousand: how many wolves are there currently in Italy?', *Mammal Research*, vol 61, pp13–24.

Garde, L. (2015) 'Sheep farming in France: Facing the return of the wolf', *Carnivore Damage Prevention News*, vol 11, pp17–27.

Garde, L., Bacha, S., Bataille, J.-F., Gouty, A. N., and Silhol, A. (2007) 'Les éleveurs résidents en zone à loups: Perceptions et stratégies', in L. Garde (ed) *Loup – élevage: s'ouvrir à la complexité. Actes du séminaire des 15 et 16 juin 2006.*, CERPAM, Manosque.

Garde, L., Dimanche, M., and Lasseur, J. (2014) 'Permanence and changes in pastoral farming in the Southern Alps', *Journal of Alpine Research | Revue de géographie alpine*, vol 102, http://rga.revues.org/2416.

Garde, L., and Meuret, M. (2017) Quand les loups franchissent la lisière. Expériences d'éleveurs, chasseurs et autres résidents de Seyne-les-Alpes confrontés aux loups, CERPAM & INRA, Manosque, Montpellier.

Giovarelli, R. (1998) Land reform and farm reorganization in the Kyrgyz Republic, # 96, Rural Development Institute, Washington, DC.

Griffin, A. S., Tebbich, S., and Bugnyar, T. (2017) 'Animal cognition in a human-dominated world', *Animal Cognition*, vol 20, pp1–6.

Grooms, S. (2007) 'Ontario experiences cluster of wolf-human encounters', *International Wolf*, vol 17, pp10–13.

Hadživuković, S. (1989) 'Population growth and economic development: A case study of Yugoslavia', *Journal of Population Economics*, vol 2, pp225–234.

Hindrikson, M., Remm, J., Pilot, M., Godinho, R., Stronen, A.V., et al. (2016) 'Wolf population genetics in Europe: A systematic review, meta-analysis and suggestions for conservation and management', *Biological Reviews*.

Höchtl, F., Lehringer, S. and Konold, W. (2005) '"Wilderness": what it means when it becomes reality - a case study from the southwestern Alps', *Landscape and Urban Planning*, vol 70, pp85–95.

Hubert, B., Deverre, C., and Meuret, M. (2014) 'The rangelands of southern France: Two centuries of radical change', in M. Meuret and F. D. Provenza (eds) *The Art and Science of Shepherding*, Acres USA, Austin, TX.

Ingold, T. (2000) 'From trust to domination: An alternative history of human-animal relations', in T. Ingold (ed) *The Perception of Environment: Essays on Livelihood, Dwelling and Skill*, Routledge, London and New York.

Jacquesson, S. (2004) 'Au coeur du Tian Chan: Histoire et devenir de la transhumance au Kirghizstan', *Cahiers d'Asie Centrale*, vol 11/12, pp203–244.

Kaczensky, P. (1999) 'Large carnivore predation on livestock in Europe', *Ursus*, vol 11, pp59–72.

Kaczensky, P., Chapron, G., von Arx, M., Huber, D., Andrén, H., et al. (2012) Status, management and distribution of large carnivores – Bear, lynx, wolf & wolverine – In Europe, European Commission, Brussels.

Keçi, E., Trajçe, A., Mersini, K., Bego, F., Ivanov, G., et al. (2008) 'Conflicts between lynx, other large carnivores, and humans in Macedonia and Albania', in *Proceedings of the III Congress of Ecologists of the Republic of Macedonia with International Participation*, 06–09 October 2007 (pp. 257–264), Struga, Macedonia.

Kellert, S. R., Black, M., Reid Rush, C., and Bath, A. J. (1996) 'Human culture and large carnivore conservation in North America', *Conservation Biology*, vol 10, pp977–990.

Latour, B. (1993) *We Have Never Been Modern*, Harvard University Press, Cambridge, MA.

Léger, F., Meuret, M., Bellon, S., Chabert, J.-P., and Guérin, G. (1996) 'Elevage et territoire : quelques enseignements des opérations locales agri-environnementales dans le sud-est de la France', *Rencontre Recherches Ruminants*, vol 3, pp13–20.

Lescureux, N. (2006) 'Towards the necessity of a new interactive approach integrating ethnology, ecology and ethology in the study of the relationship between Kirghiz stockbreeders and wolves', *Social Science Information*, vol 45, pp463–478.

Lescureux, N. (2007) *Maintenir la réciprocité pour mieux coexister. Ethnographie du récit kirghiz des relations dynamiques entre les hommes et les loups*, PhD thesis. Muséum National d'Histoire Naturelle, Paris (France).

Lescureux, N., and Linnell, J. D. C. (2010) 'Knowledge and perceptions of Macedonian hunters and herders: The influence of species specific ecology of bears, wolves, and Lynx', *Human Ecology*, vol 38, pp389–399.

Lescureux, N., and Linnell, J. D. C. (2013) 'The effect of rapid social changes during post-communist transition on perceptions of the human – Wolf relationships in Macedonia and Kyrgyzstan', *Pastoralism: Research, Policy and Practice*, vol 3, p4.

Lestel, D., Brunois, F., and Gaunet, F. (2005) 'Etho-ethnology and ethno-ethology', *Social Science Information*, vol 45, pp155–177.

Linnell, J. D. C., Solberg, E. J., Brainerd, S. M., Liberg, O., Sand, H., et al. (2003) 'Is the fear of wolves justified? A Fennoscandian perspective', *Acta zoologica Lituanica*, vol 13, pp34–40.

Linnell, J. D. C., Swenson, J. E., and Andersen, R. (2001) 'Predators and people: Conservation of large carnivores is possible at high human densities if management policy is favourable', *Animal Conservation*, vol 4, pp345–349.

Löe, J., and Røskaft, E. (2004) 'Large carnivores and human safety: A review', *Ambio*, vol 33, pp283–288.

Lopez, B. H. (1978) *Of Wolves and Men*, Charles Scribner's Sons, New York.

MAFWE (2003) *Country Report on the State of the Animal Genetic Resources in Republic of Macedonia*, Republic of Macedonia Ministry of Agriculture, Forestry and Water Economy, Skopje.

Mauz, I. (2005) *Gens, cornes et crocs*, Inra-Quae, Paris, Grenoble.

McNay, M. E. (2002) 'Wolf-human interactions in Alsaka and Canada: A review of the case history', *Wildlife Society Bulletin*, vol 30, pp831–843.

Mech, L. D. (1995) 'The challenge and opportunity of recovering wolf populations', *Conservation Biology*, vol 9, pp270–278.

Mech, L. D. (2017) 'Where can wolves live and how can we live with them?', *Biological Conservation*, vol 210, Part A, pp310–317.

Mech, L. D., and Boitani, L. (2003) *Wolves: Behavior, ecology, and conservation*, The University of Chicago Press, Chicago.

Melovski, L., and Godes, C. (2002) 'Large carnivores in the "Republic of Macedonia" (recognised by Greece as: "the Former Yugoslav Republic of Macedonia")', in S. Psaroudas (eds) *Protected areas in the Southern Balkans – Legislation, Large Carnivores, Transborder Areas*, Arcturos and Hellenistic Ministry of the Environment, Physical Planning, and Public Works, Thessaloniki, Greece.

Meuret, M., Garde, L., Moulin, C.-H., Nozières-Petit, M-O., and Vincent, M. (2017) 'Élevage et loups en France: historique, bilan et pistes de solution', *INRA Productions animales*, vol 30, pp465–478.

Miklósi, Á., and Kubinyi, E. (2016) 'Current trends in canine problem-solving and cognition', *Current Directions in Psychological Science*, vol 25, pp300–306.

Ministry of Environment and Physical Planning (2003) *Biodiversity Strategy and Action Plan for the Republic of Macedonia*, Skopje.

Moriceau, J.-M. (2007) *Histoire du méchant loup. 3 000 attaques sur l'homme en France*, Fayard, Paris.

Olsen, S. J. (1985) *Origins of the Domestic Dog: The Fossil Record*, The University of Arizona Press, Tucson, Arizona.

Oriol-Cotterill, A., Valeix, M., Frank, L. G., Riginos, C., and Macdonald, D. W. (2015) 'Landscapes of Coexistence for terrestrial carnivores: The ecological consequences of being downgraded from ultimate to penultimate predator by humans', *Oikos*, vol 124, pp1263–1273.

Packard, J. M. (2003) 'Wolf behavior: Reproductive, social, and intelligent', in L. D. Mech and L. Boitani (eds) *Wolves: Behavior, Ecology and Conservation*, The University of Chicago Press, Chicago and London.

Pardini, A., and Nori, M. (2011) 'Agro-silvo-pastoral systems in Italy: Integration and diversification', *Pastoralism: Research, Policy and Practice*, vol 1, p26.

Peillon, V., and Carbonne, G. (1993) 'Bienvenue aux loups', *Terre Sauvage*, vol 73, pp23–42.

Peterson, R. O., and Ciucci, P. (2003) 'The wolf as a carnivore', in L. D. Mech and L. Boitani (eds) *Wolves: Behavior, Ecology, and Conservation*, The University of Chicago Press, Chicago.

Petkovski, S., Smith, D., Petkovski, T., and Sidorovska, V. (2003) *Study on Hunting Activities in Macedonia: Past, Present and Future*, Society for the Investigation and Conservation of Biodiversity and the Sustainable Development of Natural Ecosystems (BIOECO), Skopje.

Poudyal, N., Baral, N., and Asah, S. T. (2016) 'Wolf lethal control and livestock depredations: Counter-evidence from respecified models', *PLoS ONE*, vol 11, ppe0148743.

Rajpurohit, K. S. (1999) 'Child lifting: Wolves in Hazaribagh, India', *Ambio*, vol 28, pp162–166.

Sablin, M.V., and Khlopachev, G. A. (2002) 'The earliest ice age dogs: Evidence from Eliseevichi 1', *Current Anthropology*, vol 43, pp795–799.

Savini, I., Landais, E., Thinon, P., and Deffontaines, J.-P. (2014) 'Taking advantage of an experienced herder's knowledge to design summer range management tools', in M. Meuret and F. D. Provenza (eds) *The art and Science of Shepherding*, Acres USA, Austin, TX.

Sidorovich, V. E., Tikhomirova, L. L., and Jędrzejewska, B. (2003) 'Wolf canis lupus numbers, diet and damage to livestock in relation to hunting and ungulate abundance in northeastern Belarus during 1990–2000', *Wildlife Biology*, vol 9, pp103–111.

Silhol, A., Bataille, J.-F., Dureau, R., Garde, L., and Niez, T. (2007) 'Evaluation du schéma de protection des troupeaux en alpage: coût, travail, impact', in L. Garde (ed) *Loup – élevage: s'ouvrir à la complexité. Actes du séminaire des 15 et 16 juin 2006*, CERPAM, Manosque.

Skogen, K., Mauz, I., and Krange, O. (2008) 'Cry wolf!: Narratives of wolf recovery in France and Norway', *Rural Sociology*, vol 73, pp105–133.

Theuerkauf, J., Jedrzejewski, W., Schmidt, K., and Gula, R. (2003) 'Spatiotemporal Segregation of wolves from humans in the Bialowieza forest (Poland)', *Journal of Wildlife Management*, vol 67, pp706–716.

Treves, A., M. Krofel, and J. McManus. (2016) 'Predator control should not be a shot in the dark', *Frontiers in Ecology and the Environment*, vol 14, pp380–388.

Van Veen, T. W. S. (1995) 'The Kyrgyz sheep herders at a crossroad', *Pastoral Development Network Series*, vol 38, pp1–14.

Verginelli, F., C. Capelli, V. Coia, M. Musiani, M. Falchetti, et al. (2005) 'Mitochondrial DNA from prehistoric canid highlights relationships between dogs and South-East European wolves', *Molecular Biology and Evolution*, vol 22, pp2541–2551.

Vigne, J.-D., Carrère, I., Brios, F., and Guilaine, J. (2011) 'The early process of mammal domestication in the near east: New evidence from the pre-neolithic and pre-pottery neolithic in Cyprus', *Current Anthropology*, vol 52, ppS255–S271.

Vincent, M. (2011) *Les alpages à l'épreuve des loups*, Quae & MSH, Montpellier.

Von Uexküll, J. (1956) *Mondes animaux et mondes humains*, Denoël, Paris.

Walker, B. L. (2005) *The Lost Wolves of Japan*, University of Washington Press, Seattle.

Wielgus, R. B., and Peebles, K. A. (2014) 'Effects of wolf mortality on livestock depredations', *PLOS One*, vol 9, ppe113505.

Zeder, M. A. (2008) 'Domestication and early agriculture in the Mediterranean basin: Origins, diffusion, and impact', *Proceedings of the National Academy of Science*, vol 105, pp11597–11604.

Zeder, M. A. (2011) 'The origins of agriculture in the near east', *Current Anthropology*, vol 52, ppS221–S235.

9 Attitudes towards large carnivore species in the West Carpathians

Shifts in public perception and media content after the return of the wolf and the bear

Miroslav Kutal, Petr Kovařík, Leona Kutalová, Michal Bojda, and Martina Dušková

Introduction

The recovering populations of large carnivores in Europe have shown a remarkable ability to live in human-dominated landscapes, mostly outside large protected areas, where they share the space with people (Chapron et al., 2014). The co-existence in a cultural landscape inevitably leads to some degree of conflicts, especially with animal farming and hunting (Breitenmoser, 1998). These conflicts are not new in the Carpathians, where traditional methods such as the use of livestock guarding dogs have been applied by farmers for centuries to prevent attacks by large carnivores. However, due to the decline of large carnivore populations in some parts of the Carpathians and the introduction of significant changes in the farming system, many of these practices were abandoned. As a consequence, only 16% of Slovak farmers are now familiar with non-lethal methods to prevent attacks (Rigg et al., 2011). At the same time, sheep farming has nearly disappeared from many areas of Central Europe during recent decades (Niznikowski et al., 2006; Martinát et al., 2008) and public attitudes towards wolves have much improved, resulting also in better environmental-law enforcement (Mech, 1995). These forces and socio-demographical changes have led to divergent attitudes of rural and urban people (Kleiven et al., 2004; Blekesaune and Rønningen, 2010). The use of appropriate spatial scale – with a focus on the local level where core conflicts are most likely to occur – is therefore needed to improve our knowledge of attitudes of people living in the same areas with large carnivores (Ericsson et al., 2006). The acceptance of concerned groups is crucial especially for viable populations of brown bear (*Ursus arctos*) and grey wolf (*Canis lupus*) in the West Carpathians, at the edge of the Carpathian range of these two species.

Wolves and bears are currently regarded as flagship species of the Carpathian Mountains. By the 1930s, their populations were driven to extinction in many parts of the West Carpathian range, including the Czech Republic (Hell and Slamečka, 1999; Hell, 2003; Anděra and Červený, 2009). In recent decades, these populations have been recovering, mainly due to improved legislative protection and a decline in extensive livestock farming (Martinát et al., 2008; Chapron et al., 2014). This offers an exceptional opportunity for studying how different groups of people perceive large carnivores and how their opinions and attitudes change over time. The goals of our study were to describe: (1) the attitudes and opinions of two major groups of people – local people and visitors – towards wolves and bears, focusing on overall acceptance, fear and the level of relevant knowledge; (2) socio-demographic parameters determining these attitudes, such as age, education, occupation, relations with hunters and livestock breeders, and source of information about wolves and bears; and (3) the shift of local people's opinions and media coverage over time.

Methods

Study area

The West Carpathian Mountains, situated at the Czech-Slovak-Polish border, are the western edge of continuous wolf and bear occurrence in the Carpathians (Chapron et al., 2014; Kutal et al., 2016). On the Czech side of the border, a large forested area with altitudes ranging from 350–1324 m above sea level is designated as the Protected Landscape Area (PLA) Beskydy. Despite the 72% forest cover, the area is intensively used by people, as is typical for the Carpathian landscape. Average human density is approximately 100 inhabitants/km², with most people being concentrated in towns and villages in valleys and river basins. However, many activities are situated up in the mountains. In addition to forestry and hunting practices, there is a moderate impact of tourism, including a number of ski lifts, ski and bike routes, hiking trails, recreational cabins, and chalets. The Beskydy Mountains are also a traditional area of sheep farming in the Czech Republic. In the three districts comprising our study area (Vsetín, Nový Jičín, and Frýdek-Místek), a total of 19,896 sheep were held on 1,770 farms in 2011 (Kovařík et al., 2014). Average sheep density in the Beskydy region (6.15 sheep/km²) is more than twice the national average in the Czech Republic (3.01 sheep/km²) (Kovařík and Kutal, 2014).

By the beginning of the 20th century, wolves were exterminated from the Czech part of the West Carpathians (Anděra and Červený, 2009) and the northwest of Slovakia (Hell and Sládek, 1974). After a period of absence, wolves were again recorded in the Czech Carpathians, in the area of Beskydy, in 1994 (Bartošová, 1998), with 1–3 wolf packs being registered in each of the following years until 2003 (Anděra et al., 2004). Intensive monitoring during the period 2003–2012 did not, however, detect any reproducing pair, since only occasional wolf occurrence was recorded (Kutal et al., 2016).

Similarly, bears were wiped out in the Beskydy Mountains by the end of the 19th century. After the Second World War, rising numbers of bears in central Slovakia contributed to the expansion of the distribution of the species to mountains in western Slovakia, adjacent to the mountain ranges on the Czech side of the border. As a result, bears started to appear regularly in the Czech Carpathians, too, especially in the Beskydy and Javorníky ranges, and in the northern part of the White Carpathians. The first modern record dates back to 1946, but regular observations do not appear until the 1970s (Červený et al., 2004). Across the border, on the Slovak side of the Javorníky Mountain Range, bears were more frequently reported already in 1968 (Janík et al., 1986). Although there were some reports of bears with cubs from time to time, no evidence of bear reproduction was available until 2014 (Bojda et al., 2014), and the occurrence of bears has remained sporadic in Beskydy. Further, any damage caused by bears has been rare (Kovařík et al., 2014). The only exception was a bear, which had probably escaped or was illegally released from captivity. The origin of the animal was not confirmed, but the behaviour of the animal, later trapped and removed from the wild, indicated signs of strong habitation toward humans (Červený and Koubek, 2003). The bear caused damage of about 6500 EUR, an amount significantly higher than the total compensation paid for damages caused by bears to all local farmers in the previous 28 years. This case received considerable media attention (Pavelka and Trezner, 2001; Červený and Koubek, 2003).

Data collection and analysis

Attitude of local people and visitors towards wolf and bear presence
(2009–2010)

To investigate attitudes towards wolf and bear presence, we collected data by means of a questionnaire (see items in Annex 9.1) administered by face-to-face on-site interviews in the Beskydy Mountains in 2009–2010. There were two groups of respondents: residents of local villages (n = 158), and visitors to the Beskydy Mountains (n = 156). The survey of local residents was conducted in 21 villages randomly selected from all the 77 villages situated wholly or partly within the boundaries of the Beskydy PLA. The third author personally visited local residents, who were chosen by random selection of house numbers. The visitors were randomly approached at 13 popular sites.

We focused on four topics: (1) if respondents believed that large carnivores belonged in PLA Beskydy; (2) if they thought large carnivores were causing too much damage; (3) if they saw large carnivores as potentially dangerous to humans; and (4) how accurate was their estimation of large carnivores' numbers in the Czech Republic. Except for the estimates of wolf and bear numbers, the questionnaire consisted of multiple-choice questions. We also recorded socio-demographic variables such as age, education, size of place of residence, and respondent's relationships with sheep farming and hunting.

Shift in the attitude of local people towards wolf and bear presence
(2000; 2009–2010)

To compare the development of attitudes of local people towards wolf and bear presence over time, we used results from a survey administered by Friends of the Earth Czech Republic in 2000 in the same region, using the same method (questionnaire administered by means of face-to-face on-site interviews) and some identical questionnaire items as the study presented here. A total of 177 respondents participated in the survey that took place in 2000, following a one-year intensive awareness campaign as a part of the pilot project "Preservation and recovery of large carnivores in the West Carpathians" (Bartošová and Genda, 2001). At the time, wolves were regularly present in the survey area (Beskydy). Since then, many education and public awareness activities took place in the region, including discussion forums with local people, communication with media, distribution of educational materials and lectures at schools. For example, in 2005–2009, Friends of the Earth Czech Republic communicated its position on large carnivores through 119 media outputs (press releases, interviews, comments, etc.). In 2007–2009, it organized lectures for 119 school classes and produced approximately 50,000 copies of educational materials for distribution in the Beskydy region. The questionnaire survey that took place in 2009–2010 (described previously) was undertaken after 10 years of intensive communication work. Because the first study in 2000 involved only local residents, the presented comparison includes only answers of local residents from both time periods.

Shift in the perspective of media content (1996–2000; 2005–2009)

We searched in the media database Anopress for all available Czech articles containing the keywords "medvěd" (bear) and "vlk" (wolf) published in the years 1996–2000 and 2005–2009, which correspond to five-year periods preceding our questionnaire surveys. Anopress is one of the most comprehensive Czech databases, offering a possibility to search print, broadcast, and online media outputs dating back to 1996. The database contains both regional and national media sources, but since the Beskydy region was, at the time of survey, the only region where wolves and bears occurred, all relevant search results were related to Beskydy. After declining articles with uses of keywords not related to wildlife (e.g., surnames) and articles that informed about animals kept in captivity (e.g., zoos and circuses), we obtained 263 and 227 articles about wolves and bears, respectively, from the period 1996–2000, and 215 and 297 articles on wolves and bears, respectively, from the period 2005–2009. We evaluated how media portrayed each species on a five-point Likert scale ranging from a "purely negative" perspective (no positive aspect for large carnivore species included), through "mainly negative" (negative, unfavourable connotations dominated), and "balanced" perspectives (positive and negative aspects equally represented or no positive/negative aspects mentioned) to "mainly positive" (positive, favourable connotations dominated) and "purely positive" perspectives (no negative aspect for large carnivore species included).

Statistical analyses

We analysed data from questionnaires in JMP 5.0.1 (SAS, 2007) using the generalized linear model and likelihood ratio test to explore the relationship between responses and different socio-demographic variables such as age, education, and relationship with sheep farming and hunting. Time trends were analysed by two-sample test for equality of proportions without continuity correction in software R 3.3.0 (R Core Team, 2016).

Results

Attitudes of local people and visitors (2009–2010)

Comparison of attitude towards wolf and bear presence between local people and visitors

Local people and visitors differed significantly in their attitudes towards wolf and bear presence in the study area (data for wolves presented in Table 9.1: $\chi^2 = 50.34$, $p < 0.001$; data for bears presented in Table 9.2: $\chi^2 = 45.92$, $p < 0.001$; n = 314).

Table 9.1 Attitude of local people and visitors towards wolf presence (2009–2010)

"How do you feel about the presence of wolves in the region?"	Percentage of local people (n = 158)	Percentage of visitors (n = 156)	Difference
"It bothers me and I am not ready to accept it"	5.7	1.3	–4.4
"It bothers me but I can accept it"	14.6	4.5	–10.1
"I do not care"	8.2	2.6	–5.6
"I do not mind"	56.3	42.9	–13.4
"It makes me happy"	15.2	48.7	+33.5

Note: $\chi^2 = 50.34$; p < 0.001.

Table 9.2 Attitude of local people and visitors towards bear presence (2009–2010)

"How do you feel about the presence of bears in the region?"	Percentage of local people (n = 158)	Percentage of visitors (n = 156)	Difference
"It bothers me and I am not ready to accept it"	4.4	0.6	–3.8
"It bothers me but I can accept it"	17.1	7.7	–9.4
"I do not care"	8.9	1.9	–7.0
"I do not mind"	54.4	43.6	–10.8
"It makes me happy"	15.2	46.2	+31

Note: $\chi^2 = 45.92$; p < 0.001.

Although the highest percentages for both target groups were recorded for neutral responses, visitors endorsed the presence of large carnivore species (48.7% and 46.2% for the wolf and the bear, respectively, for the item "It makes me happy") more than local people (15.2% for the both wolf and the bear, for the item "It makes me happy"). How long had local people lived in the Beskydy region also proved to be a significant factor. Local residents who had lived in the area longer were more likely to have a negative attitude towards large carnivores (wolves: $\chi^2 = 32.53$, $p < 0.001$; bears: $\chi^2 = 29.54$, $p < 0.01$; n = 158).

Illegal killing of large carnivores

Local people opposed illegal hunting less often than visitors (wolves: $\chi^2 = 25.02$, $p < 0.001$; bears: $\chi^2 = 20.72$, $p < 0.001$; n = 314) and showed lower willingness to report these cases to competent authorities (wolves: $\chi^2 = 8.41$, $p < 0.01$; bears: $\chi^2 = 7.72$, $p < 0.01$; n = 314). It is noteworthy that respondents holding a negative or indifferent opinion on the presence of large carnivores would seem to endorse illegal killing of large carnivores (response "I would be glad" for the item "If you learned that someone you know has illegally killed a wolf or bear, how would you react?") (for wolves: $\chi^2 = 32.19$, $p < 0.001$; for bears: $\chi^2 = 35.59$, $p < 0.001$; n = 314). At the same time, these respondents showed less interest in getting more information about large carnivores than the rest of the sample (wolves: $\chi^2 = 32.19$, $p < 0.001$; bears: $\chi^2 = 35.59$, $p < 0.001$; n = 314).

Damage caused by large carnivores

Respondents commonly held the view that large carnivores did not belong in the Beskydy Mountains, and they were believed to be causing too much damage to livestock and wild game. These attitudes were particularly pronounced in relation to wolves, where 67% of local people and 44% of visitors believed that wolves caused a lot of damage to livestock. Respondents' attitudes did not seem to align with actual figures for damages, though: On average, there were only 16 sheep killed by wolves per year (2001–2009), which is negligible as a magnitude of losses considering that there were nearly 20,000 sheep in the Beskydy region. In addition, the number of attacks decreased considerably since 2001 (Kovařík et al., 2014). Damage caused by bears was perceived as being significantly lower: Only 30.4% of locals and 17.3% of visitors agreed with the previously mentioned statement in relation to bears. Despite the lower percentages as compared to wolves, respondent attitudes once again stood in contrast to actual damage recorded, with just four sheep killed during four documented bear attacks during the period 2001–2009 (Kovařík et al., 2014).

Fear of large carnivores

Another major issue appeared to be the deep-rooted perception of large carnivores as a threat to humans. Many respondents regarded an encounter with large carnivores, bears in particular, as dangerous for humans (69.1% and 89.5% for wolves and bears, respectively; n = 314). Interestingly, those who relied mainly on TV for information were significantly more likely to consider an encounter with a wolf or bear to be slightly or highly dangerous (wolf: χ^2 = 8.14, p < 0.05; bear: χ^2 = 9.82, p < 0.05; n = 314). A similar relationship was not found among respondents who relied mainly on other sources of information (newspapers, radio broadcast).

Estimates of large carnivore numbers

Fear of possible attacks on humans might have been amplified by the fact that most respondents overestimated actual numbers of large carnivores in the region. Therefore, most respondents may have exaggerated the potential impact of large carnivores on human safety. Indeed, respondents who overestimated the numbers of wolves and bears were more likely to consider the encounter with these animals to be highly dangerous (wolves: χ^2 = 16.92, p < 0.01; bears: χ^2 = 20.13; p < 0.01; n = 314). On the contrary, people who realistically estimated the numbers of large carnivores were more likely to express tolerance towards their presence (wolves: χ^2 = 32.81, p < 0.01; bears: χ^2 = 21.37, p < 0.01; n = 314).

The tendency to overestimate the abundance of carnivores was especially pronounced in the case of wolves. Indeed, this has been the case, despite the fact that the number of wolves in the Czech Republic was extremely low at the time of the survey, ranging between 0–5 individuals, which have most probably crossed the border with neighbouring countries. In contrast, 31.8 % of respondents (n = 314) estimated that there were several hundred wolves in the country. Overall, 86.9% of the total sample overestimated wolf numbers. In the case of the bear, at the time of the survey, the population numbered between 0–2 individuals. Almost a half of the sample (44.6 %; n = 314) gave a more realistic estimate of up to 10 animals, as compared to wolf estimates. Still, 52.4% of all respondents overestimated bear numbers. There was no significant difference between locals and visitors in their estimates, but respondents with higher education were more likely to give a more realistic estimate of bear numbers (χ^2 = 14.08, p < 0.05; n = 314). Respondents with hunting interests of social and kin networks with hunters were also significantly better in estimating the numbers of wolves (χ^2 = 9.12, p < 0.05; n = 314).

Effect of socio-demographic characteristics on attitudes towards large carnivore presence

There was a significant variation among respondents according to the population of their place of residence. People living in large municipalities were more likely to endorse large carnivore presence than inhabitants of small villages (wolves: χ^2 = 12.97, p < 0.001; bears: χ^2 = 8.73, p < 0.01; n = 314). The effect of the size of municipality was not significant when local people and visitors were tested separately. There was a significant effect of age and education on attitudes towards large carnivore presence. Older respondents (wolves: χ^2 = 25.03, p < 0.001; bears: χ^2 = 21.86, p < 0.001; n = 314) and those with lower education had more negative attitudes (wolves: χ^2 = 71.71, p < 0.001; bears: χ^2 = 71.16, p < 0.001; n = 314). The latter were also much more tolerant of illegal hunting of these species (wolves: χ^2 = 20.86, p < 0.001; bears: χ^2 = 20.46, p < 0.001; n = 314).

Livestock breeders and their relatives and acquaintances formed a special group of respondents. These groups showed highly negative attitudes towards large carnivore presence in the region compared to other respondents. The attitudes of livestock breeders towards wolves and bears were significantly more negative than those of other respondents (wolves: χ^2 = 28.57, p < 0.001; bears: χ^2 = 26.75, p < 0.001; n = 314), and the same also applied for their relatives (wolves: χ^2 = 18.15, p < 0.001; bears: χ^2 = 15.7, p < 0.01; n = 314). In the case of people acquainted to a livestock breeder, their negative attitudes stood out in terms of wolves, but not bears (wolves: χ^2 = 11.49, p < 0.05; bears: χ^2 = 4.65, p > 0.05; n = 314). Differences between these groups were not significant when tested separately among local people and visitors, with an exception of visitors' relatives, holding more negative views towards bears, but not towards wolves (wolves: χ^2 = 5.83, p = 0.21; bears: χ^2 = 10.4, p < 0.05; n = 156).

Shift in the attitude of local people towards wolf and bear presence (2000; 2009–2010)

The comparison of results from 2000 and 2009–2010 showed that there was a significant change in the attitudes of local people towards wolf and bear presence (Tables 9.3 and 9.4, respectively). The number of local people who would not welcome wolf presence ("it bothers me and I am not ready to accept it"; Table 9.3) decreased from 13% to 5.7%. A decrease was also recorded for those who would be bothered by wolf presence but who would nevertheless accept it (from 24% in 2000 to 14.6% in 2009–2010). The same trends were also observed for bears (Table 9.4). Those who not welcome wolf presence ("it bothers me and I am not ready to accept it") dropped from 19% to 4.4%, and

Table 9.3 Shift in the attitude of local people towards wolf presence

"How do you feel about the presence of wolves in the region?"	*Percentage in 2000 (n = 177)*	*Percentage in 2009– 2010 (n = 158)*	*Change*
"It bothers me and I am not ready to accept it"	13.0	5.7	-7.3
"It bothers me but I can accept it"	24.0	14.6	-9.4
"I do not care"	7.0	8.2	+1.2
"I do not mind"	43.0	56.3	+13.3
"It makes me happy"	13.0	15.2	+2.2

Note: χ^2 = 12.23; p = 0.016.

Table 9.4 Shift in the attitude of local people towards bear presence

"How do you feel about the presence of bears in the region?"	*Percentage in 2000 (n = 177)*	*Percentage in 2009– 2010 (n = 158)*	*Change*
"It bothers me and I am not ready to accept it"	19.0	4.4	-14.6
"It bothers me but I can accept it"	21.0	17.1	-3.9
"I do not care"	9.0	8.9	-0.1
"I do not mind"	39.0	54.4	+15.4
"It makes me happy"	12.0	15.2	+3.2

Note: $\chi^2 = 20.53$; $p < 0.001$.

those who would be bothered but did not object wolf presence also dropped, from 21% to 17.1%. At the same time, there was an increase in slightly positive attitudes towards large carnivore presence ("I do not mind") for both wolves (+13.3%) and bears (+15.4%).

Shift in the perspective of media content (1996–2000; 2005–2009)

There was a substantial shift in media content about wolves and bears between the corresponding five-year periods preceding the two questionnaire surveys (i.e., 1999–2000 and 2005–2010). Specifically, the percentage of articles expressing both a purely negative and a mainly negative perspective on wolves dropped (Table 9.5; −39.2% for the "purely negative" perspective and −11.2% for the "mainly negative" perspective). Moreover, the number of articles expressing a purely positive perspective of wolves increased sharply, from 4.9% to 56.3%. Overall, trends were more pronounced in extreme positioning (i.e., purely negative or purely positing perspectives), which resulted in a complete reversal of perspectives of media content between the two periods. The main reason behind the change was probably a lower number of sheep losses (Kovařík et al., 2014), which may also mean that a lower number of human-wolf conflicts had been presented by the media. The pattern for articles about bears was very similar to what has been reported for wolves (Table 9.6).

Discussion

The survey results show that most respondents had a neutral or positive attitude towards the presence of large carnivores in the Beskydy Mountains. In addition, the increased tolerance shown towards bears as compared to wolves aligns with the general picture reported in Europe (Dressel et al., 2015). However, we cannot assume, based on these results alone, that the relation of the majority of respondents with large carnivores is (or will be) without any problems. Even respondents with a positive view of large carnivores often expressed concerns about a possible encounter or attack of large carnivores on humans or livestock.

Table 9.5 Shift in the perspective of media content for articles on the wolf

Overall perspective	Percentage of articles published 1996–2000 (n = 263)	Percentage of articles published 2005–2009 (n = 215)	Change
Purely negative (i.e., no positive aspect for large carnivore species included)	43.0	3.7	-39.2
Mainly negative (i.e., negative, unfavourable connotations dominated)	14.8	3.7	-11.1
Balanced (i.e., positive and negative aspects equally represented or no positive/ negative aspects mentioned)	26.6	28.8	+2.2
Mainly positive (i.e., positive, favourable connotations dominated)	10.7	7.5	-3.2
Purely positive (i.e., no negative aspect for large carnivore species included)	4.9	56.3	+51.3

Note: χ^2 = 199.56; p < 0.001.

Table 9.6 Shift in the perspective of media content for articles on the bear

Overall perspective	Percentage of articles published 1996–2000 (n = 227)	Percentage of articles published 2005–2009 (n = 297)	Change
Purely negative (i.e., no positive aspect for large carnivore species included)	45.4	13.1	-32.2
Mainly negative (i.e., negative, unfavourable connotations dominated)	14.5	4.7	-9.8
Balanced (i.e., positive and negative aspects equally represented or no positive/ negative aspects mentioned)	31.7	30.3	-1.4
Mainly positive (i.e., positive, favourable connotations dominated)	4.4	4.4	0.0
Purely positive (i.e., no negative aspect for large carnivore species included)	4.0	47.5	+43.5

Note: χ^2 = 148.37; p < 0.001.

This fear may shift people's attitudes towards the negative extreme in the mid- or long term.

In contrast to the endorsement of wolf presence by the majority of respondents, most of them considered an encounter with a wolf as dangerous. Such response was given by almost 70% of both local people and visitors, which is a substantially higher result than the analogous percentage (40%) for people surveyed in Estonia (Randveer, 2001). Some respondents in our survey expressed concerns during the interview; in particular, about "hungry" wolves and a possible encounter with an entire wolf pack or wolves with pups, which they considered as highly dangerous. These concerns were not specifically listed in the questionnaire, but illustrate views probably resulting from the negative cultural stereotypes related to wolves (Boitani, 2000). European literary tradition includes many reports, legends, fictional stories, and fairy tales portraying wolves in a negative way (Boitani, 2000; Bisi et al., 2007). Fear of wolves seems to be rooted very deeply in people's minds, and it seems to persist despite the fact that there has been no record of a fatal wolf attack on a human in Europe in recent decades (Linnell et al., 2002, Linnell and Alleau, 2016). This might be due to the lack of adequate information, as well as the continued popularity of old superstitions and generally negative perceptions of wolves in stories that have a negative effect on people's attitudes (Prokop et al., 2011). In our survey, the respondents who relied mainly on TV for information were significantly more likely to consider an encounter with wolves or bears to be slightly or highly dangerous. Therefore, public opinion seems to be still based largely on stereotypes rather than scientific knowledge about wolves and other carnivores. Lingering prejudices thus may continue to undermine any objective view of carnivore species (Lopez, 1979; Zibordi, 2009).

Overall, fear of bears is more common than fear of wolves. The view of bears as the most dangerous and feared wild animal also prevailed in Slovakia (Wechselberger et al., 2005). This is not surprising, in light of the fact that bear attacks on humans occasionally occur and receive high publicity in the media in both the Slovak and Czech republics (Dressel et al., 2015). However, a meta-analysis conducted in 24 European countries from 1976–2012 found that people's attitudes were more positive toward bears than wolves. Dressel et al. (2015) highlighted that one of the reasons for greater acceptance of bears, as compared to wolves, could be the widespread ability to hunt bears in Europe. European wolves, by contrast, are more often strictly protected. This could explain why our results and the results of the study conducted in Slovakia differ from overall European trend. Both wolves and bears may be legally hunted in Slovakia but are strictly protected in the Czech Republic. In a Swedish study, Eriksson et al. (2015) argued that the wolves may also have a more direct effect on people's attitudes via direct experience. In Slovakia, however, bears are about three times more abundant than wolves and encounters with bears are also much more common than encounters with wolves. Higher probability of encountering a bear, together with known attacks presented in the media, have probably led to a higher perceived threat from bears also for the Czech respondents who live close to the border with Slovakia. The latter have been probably influenced by the situation in Slovakia due to low language barrier, similar culture, and a long shared history in the federal country. Moreover, Slovak mountains have been a quite common destination for Czech tourists.

Overestimating the real numbers of wolves and bears further amplified fear. The abundance of both large carnivore species in the Czech Republic was often exaggerated, but the tendency was stronger in the case of wolves. Respondents typically estimated the numbers of wolves in Beskydy to be between 50 and several hundred animals when in fact, at the time of survey, there were only a few individuals and not a single wolf pack (Kutal et al., 2016). Given these results, we expect that the public opinion about large carnivores may become more negative should the real numbers of wolves, and resulting conflicts, increase.

People who overestimated wolf abundance also expressed more concerns about their presence, and they were more likely to show negative attitudes towards the presence of large carnivores and be less tolerant of their expansion to other areas. These results indicate that the lack of adequate information could make people more susceptible to fear and less likely to accept the return of large carnivores. In France, the fear of wolves was also influenced by the tendency to overestimate their numbers (Bath, 2000). Bath (2000) recommended that the objectivity in informing about wolf numbers should improve public acceptance of wolves among respondents where the conflict is cognitive. On the contrary, in the case of a conflict of values (Lüchtrath et al., 2015), common among hunters, more realistic estimates of carnivore numbers are not expected to correlate with a more positive attitude towards these species (Ericsson and Heberlein, 2003). In our study, respondents with hunting interests did also provide better estimates of large carnivore numbers but not more positive attitudes towards these species. Furthermore, Kaczensky (2003), reporting on a study of bears in Slovenia, argued that perceived trends of population may be probably much more important than absolute numbers of carnivores. Lack of knowledge about the spatial requirements of large carnivores often leads to overestimating their population size, whereas increasing damages are perceived as a threat and result in lower support for any further increase of large carnivore numbers.

Most local people and almost half of the sub-sample of visitors believed that wolves cause much damage on livestock, although real cases were very rare and they had been reported only sporadically during the five-year period before the survey. In contrast, perceived damage by locals and visitors on livestock in the Slovak Carpathians, an area with stable wolf and bear occurrence (Wechselberger et al., 2005), were closer to real damage compared to the results of our survey in the Beskydy Mountains. It is, therefore, possible that after a longer coexistence, perceived damage will also improve and reach real numbers also in the Czech Carpathians. However, the conflict between wolves and sheep farmers in the Beskydy area was very strong in the years 1996–2000, as we have shown in our media analysis, and that seems to have influenced the attitudes of local people even when the extent of real damage dropped. Similarly, 30%

of local people were convinced that bears cause a lot of damage on livestock, despite the fact that there were almost no conflicts since the return of the species after the 1970s. In 2000, a single bear, behaving in an unusual manner and causing a significant damage, received much public attention (Pavelka and Trezner, 2001; Červený and Koubek, 2003), which appears to still influence people's attitudes more than a decade later.

As the results of other studies show, the experience with attacks on livestock leads to more negative attitudes towards wolves (Ericsson and Heberlein, 2003; Andersone, 2004; Wechselberger et al., 2005). Given the low number of respondents who personally suffered damage to domestic animals caused by large carnivores, we were not able to confirm these results in the present study. However, we found that livestock breeders were more likely to have a negative disposition towards both wolves and bears, which indicates that the potential risk of damage alone influences attitudes towards large carnivores considerably. Farmers and livestock breeders were identified as the most negative group in relation to large carnivores in other countries, too (Wechselberger et al., 2005; Majić, 2007; Dressel et al., 2015). Another group of people with significantly more negative attitudes towards large carnivores was the hunters (Dressel et al., 2015). We were unable to investigate the attitudes of hunters, directly, because of their low representation in our randomly selected sample.

Although the majority of respondents from both target groups (70.9% of locals and 80.1% of visitors) expressed their interest in learning more about large carnivores in the Czech Republic, negative attitudes towards large carnivores were largely associated with lack of interest in getting more information about them, which is in accordance with a previous study in Slovenia (Kaczensky et al., 2004). Any awareness campaigns aimed at people who would express at least some interest thus seem to be little efficient in reaching those who are mainly opposed to large carnivores. Any possible shift in their opinions may be more readily achieved by indirect influence through people close to them, some of whom may be more prone to support the protection of large carnivores.

The visitors of the Beskydy Mountains voiced more positive attitudes towards wolves and bears than the local people, but as pointed out by Karlsson and Sjöström (2007), support for the protection of wolves tends to be overestimated by people living outside the area of their occurrence. At the same time, the inhabitants of larger municipalities were more positive towards large carnivores than people living in smaller places. Similar results have been reported by Wechselberger et al. (2005) in the Slovak Carpathians and by Kleiven et al. (2004) in Norway. A Swedish study (Karlsson and Sjöström, 2007) found that people living far from wolf territories had more positive attitudes towards wolves than those living within or close to wolf territories.

Furthermore, respondents with higher education were significantly more likely to express tolerance and positive attitudes towards large carnivores. The same trends were documented in other countries as well (Williams et al., 2002; Ericsson and Heberlein, 2003; Wechselberger et al., 2005; Røskaft et al., 2007), and in a European meta-analysis (Dressel et al., 2015). However, good knowledge about large carnivores is considered only partially important for higher tolerance of wolves (Majić, 2007), while Szinovatz and Gossow (2001) failed to find any relationship between education and attitudes towards bears. Kleiven et al. (2004) found a minor trend of tolerance of large carnivores increasing with higher education, but this was not statistically significant.

Past trends recorded, and future trends expected

Comparative analyses of data collected in 2000 and 2009–2010 showed a substantial decrease in negative or extreme negative attitudes of local people towards wolves and bears, and a substantial increase in neutral attitudes. This may indicate an increase in tolerance of large carnivore species. While sampling through a large time span, we cannot exclude a possibility that the shift was partly caused by a cohort effect (e.g., younger generations, which tend to have more positive opinions about wolves, enter the sampling frame over time, while older generations exit the sampling frame). However, the observed trend corresponds well with the overall media coverage of large carnivore issues. In that respect, there was also a significant decrease in a number of articles with negative messages and content, especially aspects connected to livestock losses, and an increase in articles with a positive perspective.

Generally, public acceptance of wolves and bears may decrease in time with increasing number of carnivores and direct or indirect experience (Eriksson et al., 2015). In our study area, wolves caused high conflict and a lot of damage during the period 1996–2000, but such conflict and damage were quite rare afterwards (Kovařík et al., 2014). The number of wolf attacks on sheep decreased over time, and conflict with bears practically disappeared after the removal of a problematic animal in 2000 (Pavelka and Trezner, 2001). Moreover, intensive education and public awareness campaigns (including work with media, communication with local stakeholders, promotion of preventive measures and the system of damage compensations; see "Methods" section of this chapter for details) have been realised continuously since 1999 by non-governmental organizations and state nature conservancy. Our results are largely in line with the fact that public attitudes may be influenced through personal contacts with other people or through media (Karlsson and Sjöström, 2007). Indeed, extreme opinions against wolves and bears in our study area have decreased since human-carnivore conflicts have practically disappeared. However, the widespread conviction that wolves and bears cause substantial damage to livestock apparently persists among local people.

An overall change in public attitudes in our study coincided with activities related to the prevention of conflicts, communication with media and public awareness campaigns. Although a low number of conflicts had been caused primarily by the sporadic occurrence of wolves and bears in 2005–2009, people were nevertheless convinced that wolves were common, as indicated by their overestimation of wolf population size. It has been underlined that attitudes towards large carnivores are not stable and could change rapidly (Majic and

Bath, 2010). Negative political and media attention, in particular, could change weak positive attitudes into stronger negative attitudes (Ericsson and Heberlein, 2003). Since successful recovery of large carnivores throughout Europe (Chapron et al., 2014) and more strict regulation of wolf hunting currently approved in Slovakia (Antal et al., 2016) could also lead to wolf recovery in the edge of the West Carpathians, the main challenge for future conservation work in the study area is to keep the number of conflicts at low levels amidst rising wolf and bear numbers. Initiatives to be undertaken should focus on support of preventive measures against wolf and bear attacks, communication with key stakeholders, and work with media focused on presenting positive stories of large carnivores' recovery.

Acknowledgment

We are grateful to Igor Genda, Vlastimil Kostkan and Emil Tkadlec for their time dedicated to consultations focused on preparation of questionnaires and statistical analyses of the data.

Annex 9.1 Questionnaire used in the survey

1. How do you feel about the presence of wolves and bears in the region where you live/which you are visiting? Please select one of the following answers:

	It makes me happy	I do not mind	It bothers me but I can accept it	It bothers me and I am not ready to accept it	I do not have an opinion on this
Wolf					
Bear					

2. I am going to read you some questions. Please express your opinion: Certainly yes = 1; Rather yes = 2; Rather no = 3; Certainly no = 4; I do not know = 0

Question	Wolf	Bear
Do the wolves/bears belong in the Beskydy Mountains?		
Is it necessary to regulate wolf/bear numbers by hunting?		
Do you think that wolves/bears are causing a lot of damage to livestock?		
Do you think that wolves/bears are causing a lot of damage to wild game?		
Do wolves/bears have an important role in regulating numbers of wild ungulates?		
Would you say that a wolf/bear is a shy animal which usually avoids people?		

3. Do you think that encountering a wild wolf or bear in nature is dangerous or not for humans? Please select one of the following answers:

	Very dangerous	Slightly dangerous	No dangerous	I do not know
Wolf				
Bear				

4. In your estimate, how many large carnivores are there in the Czech Republic?

   Wolves _____
   Bears _____

5. At the moment, where do you get information about large carnivores? Please indicate the most important source of information:

   TV
   Information materials (leaflets, brochures)
   Relatives/friends/acquaintances
   Scientific publications
   Newspapers/ magazines
   Public lectures/discussions/exhibitions
   Internet
   Excursions/guided trips
   Radio
   Books (fiction)
   School
   Other:

6. Would you like to receive more information about large carnivores in the Czech Republic?

   Certainly yes
   Rather yes
   Rather not
   Certainly not

7. If you learned that someone you know has illegally killed a wolf or bear, how would you react? (More than one answer may be selected, separate answers for each species)

   a) I would be glad
   b) I would feel sorry
   c) I would be upset
   d) I would try to explain to them that it was wrong
   e) I would report it to the police or the PLA Administration
   f) I would not care
   g) Other (please specify):

8. Would you object to further spontaneous spread of large carnivores in the Czech Republic?

	Yes	*Rather yes*	*Rather no*	*No*	*I do not know (I do not have enough information)*
Wolf					
Bear					

9. Socio-demographic characteristics of respondents

How old are you?

15–19
20–26
27–35
36–50
51–65

What is your highest achieved education?

- Basic
- Apprenticeship
- Secondary – Specify:
- Further education/University – Specify:

What is the size of your place of residence?

- Less than 2,000 inhabitants
- 2,000–5,000 inhabitants
- 5,001–10,000 inhabitants
- 10,001–20,000 inhabitants
- 20,001–500,000
- 500,001–100,000
- 100,001 or more

Your occupation or the sector you work in:

Tourism
Forestry
Manual worker (any sector)
Agriculture (animal/plant production)
Education (primary/secondary/tertiary)
Housewife
Student
Retired
Civil servant

Self-employed/private entrepreneur
Unemployed

Other:
How long have you been living in your current place of residence?/How often do you visit Beskydy?

− Less than 5 years	− This is my first visit
− 5–15 years	− Once a year
− 15–30 years	− At least twice a year
− More than 30 years	− Irregular visits

Do you have any relationship to hunting?

- I am a hunter myself: I actively hunt/I have a hunting licence but do not hunt actively
- My family member is a hunter
- My friend/acquaintance is a hunter
- I do not know any hunters
- Other (e.g., I am interested in hunting)

Do you have any relationship to livestock farming (sheep, goats, cattle, etc.)?

- I am a livestock farmer myself − what kind and how many animals do you have:
- My family member is a livestock farmer
- My friend/acquaintance is a livestock farmer
- I have no relation to livestock farmers
- I used to farm livestock but I no longer do

References

Anděra, M., and Červený, J. (2009) *Velcí savci v České republice: Rozšíření, historie a ochrana. 2. Šelmy (Carnivora)*, Národní muzeum, Praha.
Anděra, M., Červený, J., Bufka, L., Bartošová, D., and Koubek, P. (2004) 'Recent distribution of the wolf (*Canis lupus*) in the Czech Republic', *Lynx (Prague)*, vol 35, pp5–12.
Andersone, Z. O. J. (2004) 'Public perception of large carnivores in Latvia', *Ursus*, vol 15, no 2, pp181–187.
Antal, V., Boroš, M., Čertíková, M., Ciberej, J., Dóczy, J., Finďo, S., Kaštier, P., Kropil, R., Lukáč, J., Molnár, L., Paule, L., Rigg, R., Rybanič, R., and Šramka, Š. (2016) *Program starostlivosti o vlka dravého (Canis lupus) na Slovensku*, Štátna ochrana prírody Slovenskej republiky, Banská Bystrica.
Bartošová, D. (1998) 'Osud vlků v Beskydech je nejistý', *Veronica*, vol 12, no 1, pp1–7.
Bartošová, D., and Genda, I. (2001) 'Projekt Záchrana a návrat velkých predátorů v oblasti Západních Karpat – část 1', *Ochrana přírody*, vol 56, no 1, pp9–12.
Bath, A. (2000) 'Human dimensions in wolf management in Savoie and Des Alpes Maritimes, France: Results targeted toward designing a more effective communication campaign and building better public awareness materials', www.kora.ch/malme/05_library/

5_1_publications/B/Bath_2000_Human_Dimensions_in_Wolf_Management_in_ France.pdf.

Bisi, J., Kurki, S., Svensberg, M., and Liukkonen, T. (2007) 'Human dimensions of wolf (*Canis lupus*) conflicts in Finland', *European Journal of Wildlife Research*, vol 53, no 4, pp304–314.

Blekesaune, A., and Rønningen, K. (2010) 'Bears and fears: Cultural capital, geography and attitudes towards large carnivores in Norway', *Norsk Geografisk Tidsskrift – Norwegian Journal of Geography*, vol 64, no 4, pp185–198.

Boitani, L. (2000) *Action Plan for the Conservation of Wolves in Europe (Canis lupus)*, Council of Europe Nature and Environment Series, Strasbourg.

Bojda, M., Váňa, M., Kutal, M., Bartošová, D., and Krajmerová, D. (2014) 'Výskyt medvěda hnědého v letech 2003–2012 v karpatských pohořích na česko-slovenském pomezí', in M. Kutal and J. Suchomel (eds) *Velké šelmy na Moravě a ve Slezsku*. Univerzita Palackého, Olomouc.

Breitenmoser, U. (1998) 'Large predators in the Alps: The fall and rise of man's competitors', *Biological Conservation*, vol 83, no 3, pp279–289.

Červený, J., Bartošová, D., Anděra, M., and Koubek, P. (2004) 'Současné rozšíření medvěda hnědého (*Ursus arctos*) v České republice', *Lynx (Praha)*, vol 35, pp19–26.

Červený, J., and Koubek, P. (2003) 'The brown bear in the Czech Republic', in B. Kryštufek, B. Flajšman and H. I. Griffits (eds) *Living with Bears. A Large European Carnivore in a Shrinking World*, Ecological Forum LDS, Ljubljana.

Chapron, G., Kaczensky, P., Linnell, J. D. C., von Arx, M., Huber, D., Andrén, H., López-Bao, J. V., Adamec, M., Álvares, F., Anders, O., Balčiauskas, L., Balys, V., Bedö, P., Bego, F., Blanco, J. C., Breitenmoser, U., Brøseth, H., Bufka, L., Bunikyte, R., Ciucci, P., Dutsov, A., Engleder, T., Fuxjäger, C., Groff, C., Holmala, K., Hoxha, B., Iliopoulos, Y., Ionescu, O., Jeremić, J., Jerina, K., Kluth, G., Knauer, F., Kojola, I., Kos, I., Krofel, M., Kubala, J., Kunovac, S., Kusak, J., Kutal, M., Liberg, O., Majić, A., Männil, P., Manz, R., Marboutin, E., Marucco, F., Melovski, D., Mersini, K., Mertzanis, Y., Mysłajek, R. W., Nowak, S., Odden, J., Ozolins, J., Palomero, G., Paunović, M., Persson, J., Potočnik, H., Quenette, P.-Y., Rauer, G., Reinhardt, I., Rigg, R., Ryser, A., Salvatori, V., Skrbinšek, T., Stojanov, A., Swenson, J. E., Szemethy, L., Trajçe, A., Tsingarska-Sedefcheva, E., Váňa, M., Veeroja, R., Wabakken, P., Wölfl, M., Wölfl, S., Zimmermann, F., Zlatanova, D., and Boitani, L. (2014) 'Recovery of large carnivores in Europe's modern human-dominated landscapes', *Science*, vol 346, no 6216, pp1517–1519.

Dressel, S., Sandström, C., and Ericsson, G. (2015) 'A meta-analysis of studies on attitudes toward bears and wolves across Europe 1976–2012', *Conservation Biology*, vol 29, no 2, pp565–574.

Ericsson, G., and Heberlein, T. A. (2003) 'Attitudes of hunters, locals, and the general public in Sweden now that the wolves are back', *Biological Conservation*, vol 111, no 2, pp149–159.

Ericsson, G., Sandström, C., and Borstedt, G. (2006) 'The problem of spatial scale when studying human dimensions of a natural resource conflict: Humans and wolves in Sweden', *International Journal of Biodiversity Science and Management*, vol 2, no 4, pp343–349.

Eriksson, M., Sandström, C., and Ericsson, G. (2015) 'Direct experience and attitude change towards bears and wolves', *Wildlife Biology*, vol 21, no 3, pp131–137.

Hell, P. (2003) '1. Historické rozšírenie a populačná hustota vlka na Slovensku', in P. Györgi and S. Pačenovský (eds) *Vlky a rysi v oblasti slovensko-maďarských hraníc*, WWF Hungary, Budapest.

Hell, P., and Sládek, J. (1974) *Trofejové šelmy Slovenska*, Príroda, Bratislava.

Hell, P., and Slamečka, J. (1999) *Medveď v slovenských Karpatoch a vo svete*, PaRPRESS, Bratislava.

Janík, M.,Voskár, J., and Buday, M. (1986) 'Súčasné rozšírenie medveďa hnedého (*Ursus arctos*) v Československu', *Poľovnícky zborník – Folia venatoria*, vol 16, pp331–352.

Kaczensky, P. (2003) 'Public opinion of large carnivores in the Alps and Dinaric Moutains', in B. Kryštufek, B. Flajšman and H. Griffiths (eds) *Living with Bears: A Large European Carnivore in a Shrinking World*, Ecological Forum LDS, Ljubljana.

Kaczensky, P., Blazic, M., and Gossow, H. (2004) 'Public attitudes towards brown bears (*Ursus arctos*) in Slovenia', *Biological Conservation*, vol 118, no 5, pp661–674.

Karlsson, J., and Sjöström, M. (2007) 'Human attitudes towards wolves, a matter of distance', *Biological Conservation*, vol 137, no 4, pp610–616.

Kleiven, J., Bjerke,T., and Kaltenborn, B. P. (2004) 'Factors influencing the social acceptability of large carnivore behaviours', *Biodiversity and Conservation*, vol 13, no 9, pp1647–1658.

Kovařík, P., and Kutal, M. (2014) 'Velké šelmy a chov ovcí v CHKO Beskydy', in M. Kutal and J. Suchomel (eds) *Velké šelmy na Moravě a ve Slezsku*, Univerzita Palackého, Olomouc.

Kovařík, P., Kutal, M., and Machar, I. (2014) 'Sheep and wolves: Is the occurrence of large predators a limiting factor for sheep grazing in the Czech Carpathians?' *Journal for Nature Conservation*, vol 22, no 5, pp479–486.

Kutal, M.,Váňa, M., Suchomel, J., Chapron, G., and López-Bao, J.V. (2016) 'Trans-boundary edge effects in the Western Carpathians: The influence of hunting on large carnivore occupancy', *PLoS ONE*, vol 11, no 12, p.e0168292.

Linnell, J., Andersen, R., Andersone, Z., Balciauskas, L., Blanco, J., Boitani, L., Brainerd, S., Breitenmoser, U., Kojola, I., Liberg, O., Loe, J., Okarma, H., Pedersen, H., Promberger, C., Sand, H., Solberg, E.,Valdmann, H., and Wabakken, P. (2002) 'The fear of wolves: A review of wolf attacks on humans', *Oppdragsmelding Norwegian Institute of Nature Research,*, vol 731, pp1–65.

Linnell, J. D. C., and Alleau, J. (2016) 'Predators that kill humans : Myth, reality, context and the politics of wolf attacks on people', in F. M. Angelici (ed) *Problematic Wildlife: A Cross-Disciplinary Approach*, Springer International, New York.

Lopez, B. H. (1979) *Of Wolves and Men*, Scribner, New York.

Lüchtrath, A., Schraml, U., and Lüchtrath, A. (2015) 'The missing lynx – Understanding hunters' opposition to large carnivores', *Wildlife Biology*, vol 21, no 2, pp110–119.

Majić, A. (2007) *Human Dimension in Wolf Management in Croatia: Understanding Public Attitudes Toward Wolves Over Time and Space*, Master's thesis, Memorial University of Newfoundland, Canada.

Majic, A., and Bath, A. (2010) 'Changes in attitudes toward wolves in Croatia', *Biological Conservation*, vol 143, no 1, pp255–260.

Martinát, S., Klapka, P., and Nováková, E. (2008) 'Changes of spatial differentiation in livestock breeding in the Czech Republic after 1990', in J. Bański and M. Bednarek (eds) *Rural Studies 15*, PAN Warszawa, Warszawa.

Mech, L. D. (1995) 'The challenge and opportunity of recovering wolf populations', *Conservation Biology*, vol 9, no 2, pp270–278.

Niznikowski, R., Strzelec, E., and Popielarczyk, D. (2006) 'Economics and profitability of sheep and goat production under new support regimes and market conditions in Central and Eastern Europe', *Small Ruminant Research*, vol 62, no 3, pp159–165.

Pavelka, J., and Trezner, K. (2001) *Příroda Valašska*, Český svaz ochránců přírody,Vsetín.

Prokop, P., Usak, M., and Erdogan, M. (2011) 'Good predators in bad stories : Cross-cultural comparison of children's attitudes towards wolves', *Journal of Baltic Science Education*, vol 10, no 4, pp229–242.

R Core Team (2016) *R: A Language and Environment for Statistical Computing*, R Foundation for Statistical Computing,Vienna.

Randveer, T. (2001) 'Estonians and the wolf', in L. Balčiauskas (ed) *Proceedings of BLCIE Symposium: Human Dimensions of Large Carnivores in Baltic Countries*. Šiauliai University, Šiauliai.

Rigg, R., Finďo, S., Wechselberger, M., and Gorman, M. L. (2011) 'Mitigating carnivore – Livestock conflict in Europe: Lessons from Slovakia', *Oryx*, vol 45, no 2, pp272–280.

Røskaft, E., Händel, B., Bjerke, T., and Kaltenborn, B. P. (2007) 'Human attitudes towards large carnivores in Norway', *Wildlife Biology*, vol 13, no 2, pp172–185.

SAS (2007) *JMP, Version 5.0.1*, SAS Institute, Cary.

Szinovatz, V., and Gossow, H. (2001) 'Die Akzeptanz der Bären in Österreich in Abhängigkeit von der Saison – eine Langzeitstudie', *Forest, Snow and Landscape Research*, vol 76, no 1/2, pp155–168.

Wechselberger, M., Rigg, R., and Beťková, S. (2005) *An Investigation of Public Opinion About the Three Species of Large Carnivores in Slovakia*, Slovak Wildlife Society, Liptovský Hrádok.

Williams, C. K., Ericsson, G., and Heberlein, T. (2002) 'A quantitative summary of attitudes toward wolves and their reintroduction (1972–2000)', *Wildlife Society Bulletin*, vol 30, no 2, pp575–584.

Zibordi, F. (2009) 'Large carnivores and public awareness campaigns: The role of protected areas', in M. Pavlik (ed) *Large Carnivores in the Alps and Carpathians: Living with the Wildlife*, ALPARC – Alpine Network of Protected Areas, Chambery.

10 Rural-urban heterogeneity in attitudes towards large carnivores in Sweden, 1976–2014

Göran Ericsson, Camilla Sandström and Shawn J. Riley

Introduction

Swedes traditionally exhibit utilitarian values and use wildlife resources in consumptive ways. Hunting and other uses of wildlife resources are deeply rooted, and to a high degree accepted, in Swedish society (Ljung et al., 2015). A key feature of Swedish legalisation since at least 1789, when the right to hunt on their own land was granted to all landowners regardless of class, has been to promote "good" species – ones that can be used by humans for food, fur or other products (Danell et al., 2017). Swedish governance, policy and management actions generally have disfavoured wildlife species such as large carnivores perceived to be in direct competition – and thus conflict – with human uses of resources. This view was challenged, however, during the latter part of the 19th century as globalised approaches to conservation emerged. Influences, primarily from the United States, recognised that active wildlife governance and management was necessary to regulate human uses of wildlife. The focus of early debates (1830 and on) in Sweden was on restoration of moose populations, and then brown bears (1860 and on) that could sustain hunting. Sweden consequently transformed legislation for both forestry and wildlife in the 1930s and 1940s because of the wise-use and conservation movements. The main stakeholders involved in conservation at that time were hunters, foresters and members of the public with means and time to engage in conservation (Danell et al., 2017).

Implementation of contemporary conservation policy and practices by state, private and non-governmental organizations (NGOs) included regulated hunting, state promotion of conservation education and complete protection of some socially desirable species. Consequently, ungulates thrived. Species such as moose (*Alces alces*), roe deer (*Capreolus capreolus*), red deer (*Cervus elaphus*), fallow deer (*Dama dama*) and wild boar (*Sus scrofa*) increased rapidly in absolute and relative terms by the mid-20th century (Danell et al., 2017). Although their recovery lagged in relation to the recovery of prey species, large mammalian predators made a commensurable comeback. Sweden currently contains historically high – perhaps record high – populations of large mammals and birds (Danell et al., 2017). These abundant wildlife populations create challenges and opportunities for conservation in European and North American society. Sweden serves as a functional model for planning and implementation of practices to provide sustainable populations of large mammals in the Global North based on the collaboration among governmental agencies, landowners, NGOs and the public (Singh et al., 2014).

Current attitudes towards large carnivores in Sweden are multidimensional. They are formed principally by human-wolf and human-bear interactions, heterogeneity among those people with and without direct experience large carnivores, and urban-rural human demographic differences (Dressel et al., 2015; Eriksson et al., 2015) A crucial starting point that created today's attitudinal heterogeneity among stakeholders towards large carnivores was the Green Revolution of the 1960s (Carson, 1962). Wolves (*Canis lupus*), functionally extinct from most of Sweden in the 1960s, were confined to the alpine north of the country. Abundance was assumed at that time to be fewer than 10 individuals (Chapron et al., 2014). Bears (*Ursus arctos*), lynx (*Lynx lynx*), and wolverines (*Gulo gulo*) also were scarce, and their geographical distribution restricted to boreal and mountain areas, where few people lived.

The Swedish society by the early 1960s consisted of a majority of people whose attitudes were formed by strong consumptive and utilitarian values yet by no real, first-hand, direct interactions with large carnivores. Any sporadic interactions had been limited to stories and media coverage such as occasional sightings or tracks (Ericsson and Heberlein, 2003). Swedish public television and its nature-oriented programs broadcasted flickering footage of occasional sightings of wolves in the far north, and of bears, lynx and wolverines in the low mountains. By far, wolves were the rarest animals to be reported, but also the one most likely to garner the most attention when large carnivores were missing in most peoples' everyday life, social life and culture (Danell et al., 2017).

Legalisation and governmental policies in the mid-1960s made a complete reversal in terms of overall objectives from promoting bounties to complete protection. Active control policies including bounties for wolf populations were still in effect at the end of 1965, but starting January 1, 1966, anyone killing a wolf in Sweden could be sent to jail. The abrupt U-turn in Swedish wolf conservation policy cannot be underestimated; it was the first major change in policy in nearly 100 years and involved wolves, the most controversial species, and historically most threatening to domestic animal husbandry (Danell et al., 2017). Nevertheless, it was politically easy to do because few wolves existed – no animals, no direct interaction, no impact. Breaking the ice with wolves, it was rather uncontroversial to make similar policy about other species. Wolverines were fully protected beginning in July 1969. The Swedish Environmental Protection Agency (EPA) was formed in 1967, one of many regulatory agencies established in countries of the world in the wake of the broad 1960s conservation movement. Wildlife-related issues, including restoration of all wildlife as well as regulation of hunting, have been since included in the EPA federal mandate.

Tracking large carnivore numbers and attitude trends

No wolves; large carnivores restricted to the north;
human-large carnivore interaction not salient

The protection of 1960s had no effect, in practice, because large carnivores were functionally absent from most of Sweden for another 10 years. Only sporadic predation on reindeer occurred in the far north, but stray wolves coming in from the east were implicated. Bears, wolverines and lynx were still restricted in their distribution to places where few humans lived, which also meant they were functionally absent from the minds of most of the Swedish public. Social science coincidently started slowly to emerge in most facets of governance, including wildlife conservation.

The first professionally conducted survey of Swedish attitudes toward carnivores was done in 1976, and still serves as a baseline well before large carnivores actually physically returned (Andersson et al., 1977). The survey was part of a research project led by the Swedish EPA in response to the non-recovery of the Swedish wolf population. No confirmed breeding had taken place since 1964, but the survey also was in response to confirmed wolf sightings in the border area with Norway, close to Trysil. The survey was a proactive measure to determine if wolves were indeed making a comeback. It was a three-contact mail survey to the public in four rural regions with historical occupancy of wolves (n = 1200, 72% response rate), administered to members of the nature conservation organisation Swedish Society for Nature Conservation ("Naturskyddsföreningen" in Swedish, n = 200, 91% response rate), reindeer owners (n = 200, 58% response rate), cattle farmers (n = 200, 82% response rate), and hunters (n = 200, 82% response rate). A majority of the total sample as well as of key sub-samples, including hunters and members of the Swedish Society for Nature Conservation, expressed positive attitudes towards possible existence of wolves. These latter groups were among the most influential in actively restoring wolves in the Swedish landscape. A general finding of the attitudinal survey was that most respondents supported a re-establishment of wolves in the northern mountain areas bordering Norway.

Since the rugged border area was and still is sparsely populated (fewer than 0.5 people/km²), no strong opposition was expected for wolves occurring in the far north back in 1976. One group, however, the indigenous Sámi reindeer herders, strongly opposed the occurrence of wolves in the mountainous north. Since the public in the south viewed large carnivores metaphorically as part of the vast, uninhabited northern wilderness, they supported wolf restoration. That support occurred regardless of place of residence – both urban and rural people supported wolves in the north. The works of nature writers and early nature filmmakers in the 1950s and 1960s such as Bertil Haglund and Edwin Nilsson reinforced beliefs of carnivores as wilderness-only species. Their epic black and white footage of rarely spotted wolves, bears, wolverines and lynx provided a vicarious interaction between large carnivores and predominately people in the north, to people of all backgrounds. Apart from Sámi reindeer herders, there was no rural-urban heterogeneity detected in stakeholder attitudes towards large carnivores in Sweden during the 1970s. From the perspective of Swedish carnivore conservation and its implications, it is interesting that support for taking action to establish carnivores in boreal areas of southern and central Sweden was rather low (Andersson et al., 1977).

Rewilding and initial cohort heterogeneity in Sweden

The large carnivore issue became salient in the early 1980s, when the first reproductively active wolf population was confirmed in south-central Sweden. This event appeared to be a spontaneous recolonisation, and the first one observed since 1918 in the region. Ironically, wolves colonised precisely in the area *not* preferred by any group which responded to the 1976 survey. State policy had promoted reforestation and rewilding, which had favoured growth of large carnivore populations throughout Sweden (Danell et al., 2017). Thus, the stage was set for large carnivores to re-establish themselves outside the northern peripheral areas they been forced to inhabit during the previous 200 years. Importantly, the prey base in the form of wild ungulates had also been restored to levels that could support increasing populations and range distribution of large carnivores.

A second turning policy point, although not as dramatic as the one in the late 1960s, occurred during the early 1980s, when brown bears and lynx received attention from society similar to what wolves and wolverines had received a decade earlier. This renewed attention most likely was a result of the Convention on the Conservation of European Wildlife and Natural Habitats (Bern Convention), which was adopted by European Environment Ministers in 1979. To promote recovery of the bear population, a strictly regulated licence hunt was initiated in 1981. The progression in lynx conservation roughly followed the same trajectory in terms of legislation and population recovery. Additional protection for the lynx included partial exclusion from hunting in 1986 and complete protection followed in 1991. In short, a more restrictive large conservation policy of the 1980s coincide with the peak of a major wave of urbanisation and environmental movements globally.

In 1980, a national proportional sample of 3,006 Swedes (67% response rate) from 16–65 years of age were surveyed about their attitude toward wildlife and hunting (Norling et al., 1981). At that time, the Swedish moose population peaked at unprecedented densities, estimated to be more than two individuals/km² in forested areas. Roe deer populations were also increasing due to restrictive hunting. In addition, the Sarcoptic mange (*Sarcoptes scabiei*) had severely reduced red fox (*Vulpes vulpes*) populations, which had also favoured roe deer. The 1980 survey was conducted just around the peak rate of urbanisation in Sweden, 18 years after Rachel Carson's book "Silent Spring", and only 10–14 years after the abrupt changes in legislation, which had promoted large carnivore conservation instead of extermination. The origin of Sweden's heterogeneity between urban and rural attitudes toward wildlife, still prevalent today, can be traced back to this time. Attitude differences among sample cohorts were apparent, with urbanites and younger respondents being more pro-conservation. However, human-carnivore conflicts were largely absent. The main driver of these conflicts, namely the wolf, was still functionally non-existent. Although rural-urban differences had been detected, they could not have meant much for wildlife conservation and management, yet.

A key research question in two of our meta-analyses of attitudes towards large carnivores was the effect of age (Williams et al., 2002; Dressel et al., 2015). Will younger people, socialised in an era of conservation, maintain a preponderance of positive attitude towards large carnivores? Can this trend be compromised with increasing social interaction, experience and age? Initially, older people expressed negative attitudes towards large carnivores, probably because those people had spent their formative years in an era when utilitarian values prevailed. However, the meta-analyses revealed the effects of direct experiences with carnivores, such as notable depredation events, as important drivers for less positive attitudes. Norling et al. (1981) had already pointed out that place of residence (a reliable proxy of direct experience of wildlife) was a more important driver (predictor) of valence (positive/negative) of attitudes than age (cohort) or upbringing (socialising). This is how Norling et al. (1981) summarised this (quote translated by authors from Swedish, emphasis added):

> in general, the Swedish public expressed a strong interest in wildlife; most species interested a majority of Swedes, typically 60–70%. . . . The interest for wildlife differed among cohorts, however. Older people expressed a strong interest for common species like moose, roe deer, hare and birds (but not birds of prey). For between every second and every third respondent over 55 years of age, there was an expression of increased interest for those common species, which was 2–3 times higher than interest expressed by younger people. *With respect to birds of prey and large carnivores, there was no general age difference, but there were differences with respect to gender and place of residence.* The largest interest for those species [Authors' note: birds of prey and large carnivores] was expressed by men as compared to women and urban citizens as compared to rural residents. . . . Groups with the highest expressed interest for those species were members of the Swedish Society for Nature Conservation and hunters; however, the hunters' interest was more restricted to game species.

Please note only 3% of respondents in this survey expressed that there were too few wolves, bears, lynx and wolverines in relation to what they had known of the population sizes of these species at that time (Norling et al., 1981). In reality, however, wolves were almost absent and the other three carnivore species' populations were critically low in number as compared to their current population. That only 3% of respondents expressed concern for large carnivore species is not surprising; why would the public miss something that had been absent for

100–150 years for most of Sweden, and had been thus absent in culture, media and everyday life?

Despite the lack of support for more large carnivores, Norling et al. (1981) documented a greater occurrence of naturalistic/ecological value orientations among the public than was anticipated from previous work in Sweden and elsewhere. Yet, as expected, greater occurrence of utilitarian value orientation about nature and animals was expressed in regions dominated by forestry and agriculture. There was no difference at that time in attitudes towards large carnivores between rural and urban respondents. Our conclusion is that around 1980, the fraction of people who had an urban upbringing with no experience of the countryside was rather low, as was the fraction of the population with experience of large carnivores. Therefore, attitudes toward large carnivores were largely homogenised within the general Swedish population.

The comeback of large carnivores led to increasing attitude heterogeneity

When wolves re-entered the scene in the region of Värmland, Swedes' attitudes were rather homogenous. From studies reviewed, we assume that the first reports of wolf sightings in the early 1980s induced positive attitudes towards restoration of *all* large carnivores among the general public (sensu Williams et al., 2002). We also presume, given reports in papers and magazines, that attitudes began to shift when people in Värmland started to report negative experiences with wolves from 1982 onward. We assume that the early 1980s was the starting point of the conflict depicted by attitudes towards large carnivores between a handful of people in the areas recolonised by the wolves (a minority with direct experience from human–carnivore interaction) and the wider public (the majority of people in the country, without such a direct experience; Williams et al., 2002). By 2000, there were nearly 100 wolves, 1,000 lynx, 2,000 bears, and 200 wolverines in Sweden, all still spatially restricted to rural areas with low human population densities.

The few attitude surveys performed until around 2000 were based on national proportional sampling (Williams et al., 2002). They signalled that a majority of Swedes supported the presence of wolves and other carnivores, and disclosed no major reluctance and opposition towards them, when data were at a *national scale*. That finding was challenged repeatedly by the early 2000s by vocal opposition at local, regional and national levels, including letters to the editor of all major local, regional and national newspapers, on radio and TV. In rural Sweden, these voices were particularly pronounced, but as the national, proportional surveys continued to show no strong opposition to large carnivores, concerns were mostly dismissed as non-relevant by the media in urban areas (Sandström and Lindvall, 2006). Thus, in the first years of the 2000s, there was a growing mistrust towards attitude surveys, since the voices of rural opponents were too few to be detected by national surveys involving about 1,000 respondents each (Ericsson et al., 2006).

Previous studies overlooked local opposition to large carnivores

In 2001, the first Swedish study to detect heterogeneity in attitudes towards large carnivores was about attitudes toward wolves (Ericsson and Heberlein, 2003). Local oversampling was introduced to address the previous criticisms about sampling in relation to participants' experiences. Despite the fact that Swedish and Norwegian case studies had previously suggested a change in perspectives (e.g., Williams et al., 2002), it became obvious that no valid sampling of all affected subgroups had been completed previously to test the hypothesis that direct experience and knowledge influenced the valence of attitudes toward large carnivores. Ericsson and Heberlein (2003) used a scale ranging from -17 to $+17$ across nine items and contrasted four groups (general public, hunters, public and hunters in rural areas with wolves).

Based on theory and previous studies (Williams et al., 2002), rural-urban heterogeneity was predicted to increase with increasing carnivore populations. That was a key hypothesis in Ericsson and Heberlein's work of 2003. The outcome confirmed heterogeneity in local versus national public attitudes towards large carnivores for the first time in Sweden. Most positive was the national public ($+4.8$), followed by the public in areas with wolves ($+2.4$), and hunters nationwide ($+1.2$). Hunters in areas with wolves expressed the only negative attitudes (-2.0). Interestingly, another key finding of Ericsson and Heberlein (2003) was that the effect of residence on attitudes towards large carnivores was already detectable with a threshold of community population set at 10,000 people. An in-depth analysis that followed suggested attitudinal heterogeneity was to be attributed to respondents – non-hunters and hunters – in areas with large carnivores having more direct experience and knowledge, stronger utilitarian ties to the land, and being less likely to express any neutrality towards any attitudinal measurement item.

It may be widely accepted that legislation and policy at large had to be decided nationally after general elections. The cost of increasing large carnivore populations, however, was shouldered only by rural populations, which created tension and a question of legitimacy. On the other hand, members of NGOs questioned the generality of attitude heterogeneity to other geographic areas, although standard international survey methodology had been used by Ericsson and Heberlein (2003), and they suggested that findings were an artefact of the local area sampled. Our research needed to take this idea of legitimacy one step farther: Would the findings about heterogeneity hold if the survey sample population was not in the spatially restricted contested area included in Ericsson and Heberlein's (2003) sampling? Given the studies published at that time and summarised by the meta-analysis of Williams et al (2002) we had expected that findings would hold.

Ericsson and co-workers further developed, tested and confirmed the main findings about rural and urban heterogeneity in the Mountain MISTRA project (2003–2008) covering about 55% of Sweden, but only 15% of the human population of the country (e.g., Ericsson et al., 2006, 2008). Our conclusion from a string of studies and papers from 2001 and on, opened the management and public debate for better acknowledging the heterogeneity that has existed in attitudes towards large carnivores (see Table 10.1 for a comprehensive review of objectives and outcomes of these studies). Once sampling was adapted to account for large carnivore distribution as well as better reflect different segments of targeted respondent populations, human-carnivore conflicts and the causes behind the conflicting patterns among groups in society had become clearer (Ericsson et al., 2008). To arrive at these results, oversampling in rural areas with large carnivores was needed.

Knowledge and experience as major drivers of attitudes towards large carnivores

After our surveys, Sweden has continued to experience increasing numbers of wolves, bears, lynx and wolverines (Chapron et al., 2014). The increase in the wolf population is still the driver that catalyses and propels debate around large carnivores, yet an expanding bear population has been another important aspect (Dressel et al., 2015; Eriksson et al., 2015). Growth in large carnivore populations and commensurate human-carnivore interactions have correlated with attitude change – mostly perceptions of increasing risk and increasing negative disposition (Dressel et al., 2015). The role of various direct experiences such as livestock depredation, finding remains of various wild or domesticated animals killed by carnivores, and seeing or hearing large carnivores are receiving new interest as the distribution of large carnivore expands toward densely human-populated areas such as the Stockholm region.

Every time a new wolf territory is established or bear populations expand into a new region, societal as well as scientific debate related to coexistence with large carnivores is renewed. This debate tends to focus on similar issues again and again: perceived competition between hunters and large carnivores for game animals, livestock depredation and contrasting views about human's role in nature. Central to these discussions was, and still is, the role of direct experiences, either positive or negative, with large carnivore species (Williams et al., 2002, Ericsson and Heberlein, 2003; Dressel et al., 2015). Rural people and hunters are always central to this discussion, since they are always believed to hold negative attitudes toward wolves and other large carnivores, mainly due to carnivores preying on domestic and game animals. We tested this assumption with data from national, regional and local surveys between 2001 and 2004. In all our studies, place of residence proved to be a good proxy for experience and knowledge of large carnivores (see Table 10.1 for a comprehensive review of the main objectives and outcomes of these studies).

Experiences incurred by living in areas with large carnivores influenced people, even those without direct connection to hunting, to express less positive towards large carnivores. Similar patterns were revealed for hunters throughout Sweden, whether they lived in an area with carnivores or not. Perceived competition over game, nonetheless, was not the driver behind disagreements over

Table 10.1 Background, main objective and main outcomes of selected human dimension studies performed in Sweden since 1976 on large carnivores

Study	Background of the study	Main objective	Main outcomes
Andersson et al. (1977)	• Research project led by the Swedish Environmental Protection Agency in response to the non-recovery of the Swedish wolf population • Confirmed wolf sightings in the border area with Norway (Trysil)	Examine attitudes towards wolves	• A majority of the total sample (n = 1200) expressed positive attitudes towards wolves • Sami reindeer herders strongly opposed the occurrence of wolves in the mountainous north • Support for taking action to establish carnivores in boreal areas of southern and central Sweden was rather low
Norling et al. (1981)	Increase of wild ungulates (moose; roe deer)	Examine attitudes towards wildlife	Only 3% of respondents (n = 3006) expressed that there were too few wolves, bears, lynx and wolverines in relation to what they had known of the population sizes of these species at that time
Williams et al. (2002)	• Early 1980s: First signs of a rural minority opposing wolf recovery • Early 2000s: Vocal opposition against wolf recovery; mistrust towards surveys, since the voices of opponents are hardly detected	Meta-analysis of wolf attitudes reported in 38 quantitative surveys between 1972 and 2000	The few attitude surveys performed until around 2000 were based on national proportional sampling and depicted a clear majority of Swedes supporting the presence of large carnivores, without disclosing any major reluctance or opposition
Ericsson and Heberlein (2003)	• Rural–urban heterogeneity was predicted to increase with increasing carnivore populations • Criticism to previous research addressing sampling methods	• Examine experiences, knowledge, and attitudes toward wolves • Local oversampling introduced to address criticisms about sampling	• First Swedish study to detect heterogeneity in attitudes towards large carnivores (n = 1734) • Endorsement of large carnivores more pronounced for the general public and hunters as compared to public and hunters in rural areas • An in-depth analysis suggested that attitudinal heterogeneity was to be attributed to respondents – both non-hunters and hunters – in areas with large carnivores having more direct experience and knowledge, stronger utilitarian ties to the land, and being less likely to express any neutrality towards any attitudinal measurement item

Source			
Dressel et al. (2015)	Ranges of wolves and bears have expanded across Europe considerably, necessitating an assessment of public attitudes	• Extend results of individual surveys over temporal and geographical scales • Identify attitude patterns and shifts	• Respondents (105 quantitative surveys involving 24 European countries; 1976–2012) were more favourable towards bears than wolves; attitudes toward bears improved over time • Attitudes toward wolves became less favourable the longer people have coexisted with the species • Wolf attitudes differed between groups, with farmers and hunters being less favourable; no differentiation was detected for bears
Johansson et al. (2016)	To inform wildlife management and allocate management resources in an efficient way, fear of wolves needs to be explored	Examine psychological antecedents of fear of wolves	• Self-reported fear of wolves addressed via two different paths: (1) at an individual level, with a concentration on potential human–wolf encounters, and (2) at a collective level, by facilitating trust building between the public and competent authorities • There was a moderating effect of geographical location, where the social path (management-related) was more important among respondents who lived closer to the animals (national sample, n = 545; regional sample, n = 1892)
Eriksson (2017)	The return of the wolf can be assumed to aggravate pre-existing urban-rural tensions, resulting in conflicts over wolf policy	Examine the attitude trend from 2004–2014	• Significant decrease of wolf support, overall, as well as in counties with continuous and increasing wolf presence • Attitudinal change associated with political alienation • Wolf policy symbolically attracting rural opposition against urban interests

large carnivore conservation. Instead, the main drivers were concrete actions to be taken or not, such as the right to protect hunting dogs. The common denominator in why majorities or respondent samples, whether urban or rural, accepted hunting of wolves and other large carnivores was protection of domestic animals, hunting dogs and human safety, regardless of whether respondents were hunters or non-hunters. Even without direct experience or knowledge, people supported the right to protect property and human lives at the expense of large carnivores.

Attitude trends between 2004 and 2014

A finding in our work is that an increase of large carnivores contributes to more heterogeneous urban *and* rural samples in terms of attitudes (Johansson et al., 2016; Eriksson, 2017). A majority of Swedes (74%) in 2004 expressed positive attitudes toward the presence of wolves in Sweden, which decreased to 71% in 2009 and 66% in 2014. The pairwise difference of 8% between 2004 and 2014 was statistically significant ($t = 3.98$, $df = 1000$, $p < 0.001$). Among counties with continuous and increasing presence of large carnivores, we revealed a decrease ($t = 3.75$, $df = 1050$, $p < 0.001$) in approval for wolves in Dalarna, Norrbotten, Jämtland and Gävleborg. Despite mitigation effects and other actions, the average approval for wolves dropped from nearly 60% to 50% per county between 2004 and 2014. The trends in the other counties are unknown, as they have never been sampled.

In the 2014 survey, we included the south-central county Värmland as a sample. Värmland is the only county with uninterrupted wolf presence since the beginning of the 1980s, when the first pair of breeding wolves was established there, most likely in 1982–1983. In 2014, Värmland had a 49% approval of wolves, which is comparable with other counties (e.g., Dalarna and Gävleborg) that experienced long permanent presence of wolves. Moving from rural to urban settings, respondents from the Stockholm County, who were included the studies of 2009 and 2014, expressed the most positive endorsement of wolves (78% wolf approval in 2009; 74% in 2014). Even though finding such a strong support may well fit the stereotype of pro-wolf urbanites, there was a marked decrease in wolf approval of 4%, which was statistically significant ($t = 4.14$, $df = 3900$, $p < 0.001$).

Regardless of county or presence of carnivores, the public expressed more positive attitudes toward bears than wolves. National support, overall, for brown bears was also higher than the wolf, but there was a decrease from 79% in 2004 to 75% in 2009, and further down to 73% in 2014 (pair-wise difference between 2004 and 2014 proved to be significant; $t = 3.20$, $df = 1000$, $p < 0.01$). In parallel with wolves, an analogous decrease in positive attitudes toward bears was also observed between 2004 and 2009 in the northern counties of Norrbotten, Västerbotten and Dalarnas. For lynx and wolverines, attitudes were once again more positive than those toward wolves, and there was no significant reduction in support through time.

Implications for large carnivore conservation and management

Easier said than done: governance challenges for large carnivore conservation and management

Wildlife governance focused on addressing interactions between people and wildlife commensurate with abundance of wildlife are generally most effective. Large carnivores are no exception to that rule. At the core of wildlife governance in Sweden, there are species harvested for meat, non-consumptive uses such as wildlife-related viewing and recreation, and species such as wolves, bears, wolverines and lynx that affect humans, human activities and human interests in nature. The current governance regime is challenged by demographic changes such as internal migration from rural to urban areas, external immigration of people from cultures with different values and little direct experience with wildlife, and from the effects of urban sprawl. All these phenomena are likely to affect resident values and publicly expressed attitudes and thus influence the character of future wildlife governance and management.

The paradigm shift in Swedish large carnivore policy in the 1960s, from management to conservation, has turned out 50 years later to be a conservation policy success (Chapron et al., 2014). As we have shown in this chapter, the increasing number of carnivores has resulted in more diversity and heterogeneity in values towards wildlife and attitudes towards large carnivores and their management in particular. This currently engages stakeholders with diverse values in seeking effective and legitimate solutions to these challenges. Sweden's case may probably be a forerunner of what will occur in other European countries, since there is a recovery of large carnivore species across the whole continent (Chapron et al., 2014; Dressel et al., 2015). Key governance challenges will be twofold: to track large carnivore species, as well as empower those affected to manage human-carnivore conflicts (Riley and Sandström, 2016).

There has been an extensive institutional change in how natural resources and wildlife are governed in Sweden, gradually shifting to more inclusive processes with respect to stakeholders affected by the return of large carnivores. After all, an integrated approach, drawing on knowledge and insights from different disciplines, is more likely to result in meeting societal needs (Riley et al., 2002). For many environmental managers, much of their time is allocated to dealing with human responses to governance and management actions. For instance, conservation and management of increasing populations of large carnivores thus has become more of a political-institutional and socio-cultural challenge than a purely biological matter. People are responding to the frequency and quality of interactions they have with wildlife in various ways, where abundance of species is not the sole driver (Riley et al., 2002; Sjölander-Lindqvist et al., 2015). Because costs and benefits of wildlife presence are distributed unevenly, and because costs may intrude on local livelihoods, striking an acceptable balance in interventions to support wildlife management is difficult (Sjölander-Lindqvist et al., 2015).

Since public involvement is a mandate of contemporary democratic decision-making, however, within the frame of natural resource governance (Sjölander-Lindqvist et al., 2015), governance approaches that accommodate and bridge between diverse values and attitudes will increase legitimacy. Thus, these approaches are highly likely to result in sustainable decisions, which improves chances to achieve policy objectives. Approaches associated with this sort of governance include building on shared vision and increased local acceptance of state initiatives (e.g., in the implementation of large carnivore policy). In short, concepts such as social trust, fair representation, and inclusion of different knowledge types, leadership and communication, play an important role in achieving legitimacy (Sjölander-Lindqvist et al., 2015).

A major challenge is how to best incorporate voices and concerns that transcend spatial scales and may be disproportionally represented in different localities. Most people do not have direct experiences with large carnivores. Their perceptions towards these species are influenced more by media and stories conveyed through social networks. The specific context and characteristics of human-carnivore interactions has predictable consequences for how people react to those interactions. In the same direction, it has been underlined that direct experience usually solidifies attitudes, either positive or negative, toward the object of the attitude. Riley and Sandström (2016, p. 63) introduced a hierarchy of responses, whereby people reacted to various risks, including risks related to large carnivores. Although conflicts related to livestock depredation are perhaps most common, human health and safety may lay a greater role in determining tolerance for large carnivores as compared to other species (Inskip et al., 2016).

New roles to be facilitated

Scientific knowledge does not present a limiting factor for the effectiveness of large carnivore conservation, in most cases, and this is because scientific knowledge can never be enough for addressing human-carnivore conflicts when the issue is principally one of overcoming value differences in society. We suggest that consideration of the biological dimension (e.g., density of large carnivores and their prey species) may be necessary but not sufficient for sustained large carnivore conservation and management. Monitoring the human dimension is also needed to shed light on causal factors affecting values, attitudes and behaviours towards carnivores and inform large carnivore policy from the perspective of complex socio-ecological systems. It is crucial to track indicators of wildlife value orientations (Patterson et al., 2003), and how these are affecting socially acceptable ways wildlife can be managed. These changes may be caused or influenced by broader societal shifts, including a society's socio-economic and cultural development, but also by shifts in socio-demographic variables and behaviour at the individual level such as income, education, migration and outdoor recreation patterns.

All knowledge and concerns exchanged will be subject to circulating discourses, multiple contestations and regimes of power. The comeback of large carnivores was accompanied by the realisation of a general deficit in trust among stakeholders, which presents an important barrier to effective coordination in large carnivore conservation and management (Eriksson, 2017). To be effective, any planned institutional transition will need to rely on enabling social relationships. Building adaptive management's capacity to pool resources, expertise and influence (Sandström et al., in this volume), will require development of new roles for the public sector that establish collaborative arrangements, common values and trust between and among the public and the government. We would single out trust as a variable that may prove decisive in several contexts: Recent research revealed that strengthening trust between the public and authorities may be instrumental in addressing fear of wolves (Johansson et al., 2016).

The new roles that need to be facilitated, often addressed as "New Public Service", imply that public authorities seek solutions to societal problems in collaboration with private and civil society organisations. In contrast to Traditional Public Administration, the new steering mode provides a social arena to enable citizens to articulate their shared interests to promote collective action. The focus is thus on citizen and stakeholder involvement and on negotiating interests among actors (Denhardt and Denhardt, 2007). In this process, the role of the government is transformed from one of mainly prescribing and controlling to one of bringing various actors to the table and brokering efforts to figure out what are the desired directions to be taken, prioritise them and put solutions into practice (Denhardt and Denhardt, 2007; Mårald et al., 2017).

Conclusion

In the absence of wolves and other large carnivores, attitudes expressed by various groups in Swedish society towards these species may have been quite homogeneous. With increasing numbers of wolves, bears, lynx and wolverines, different segments of society experienced and interpreted different sorts of interactions with these unique species that had been functionally extinct for several human generations. A main conclusion of our work since 2001 is that the presence of large carnivores creates opportunities for direct and indirect experiences of humans with these species. Heterogeneity in lived and interpreted encounters, and how these interpretations may shape knowledge, will fuel any controversy over the best way to manage controversial species such as wolves and bears.

As soon as the number of carnivores increased, human–carnivore conflicts increased and social tension emerged. So did conflicts between the Swedish state and affected rural minorities; for instance, Sámi people. Contemporary wildlife management systems rely heavily on development of inclusive input from stakeholders and interdisciplinary management approaches to achieve conservation targets. Rural opposition, however, seems to imply that rural people believe they are deprived of their former rights and roles. Consideration of scale will improve chances for successful large carnivore conservation in Sweden. Success will require attention to human-carnivore interactions at the local scale, where people and animals co-exist with direct negative and positive impacts, as well as at the regional and national scale that needs to address binding international conventions. Although large carnivores will continue to exist in areas with lower human population density, heterogeneity of attitudes – exacerbated by increased urbanisation and differences in carnivore abundance – will provide the social background for any future decision, and therefore, a continuous monitoring of human dimensions will be required.

References

Andersson, T., Bjärvall, A., and Blomberg, M. (1977) *Inställning till varg i Sverige – en intervjuundersökning.* SNV PM 850, Naturvårdsverket (Swedish EPA), Sweden.

Carson, R. (1962) *Silent Spring*, Greenwich, Connecticut.

Chapron, G., Kaczensky, P., Linnell, J. D. C., et al. (2014) 'Recovery of large carnivores in Europe's modern human-dominated landscapes', *Science*, vol 346, no 6216, pp1517–1519.

Danell, K., Bergström, R., Mattson, L., and Sörlin, S. (2017) *Jaktens historia i Sverige (The History of Hunting in Sweden)*, Liber, Stockholm, Sweden.

Denhardt, J.V., and Denhardt, R. B. (2007) *The New Public Service: Serving, Not Steering*, M. E. Sharpe, New York.

Dressel, S., Sandström, C., and Ericsson, G. (2015) 'A meta-analysis of attitude surveys on bears and wolves across Europe 1976–2012', *Conservation Biology*, vol 29, no 2, pp565–574.

Ericsson, G., Bostedt, G., and Kindberg, J. (2008) 'Wolves as a symbol for people's willingness to pay for large carnivore conservation', *Society and Natural Resources*, vol 21, no 4, pp249–309.

Ericsson, G., and Heberlein, T. (2003) 'Attitudes of hunters, locals and the general public in Sweden now that the wolves are back?', *Biological Conservation*, vol 111, no 2, pp149–159.

Ericsson, G., Sandström, C., and Bostedt, G. (2006) 'The problem of spatial scale when studying human dimensions of a natural resource conflict: Humans and wolves in Sweden', *International Journal of Biodiversity Science and Management*, vol 2, no 4, pp343–349.

Eriksson, M. (2017) *Changing Attitudes to Swedish Wolf Policy Wolf Return, Rural Areas, and Political Alienation*, PhD thesis, Department of Political Science, Umeå University, Sweden.

Eriksson, M., Sandström, C., and Ericsson, G. (2015) 'Direct experience and attitude change towards bears and wolves', *Wildlife Biology*, vol 21, no 3, pp131–137.

Inskip, C., Carter, N. H., Riley, S. J., Fahad, Z., Roberts, T. M., and Macmillan, D. C. (2016) 'Toward human-carnivore coexistence: Understanding tolerance for tigers in Bangladesh', *PLOS One*, vol 11, no 1, pp1–20.

Johansson, M., Sandström, C., Pedersen, E., and Ericsson, G. (2016) 'Factors governing human fear of wolves: Moderating effects of geographical location and standpoint on nature protection', *European Journal of Wildlife Research*, vol 62, pp749–760.

Ljung, P. E., Riley, S. J., and Ericsson, G. (2015) 'Game meat consumption feeds urban support of traditional use of natural resources', *Society and Natural Resources*, vol 28, no 6, pp657–669.

Mårald, E., Sandström, C., and Nordin, A. (2017) *Forest Governance and Management across Time: Developing a New Forest Social Contract*, Earthscan, Routledge, London.

Norling, I., Jägnert, C., and Lundahl, B. (1981) 'Viltet och allmänheten', *I Vilt och Jakt. Sociala och ekonomiska värden.* Jordbruksdepartementet. Ds Jo 1981 vol 5, pp9–79.

Patterson, M. E., Montag, J. M., and Williams, D. R. (2003) 'The urbanization of wildlife management: Social science, conflict, and decision making', *Urban Forestry and Urban Greening*, vol 1, no 3, pp171–183.

Riley, S. J., Decker, D. J., Carpenter, L. H., Organ, J. F., Siemer, W. F., Mattfeld, G. F., and Parsons, G. (2002) 'The essence of wildlife management', *Wildlife Society Bulletin*, vol 30, no 2, pp585–593.

Riley, S. J., and Sandström, C. (2016) 'Human dimensions insights into reintroduction of wildlife populations', in D. S., Jachowski, J. J. Millspaugh, P. L. Angermeir, and R. Slotow (eds) *Reintroduction Of Fish and Wildlife Populations*, University of California Press, Davis, CA.

Sandström, C., Sjölander-Lindqvist, A., Pellikka, J., Hiedanpää, J., Krange, O., and Skogen, K. (in press) 'Between politics and management: Governing large carnivores in Fennoscandia', in T. Hovardas (ed) *Large Carnivore Concervation and Management: Human Dimensions*, Routledge, London.

Sandström, C., and Lindvall, A. (2006) Regional förvaltning av rovdjur i Västerbotten och Norrbotten – om likheter och skillnader ur ett samförvaltningsperspektiv, FjällMistrarapport, Rapport nr: 18, ISSN 1652-3822, Umeå, Sweden.

Singh, N. J., Danell, K., Edenius, L., and Ericsson, G. (2014) 'Tackling the motivation to monitor: Success and sustainability of a participatory monitoring program', *Ecology and Society*, vol 19, no 4, p7.

Sjölander-Lindqvist, A., Johansson, M., and Sandström, C. (2015) 'Individual and collective responses to large carnivore management: The roles of trust, representation, knowledge spheres, communication and leadership', *Wildlife Biology*, vol 21, no 3, pp175–185.

Williams, C. K., Ericsson, G., and Heberlein, T. A. (2002) 'A quantitative summary of attitudes toward wolves and their reintroduction (1972–2000)', *Wildlife Society Bulletin*, vol 30, no 2, pp575–584.

11 Challenging the false dichotomy of Us vs. Them

Heterogeneity in stakeholder identities regarding carnivores

Michelle L. Lute and Meredith L. Gore

Controversy in carnivore management

Sociopolitical controversy over rules governing natural resource use and wildlife conservation is globally ubiquitous. Carnivore conservation can be particularly contentious, and decisions about whether or not to lethally remove carnivores are often framed as binary options representing polarized pro- and anti-protectionist sides where each is caricatured as universally and singularly supporting or opposing a particular policy option. Lethal control of carnivores is one of the most commonly debated policy options; policy debate over lethal control is geographically indiscriminate (Bruskotter, 2013). Polarization over policy options such as lethal control can result in a number of negative policy outcomes, including stymied decision-making, increased intolerance, and retaliatory killing of carnivores that are widely reported and denigrated. It is understandable that – given how widespread polarization is, how negatively polarization is portrayed in the mass media, and how common the caricatures of binary options about lethal control are – decision-makers may assume dichotomized public opinion is the norm and to be expected. In practice, most decision-makers know stakeholders are more diverse than an oversimplified dichotomy suggests; they work to leverage principles and practices from the knowledge base, such as community-based carnivore management, public participation, and risk communication (Lichtenfeld et al., 2015). Even when policy is science-based, the dichotomy can persist (Triezenberg, 2010).

The question then becomes, why cannot the negative effects of distorted stakes be overcome? Social science provides rich understanding about human-wildlife interactions, what human dimensions underlie such assumptions and what factors may be driving the tendency to oversimplify a dichotomy in carnivore conservation. The social science evidence base tells us stakeholder positions are rooted in complex social identities that stakeholders use to organize in groups (Marchini and Macdonald, 2012; Lute et al., 2014; Kreye et al., 2016). Deeper understanding about these group dynamics and tendencies can offer innovative ways of thinking about the negative effects of conflict over carnivore management, introduce alternative means for resolving problems, and harmonize the relationship between science and policy.

This chapter explores how the concept of social identity can be used to explain key drivers of stakeholder conflict in carnivore management. First, we introduce a review of the concept as it relates to carnivore management, paying particular attention to the implications of considering group identity as a defining characteristic or attribute during decision-making. We then discuss how social identity can be measured to optimize the efficacy and efficiency of decision-making about carnivores. Next, we explore how social identity can be incorporated into carnivore management. Our review is anchored on the case of wolf management in Michigan, and draws upon a substantial historical and social science knowledge base about the context.

The concept of social identity

Identity is the component of one's self-concept that is derived from group membership and the value and emotion attached to that membership. Social identity theory (SIT) explains how language, appearance, and other signals cue group affiliation, which drives many perceptions and behaviors relevant to social and political spheres including conservation. In response to opposition, different groups can affiliate with each other when their positions have sufficient overlap. Importantly, such affiliations do not necessarily constitute a homogenous group with the same values and perceptions. Yet outsiders often assume just that, stereotyping members of a group or affiliated groups. In SIT terminology, this is referred to as ingroup bias, whereby individuals cognitively seek to increase positive ingroup characteristics and negative aspects of outgroups in such a way as to result in a preconceived judgment about outgroup members (Sherif, 1967; Labianca et al., 1998).

Consider, for instance, initiatives to ban trapping in various western U.S. states. Arizona, California, Colorado, and Washington (as well as a number of eastern states and other countries) have passed legislation banning certain types of, or all, traps while attempts have been made in Montana, New Mexico, Oregon, and other states. In many of these cases, livestock ranchers and hunters, each as distinct identity groups, become allies in defense of wildlife trapping, whereby trapping is considered a consumptive use. Likewise, animal welfare groups sometimes align with environmentalists (i.e., those focused on natural resources protection over use) over initiatives to ban trapping. Both coalitions unite over a position that trapping (e.g., of a particular species, on public lands) is or is not an appropriate policy. Each group may have similar notions regarding how humans ought to interact with animals. Yet, animal welfare advocates focus more on the welfare of individual animals while environmentalists are often concerned more about protecting a wild population (Ramp and Bekoff, 2015). Livestock ranchers may be motivated to defend livelihoods and hunters defend trapping to obtain materials (e.g., meat, pelts) or maintain tradition and recreation. It is important to note that environmentalists or hunters as stakeholder groups may also care about individual wellbeing and animal welfare advocates can be concerned about others' livelihoods. Although there may be some overlap in secondary dimensions, primary aspects likely determine the alliances that can lead to intergroup conflict.

It is widely assumed that affiliated groups have the same motivating values and attitudes underlying a similar position, or that taking a certain position means there are no shared values with the opposition, as in the previous example or the many social conflicts in which controversy is presented as an "Us versus Them" two-sided debate (e.g., in the United States: climate change, early childhood vaccinations, genetically modified organisms, evolution/science education). Although common, such assumptions can result in the so-called "predator pendulum" observed most notoriously with gray wolf conservation in the United States, where a carnivore species – in this case wolves – is managed according to alternatives that solely eradicate or protect the species. The pendulum swings from one extreme to the other depending on the identity group holding political decision-making power at the time a particular decision is made (Bruskotter, 2013). Identity conflict such as this case is at its root about inequalities between groups or perceptions of inequality, which leads to competition over a resource (Labianca et al., 1998; Brown, 2000; Hornsey, 2008). Although the conflict over how to manage carnivores is in part based on foundational values about how to treat non-human lives (Nie, 2003a, 2003b), carnivores or access to the legislators that make decisions about carnivores may also be a symbolic contested resource.

Identity conflict over carnivores, at least in the North American context, is rooted in historically unequal influences of social groups on wildlife management decisions. Traditionally, hunting, ranching, and rural interests dominated the decision space around wildlife in North America (Chase et al., 2011). The North American Model of Wildlife Management, long hailed as the epitome of conservation ethos, holds hunting as a central tenet in management. Contemporary critique of the model suggests this focus has, over time, had the possibly unintended outcome of disenfranchising non-hunting stakeholders (Nelson et al., 2011; Peterson and Nelson, 2016). As societal worldviews have shifted toward post-materialism and away from consumptive uses, "non-traditional" wildlife interests – such as those coming from urban, animal welfare, and wildlife advocate groups – vie for a greater voice in decision processes (Inglehart, 1977; George et al., 2016; Lute and Attari, 2016; Lute et al., 2016). In many ways, this historical shift in wildlife-related perceptions parallels other societal issues in which communities and policies are becoming more inclusive of diverse groups (e.g., marriage equality in the United States).

Thus, as the predator pendulum swings from one ingroup having power to another, the shift upsets the status quo and intergroup conflict over carnivore conservation flares. These upticks in controversy can be marked with the vitriol, misunderstanding, assumptions and power struggles typical of any identity conflict (Sherif, 1967; Triezenberg et al., 2011). These struggles often play out in courtrooms, legislative halls, and social media where groups retreat to hardened positions, appear to forget that groups are comprised of diverse individuals, and assume all their opponents are the same uneducated, wrongly-motivated stereotype. In places where enforcement of the rule of law is weak (e.g., Cameroon, Venezuela, Madagascar), such power struggles can manifest in poaching and retaliatory killing of carnivores and other wildlife (Liberg et al., 2012; Kahler and Gore, 2015; Solomon et al., 2015). When power struggles related to carnivore conservation become part of broader social conflicts (e.g., related to land tenure, recourse allocation), conflicts can turn violent or destructive (Raik and Decker, 2007). Importantly, these social conflicts do not occur in isolation; Carter et al. (2017) discussed the coupling of social conflicts over conflict with biological processes and the emergent properties of such coupling. Without considering the micro and macro level dimensions of the issue, it is nearly impossible for an accurate picture of the problem and solution set to be made (Gore et al., 2007).

Implications of bifurcating stakeholders

Ignoring nuances in stakeholder identities and policy positions can result in diverse implications. First, oversimplifying the diversity of policy perspectives reduces the number of program options and the possibility those options will be adopted as efficient and effective policies. For example, when a decision-maker such as a wildlife manager moves forward in a decision-making process with the assumption that livestock ranchers' position is dominated by care for livelihood preservation, they essentially eliminate the possibility that ranchers might also hold values for animal welfare which could be applied to encourage non-lethal tactics for preventing carnivore attacks. Indeed, ranchers work closely with domestic and wild animals on a daily basis, and may defend their actions and positions based on care for the animals in their charge. Dismissing ranchers' concerns for animal lives unnecessarily reduces the suite of strategies available.

The Blackfoot Challenge in Montana is a conservation partnership designed to promote natural resource stewardship that benefits the Blackfoot watershed. Directed by a volunteer board of private landowners, federal and state land managers, and local government officials, the program follows a consensus-based model and works on topics including forestry, wildlife, conservation, water, weeds, and education. It aims to foster more cohesive rural communities through cooperative conservation and members' shared identity as stewards of the watershed and its inhabitants. Grizzly bears, wolves, elk, and other wildlife live within the Blackfoot Watershed. One of the program's activities involves grizzly bear conservation. The program works with ranchers in particular to help them proactively implement non-lethal management and prevent wolf and grizzly bear attacks on livestock (e.g., installing electric fences to protect calving areas and beehives, guarding herds with full- and part-time range riders or guard dogs). According to program metrics, grizzly bear conflicts with livestock deceased 93% since 2003. The collaboration would never have come to fruition and the success metrics never achieved if those involved stood by dichotomous assumptions about each other. Ranchers proved to be open to new methods of mitigating human–bear conflict because they saw the value of carnivores on their landscapes. Conservationists proved to be open-minded to working with a relatively new group of partners and finding pragmatic solutions. Success was achieved in part because collective work toward a common cause increased the salience of a shared identity among diverse stakeholders in the Blackfoot Watershed (Ashforth and Johnson, 2001; Giannakakis and Fritsche, 2011).

Measuring social identities

Recognizing the heterogeneity of stakeholder identities and associated values and perceptions may be a critical key to minimizing ingroup bias, addressing the roots of conflict, and overcoming controversy in carnivore conservation. Including social identities along with the more commonly measured sociodemographic characteristics may strengthen predictive models of intergroup conflict in conservation (Naughton-Treves et al., 2003; Lute et al., 2016; Manfredo, 2008). By going beyond standardized sociodemographic variables that are known by social scientists to predict public opinion to a varying degree, such as education level, political affiliation, and employment status (e.g., Bright et al., 2000), researchers can measure social identity as well. Doing so, research enhances understanding of *why* relevant sociocultural factors explain why a certain group of people so strongly conflicts with another.

Oftentimes, identities are quantified to aid public engagement because, through socialization within a cultural group, identity reflects deeply held, value-laden perceptions that ultimately influence behaviors towards wildlife (Kaltenborn et al., 1999; Naughton-Treves et al., 2003). Strength of group identification increases positive emotions for members who act in accordance with group norms, which may in turn strengthen group identification further (Christensen et al., 2004). This positive catalysis may reflect some of the difficulty in voicing concern with regard to ingroup positions. Along with insight about norms and social pressures, utilizing SIT may inform understanding of influences on behavioral intentions (Christensen et al., 2004) and best strategies for public involvement and communication in wildlife decision-making. Strategies that incorporate SIT in public participation include disseminating information through trusted thought leaders within an ingroup rather than from an untrusted outgroup member or removing signals that display differing identities (i.e., using neutral language, not wearing a certain uniform or even passing out t-shirts to all participants; Ford et al., 2012).

Social identity theory explores how one's self concept relates to group affiliation (Tajfel and Turner, 1979; Tajfel, 1982). According to SIT, individuals locate ingroups consisting of like-minded individuals through self-categorization, which occurs when an individual enters a situation they believe relevant to a certain social group for which s/he is a member (Tajfel et al., 1971; Tajfel and Turner, 1979). The individual views himself or herself as a representative of that group and acts according to group social norms (Jetten et al., 1996; Christensen et al., 2004). Ingroups are cohesive because of a shared desire for positive social identity (i.e., high self-esteem), which is attained by comparisons of their ingroup to relevant outgroups (Labianca et al., 1998; Brown, 2000). Comparisons that reveal perceived inequalities in status (e.g., based on socioeconomic levels, power) result in intergroup competition and ingroup bias (Sherif, 1967; Labianca et al., 1998).

Identity is measured by group salience (i.e., the extent to which an individual finds groups membership salient; Jetten et al., 2006), which consists of accessibility (i.e., how readily evident a social categorization is over time and across contexts) and fit (i.e., the extent to which similarities and differences among people are attributable to social categorization; Hogg and Reid, 2006). Identity can be tied to relationships with specific group members (e.g., "my father taught me to be a good hunter") or to a broad affiliation with the group as a whole (e.g., "I consider myself a representative of the hunting community"; Karasawa, 1991). All these measures address degree of identification (Perreault and Bourhis, 1998), and make up cognitive (as opposed to emotional) assessments of identity (e.g., cohesiveness, commitment, importance, similarity; Hogg and Reid, 2006). But the emotional quality of identity also applies (Perreault and Bourhis, 1998) and consists of affective components such as happiness, liking, and sense of fitting in (Karasawa, 1991). Examples of how to measure identity and the various relevant concepts are summarized in Annex 11.1.

Methodological considerations for identity research

Whether one chooses qualitative or quantitative measures of identity depend on what is already known about the identities involved in the social conflict in question. Qualitative inquiry is often best for initial exploratory phases of inquiry where relevant identities are not well understood or the research context is novel, and has the advantage of allowing stakeholders to self-identify. On the other hand, quantitative inquiry may be advantageous when there is more knowledge about the context of the conflict and actors involved. Quantitative measures in the form of internet surveys provide multiple benefits and reduced transaction costs. For example, internet surveys help simplify data sampling and collection (e.g., no printing hard copies, affixing stamps and mailing paper, manual recording of respondents and nonrespondents) and can increase sample size by extending reach, which may be necessary to precisely measure heterogeneity among diverse groups.

One of the challenges of using a structured, close-ended survey for measuring identity is that it requires 1) knowing all germane identity groups and 2) deciding how specific to be in outlining borders of group membership. For example, important differences may occur within hunting groups (e.g., deer versus bear hunters) or between animal rights, animal welfare, and other wildlife advocates. A survey item that asks a respondent to self-identify as only one type of "hunter" may fail to capture important differences among hunters' motivations, for example, thus compromising the study's ability to predict and understand intergroup conflict over what is and is not an appropriate method or manner of take. Specificity may be useful and important to measure, but it can also be impractical to include a long laundry list of possible groups and affiliations. Researchers should exercise care and acknowledge important differences in groups to avoid cultural insensitivity and indicating ignorance of the groups being studied. Such differences may be revealed through thorough literature review for more studied conflicts. For novel or understudied cases, exploratory (e.g., qualitative or mixed methods) research can be carefully designed to explore salient identities of involved stakeholders. For instance, treating all tribal and First Nation groups as the same group would not do justice to or allow capture of the varied perspectives they hold in regard to natural resources (Shelley et al., 2011). While some tribes consider the wolf a brother who shares their fate, others are open to hunting wolves (Johnson, 2013). Assuming both groups share a common identity would be a mistake.

Once relevant groups and the appropriate level of specificity have been identified, researchers can assess heterogeneity in stakeholder identity by measuring one or more of the identity concepts in Annex 11.1 and use standard statistical techniques to explore how they relate to perceptions, norms and behaviors relevant to the case study or conservation challenge. Convergence among identity groups indicates, at a minimum, common ground and a launching point for collaborative decision-making or conflict facilitation. On the other hand, if degree of identification with a particular group strongly predicts a high perception of risk related to black bears, for example, then specific interventions can be aimed at that group (Gore et al., 2007; Slagle et al., 2013).

The interplay between identity and group versus personal norms may be particularly fruitful areas to explore. Group and personal norms should be first treated as independents constructs, although analysis may reveal that they correlate. Therefore, we suggest considering the interplay between the two kinds of norms and exploring whether they align. If personal and group norms align, this may indicate strong identification with the group; this can be measured within the individual or at the group level. If personal and group norms are somewhat at odds, within-group heterogeneity may be a place for intervention. Consider, for example, a hunting organization that wants to partner with an environmental group to conserve large carnivore habitat. But the environmental group has a policy (i.e., indication of a group norm) that they do not support any form of hunting (i.e., with guns, traps, poisons, aircraft) and is reluctant to partner with the hunting organization because of their ascription to the norm. If thought leaders (i.e., influential and authoritative members) of the group hold personal norms that hunting prey species for food is acceptable, they may be able to convince the group to ease the policy around hunting and collaborate with what was once considered an opposing group (see Triezenberg et al., 2011). Additionally, such diversity in intragroup personal norms, and the behaviors they may influence, can indicate areas of disagreement (perhaps healthy level, because disagreement can be positive and progressive) within a stakeholder group or evolution in the group's thinking or in the thought leaders' thinking (Triezenberg, 2010). Differing personal and group norms may also present opportunities to shift group norms in support of a conservation goal. For example, if a key thought leader in a woolgrowers' association holds personal norms that support environmental stewardship, s/he may be able to influence the entire association's approach to nonlethal methods to reduce risks associated with coyote attacks on sheep (e.g., stress, injury, death).

Importantly, a single individual can, and likely usually does, hold "multiple identities" or express a particular identity over another during a decision-making process. Indeed, everyone identifies with various groups, such as a parent, sports fan, movie lover, or comic book nerd, depending on context. A stakeholder supporting a particular carnivore-related policy may come into a decision-making process as a representative of a local Humane Society chapter, but also identify as an outdoor recreationalist. Assuming this individual is a "prototypical" animal welfare advocate and failing to recognize multiple identities in the issue can undermine, and even counteract, the ability to understand the motivations and perceptions influencing stakeholder behavior and decision-making. Sometimes varying or contested definitions of terms can lead to confusion in measuring identity. For instance, "coexistence" could be defined as land sparing (i.e., allowing carnivores to exist in protected areas) or land sharing (i.e., people and carnivores living on the same landscapes; Carter et al., 2012; Carter and Linnell, 2016). Even the word "conservation" can be, and usually is, defined differently; some equate it with wise use while others consider it synonymous with environmentalism and protectionism (Lute and Gore, 2014a, 2014b). Therefore, it often useful to 1) measure an individual's degree of identification with all relevant identities that are clearly defined and 2) ask them to choose the most important group. Each measure can answer different and complementary questions about the role of identity in the issue at hand.

Leveraging social identity for more effective conservation

Understanding an individual's identity is a critical first step in relating to anyone, and relating to someone is a requisite for effective wildlife conservation policy outcomes. Identity helps us understand where a person is coming from, how they may view the world, and what motivates them to participate or have an interest in carnivore conservation or management. Although identity provides us with a quick and easy heuristic for relating to someone, when taken too far and the individual is seen only as a representative of the relevant ingroup, the heuristic leads to stereotypes. With this caution in mind, an analysis of social identity can be a useful step to understand human dimensions of environmental controversies or facilitate community entry prior to launching research or community conservation projects. Understanding the role of identity in wildlife conservation can also aid in objectives of increasing inclusion and diversity.

As public participatory processes and community-based conservation initiatives become increasingly common in wildlife management (Bright et al., 2000; Berkes, 2004), diverse stakeholder interests will become involved and can be considered in decision-making. Social identity may serve as one tool to help facilitate collaboration among these many groups by providing a framework to recognize, understand, and respect diversity at group and individual levels. A first step in recognizing diversity is recognizing the assumptions and stereotypes into which one can slip. Ingroup bias may come into play in many public participatory processes. For example, qualitative research on stakeholders involved in public participation regarding wolf management decisions in Michigan indicated that ingroup bias was at play and interviewed stakeholders held common assumptions about each other (Lute, 2014; Lute and Gore, 2014b). Stakeholders identifying with a wise use conservation model (similar to – or even synonymous with – the conservationist definition used in Annex 11.1) believed that, in general, those opposed to some form of hunting were opposed to all forms (e.g., trapping, hunting with hounds). Yet stakeholders opposed to wolf hunting in Michigan focused their social identity on particular aspects of the wolf hunting policy decision making process that they felt were not justified; for example, killing carnivores or hunting for recreation or trophies. Such individuals were often supportive of hunting, in various forms, for sustenance, a reason many Americans have indicated as acceptable for taking animal life (Lute and Gore, 2014b).

Similar assumptions can occur across carnivore management groups and contexts. Animal rights or welfare advocates often assume that proponents of hunting did not weigh moral considerations and were prone to angry overreactions to the presence of wolves. Additionally, various stakeholders can make assumptions about wildlife managers (e.g., that they are either pro- or anti-hunting based on perceived favoring of various stakeholder groups). Although there are measurable differences between stakeholder groups conflicting over carnivores (Figure 11.1), ingroup bias may not fairly assess those differences but instead exploits perceived or real inequalities to serve the competitive nature of intergroup conflict. Minimizing ingroup bias is not without challenges, but working to do so may help facilitate maximally collaborative public participatory processes by helping to overcoming stereotypes and creating an open space for equal footing among multiple interests.

Minimizing ingroup bias requires interpersonal interactions among stakeholders, rather than intergroup interactions where individuals are viewed simply as prototypical representatives of their respective groups (Tajfel and Turner, 1979; Hornsey, 2008). Explicitly recognizing and addressing assumptions in socially safe discursive settings may also ease intergroup conflict and aid inclusive decision-making processes (Labianca et al., 1998; Kahan, 2010; Sponarski et al., 2013). Highlighting diversity within identity groups may facilitate healthy debate and dialogue among ingroup members, as well as increase accurate perceptions between groups. Social scientists have long noted that contact alone is usually not adequate to assuage negative stereotypes and associated conflict (Tyerman and Spencer, 1983; Labianca et al., 1998). Inclusive public participatory processes are ideally carefully designed to facilitate consistent interpersonal interactions (Lute et al., 2017). Research identifying and describing ingroup

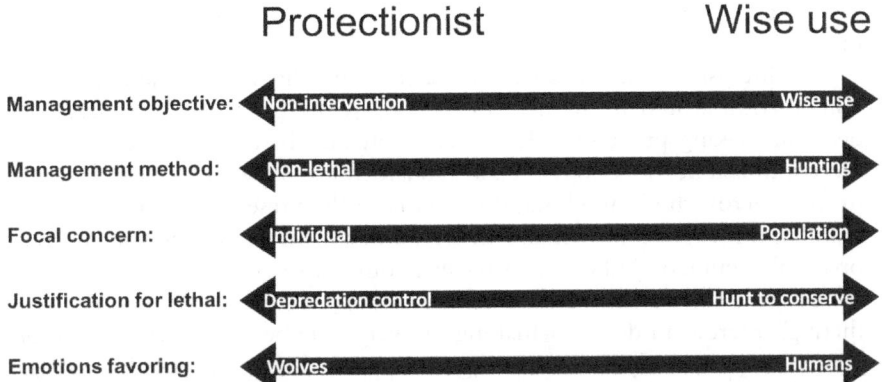

Figure 11.1 Several human dimensions of wolf management studies in Michigan (Lute and Gore, 2014a, 2014b; Lute et al., 2016) revealed individuals and identity groups could fall along a spectrum of protectionist or wise use principles, preferences and emotions. Emergent themes from qualitative studies (Lute and Gore, 2014a, 2014b) first delineated the four dimensions of objective (i.e., goals of management), method (i.e., strategies to attain objective), focal concern (i.e., target of management), and justification for lethal control (i.e., reasons used to determine whether lethal control is appropriate). The fifth dimension of emotions was revealed by quantitative research that found stakeholders' emotions fell on a spectrum of strongly favoring people to strongly favoring wolves (Lute et al., 2016).

bias may be leveraged in such processes to promote collaboration by comparing biased expectations (e.g., "most hunters think this way") to actual data (e.g., "25% of deer hunters think this way and 25% are undecided, while 50% of bow hunters … ").

Focusing on common ground and shared values may counteract multiple negative assumptions and stereotypes related to ingroup bias. For example, in the previously mentioned human dimensions studies of stakeholders involved in Michigan wolf management, diverse stakeholders identified with a shared concept of being a "wolf steward" (Lute and Gore, 2014b) that included taking actions, such as educating others about depredation prevention, to benefit decision-making, wolves, and ecosystems. Public participation processes may benefit from encouraging diverse stakeholders to identify as stewards and enfranchising groups to collaborate on shared responsibilities to nature (Benson, 1998). A "steward" identity might blend both ingroups' ideas regarding human relationships with the natural world. To maximize cooperation, managers may encourage shared values and actions that transcend separate group identity and redefine a common identity as wolf stewards with shared ultimate goals (Sherif, 1967; Tyerman and Spencer, 1983). For example, diverse stakeholders might be primed to consider a shared identity during public meetings or other participatory processes by pre-polling participants on topics on which a manager or meeting facilitator knows they agree (from prior conversations or research).

Any discussion of inclusion and diversity in wildlife management and nature conservation would be remiss not to address the groups disenfranchised by decision-making processes. More than likely an observer of state level wildlife commissioner meetings (i.e., a standard form of wildlife-related decision making) across the United States would note the presence of more men than women, more older individuals than youth, and more whites than minority or tribal members. Although managers, commissioners, and legislators might believe that the transactional costs of including even more (potentially conflicting) interests in decision-making outweigh the benefits, participation processes that proactively include, engage, and satisfy diverse stakeholders may lead to higher quality decisions that serve a broader suite of interests, are in line with public opinion, and avoid the "predator pendulum." Such processes may also result in higher acceptance of final decision outcomes (Webler and Tuler, 2006). To achieve more inclusive and diverse conservation communities, current leaders can actively seek collaboration and input from members of feminist, immigrant, LGBTQ (i.e., lesbian, gay, bisexual, transgender and queer), First Nations/tribal, and other minority communities. Creating environments where respectful debate is encouraged as part of transparent and fair processes is no small task, but essential to engaging and sustaining public participation and input. If current trends in de-democratization – the erosion of democracy often associated with a range of phenomena including declining civil freedoms and dissatisfaction with democratic institutions (Cassani and Pellegata, 2015) – continue (particularly in the United States, select European countries and countries where democracy has been limited or nonexistent), identity politics may increasingly influence public engagement in decision-making. In these cases, understanding of social identity may be even more relevant and critical to anticipating and influencing decision-makers.

Conclusion

In this chapter, we explored how social identity can be measured to enhance its contributions to carnivore conservation. As diverse stakeholders become more involved in decisions regarding carnivores, public participatory processes will remain one tool for encouraging cooperation by recognizing the role of social identities in influencing perceptions and behaviors. Failing to address assumptions about different groups resulting from ingroup bias can result in ineffective communication, entrenched positions, and stymied decision-making. Power struggles between identity groups helps explain the predator pendulum observed in the ever-changing endangered status of gray wolves and potentially grizzly bears in North America (Bruskotter et al., 2016). We discussed approaches and methodological considerations for measuring identity differences among stakeholders and creating more inclusive and diverse conservation communities. The unique insights which managers, decision-makers and public participants can gain from considering social identity has the potential to help overcome conflict regarding not only carnivores but also other contentious environmental issues, such as global climate change, environmental injustice, multi-use landscape management, public and private lands access, and sovereignty of tribal lands.

Annex 11.1 Example measures of general and carnivore-specific identities, perceptions of outgroups, and identity strength, degree, and quality for interviews and surveys

Measure	Questions and follow-up statements	Response choice	Reference
General identity with an ingroup (i.e., individuals affiliating with the same identity)	• What role or group related to [research problem] do you identify with? For example, I strongly identify as a [group] and I feel it is something that defines who I am and how others perceive me. • Why is this role or group important to you? • How strongly do you identify with this group: strongly, moderate or weakly identify? • Are there other roles or groups that you are aware of but do not identify with? If so, what are they and why are they important?	Open-ended response providing qualitative/ categorical measure(s)	Lute and Gore, 2014b
Perceptions related to outgroups	• Who are the groups that do not see eye to eye with your group? • Who is in greatest disagreement? • Why do you believe they disagree with your role or group?	Open-ended response providing qualitative/ categorical measure(s)	Lute and Gore, 2014b
Carnivore relevant identity groups	To what extent do you identify yourself as: • Animal rights advocate • Animal welfare advocate • Conservationist* • Environmentalist** • Farmer • Gun rights advocate • Hunter • Rancher • Wildlife advocate • Other (please specify)	Likert scale (e.g., strongly disagree/ strongly agree)	Lute et al., 2016

Strength of identity	To what extent do you agree with the following statements: • It would be accurate if someone described me as a typical member of my group. • I often acknowledge the fact that I am a member of my group. • I would feel good if I were described as a typical member of my group. • I often refer to my group when I introduce myself.	Likert scale (e.g., strongly disagree/ strongly agree)	Karasawa, 1991
Degree of identity	At this moment, to what extent do you think of yourself as a member of [group] or as an individual?	Likert scale: Mainly think of myself as an individual –Mainly think of myself as a member of [group]	Jetten et al., 2006
Degree of identity	How much do you identify yourself as a member of [group]?	Likert type scale: not at all/very much	Perreault and Bourhis, 1998
Quality of social identity	• How happy do you feel as a member of [group]? • How satisfied do you feel as a member of [group]? • How comfortable do you feel as a member of [group]? • How confident do you feel as a member of [group]? • How much do you like being a member of [group]?	Likert scale: not at all/ very much	Perreault and Bourhis, 1998

* Because different groups may claim this name but intend different meanings, we defined conservationist as "a person who is concerned with conserving natural resources."

** Likewise, we defined an environmentalist as "a person who is concerned with protection of the environment."

References

Ashforth, B. E., and Johnson, S. A. (2001) 'Which hat to wear? The relative salience of multiple identities in organizational contexts', in M. A. Hogg and D. J. Terry (eds) *Social Identity Processes in Organizational Contexts* Psychology Press, Philadelphia.

Benson, D. W. (1998) 'Enfranchise landowners for land and wildlife stewardship: Examples from the Western United States', *Human Dimensions of Wildlife*, vol 3, no 1, pp59–68.

Berkes, F. (2004) 'Rethinking community-based conservation', *Conservation Biology*, vol 18, no 3, pp621–630.

Bright, A. D., Manfredo, M. J., and Fulton, D. C. (2000) 'Segmenting the public: An application of value orientations to wildlife planning in Colorado', *Wildlife Society Bulletin*, vol 28, no 1, pp218–226.

Brown, R. (2000) 'Agenda 2000 social identity theory: Past achievements, current problems and future challenges', *European Journal of Social Psychology*, vol 30, pp634–667.

Bruskotter, J. T. (2013) 'The predator pendulum revisited: Social conflict over wolves and their management in the western United States', *Wildlife Society Bulletin*, vol 37, no 3, pp674–679.

Bruskotter, J. T., Szarek, H., Karns, G., Heeren, A., Toman, E., and Wilson, R. S. (2016) *To List or not to List? Experts' Judgments about Threats to Greater Yellowstone Grizzly Bears*, Ohio State University, Columbus.

Carter, N. H., and Linnell, J. D. C. (2016) 'Co-adaptation is key to coexisting with large carnivores', *Trends in Ecology and Evolution*, vol 31, no 8, pp575–578.

Carter, N. H., López-Bao, J. V., Bruskotter, J. T., Gore, M., Chapron, G., Johnson, A., . . . Treves, A. (2017) 'A conceptual framework for understanding illegal killing of large carnivores', *Ambio*, vol 46, no 3, pp251–264.

Carter, N. H., Shrestha, B. K., Karki, J. B., Man, N., Pradhan, B., and Liu, J. (2012) 'Coexistence between wildlife and humans at fine spatial scales', *Proceedings of the National Academy of Sciences of the United States of America*, vol 109, no 8, pp15360–15365.

Cassani, A., and Pellegata, A. (2015) 'The other way around: Investigating the reverse of de-democratization hypothesis', in *XXIX Annual Conference of the Italian Political Science Association*, Universita della Calabria, Arcavacata di Rende.

Chase, L. C., Schusler, T. M., and Decker, D. J. (2011) 'Innovations in stakeholder involvement: What's the next step?' *Wildlife Society Bulletin*, vol 28, no 1, pp208–217.

Christensen, P. N., Rothgerber, H., Wood, W., and Matz, D. C. (2004) 'Social norms and identity relevance: A motivational approach to normative behavior', *Personality and Social Psychology Bulletin*, vol 30, no 10, pp1295–1309.

Ford, J., O'Hare, D., and Henderson, R. (2012) 'Putting the "We" into teamwork: Effects of priming personal or social identity on flight attendants' perceptions of teamwork and communication', *Human Factors*, vol 55, no 3, pp499–508.

George, K. A., Slagle, K. M., Wilson, R. S., Moeller, S. J., and Bruskotter, J. T. (2016) 'Changes in attitudes toward animals in the United States from 1978 to 2014', *Biological Conservation*, vol 201, pp237–242.

Giannakakis, A. E., and Fritsche, I. (2011) 'Social identities, group norms and threat: On the malleability of ingroup bias', *Personality and Social Psychology Bulletin*, vol 37, no 1, pp82–93.

Gore, M. L., Knuth, B. A., Curtis, P. D., and Shanahan, J. E. (2007) 'Factors influencing risk perception associated with human – Black bear conflict', *Human Dimensions of Wildlife*, vol 12, no 2, pp133–136.

Hogg, M. A., and Reid, S. A. (2006) 'Social identity, self-categorization and the communication of group norms', *Communication Theory*, vol 16, no 1, pp7–30.

Hornsey, M. J. (2008) 'Social identity theory and self-categorization theory: A historical review', *Social and Personality Psychology Compass*, vol 2, no 1, pp204–222.

Inglehart, K. (1977) *The Silent Revolution*, Princeton University Press, Princeton, NJ.

Jetten, J., McAuliffe, B. J., Hornsey, M. J., and Hogg, M. A. (2006) 'Differentiation between and within groups: The influence of individualist and collectivist group norms', *European Journal of Social Psychology*, vol 36, no 6, pp825–843.

Jetten, J., Spears, R., and Manstead, A. S. (1996) 'Intergroup norms and intergroup discrimination: Distinctive self-categorization and social identity effects', *Journal of Personality and Social Psychology*, vol 71, no 6, pp1222–1233.

Johnson, B. (2013) 'On eve of Michigan's first wolf hunt, Saginaw Chippewa Indian Tribe hosts vigil to honor animal called "brother."' *MLive*, 14 November, online.

Kahan, D. (2010) 'Fixing the communications failure', *Nature*, vol 463, no 7279, pp296–297.

Kahler, J. S., and Gore, M. L. (2015) 'Beyond the cooking pot and pocket book: Factors influencing noncompliance with wildlife poaching rules', *International Journal of Comparative and Applied Criminal Justice*, vol 36, no 2, pp103–120.

Kaltenborn, B. P., Bjerke, T., and Vitterso, J. (1999) 'Attitudes toward large carnivores among sheep farmers, wildlife managers and research biologists in Norway', *Human Dimensions of Wildlife*, vol 4, no 3, pp37–41.

Karasawa, M. (1991) 'Toward an assessment of social identity: The structure of group', *British Journal of Social Psychology*, vol 30, pp293–307.

Kreye, M. M., Pienaar, E. F., and Adams, A. E. (2016) 'The role of community identity in cattlemen response to Florida panther recovery efforts', *Society and Natural Resources*, vol 30, no 1, pp79–94.

Labianca, G., Brass, D. J., and Gray, B. (1998) 'Social networks and perceptions of intergroup conflict: The role of negative relationships and third parties', *The Academy of Management Journal*, vol 41, no 1, pp55–67.

Liberg, O., Chapron, G., Wabakken, P., Pedersen, H. C., Hobbs, N. T., and Sand, H. (2012) 'Shoot, shovel and shut up: Cryptic poaching slows restoration of a large carnivore in Europe', *Proceedings of the Royal Society Society of London Series B*, vol 279, no 1730, pp910–915.

Lichtenfeld, L. L., Trout, C., and Kisimir, E. L. (2015) 'Evidence-based conservation: Predator-proof bomas protect livestock and lions', *Biodiversity and Conservation*, vol 24, no 3, pp483–491.

Lute, M. (2014) *Influence of Morality, Identity and Risk Perception on Conservation of a Recovered Carnivore*, PhD thesis, Michigan State University at East Lansing.

Lute, M., Bump, A., and Gore, M. L. (2014) 'Identity-driven differences in stakeholder concerns about hunting wolves', *PloS One*, vol 9, no 12, e114460.

Lute, M. L., and Attari, S. Z. (2016) 'Public preferences for species conservation: Choosing between lethal control, habitat protection and no action', *Environmental Conservation*, vol 44, no 2, pp139–147.

Lute, M. L., Gillespie, C., Martin, D. R., and Fontaine, J. J. (2017) 'Landowner and practitioner persepctives on private land conservation programs', *Society and Natural Resources*, in press, pp1–14.

Lute, M. L., and Gore, M. L. (2014a) 'Knowledge and power in Michigan wolf management', *Journal of Wildlife Management*, vol 78, no 6, pp1060–1068.

Lute, M. L., and Gore, M. L. (2014b) 'Stewardship as a path to cooperation? Exploring the role of identity in intergroup conflict among Michigan wolf stakeholders', *Human Dimensions of Wildlife*, vol 19, no 3, pp267–279.

Lute, M., Navarrete, C. D., Nelson, M. P., and Gore, M. L. (2016) 'Assessing morals in conservation:The case of human-wolf conflict', *Conservation Biology*, vol 30, no 6, pp1200–1211.

Manfredo, M. (2008) *Who Cares About Wildlife?* Springer, New York.

Marchini, S., and Macdonald, D.W. (2012) 'Predicting ranchers' intention to kill jaguars: Case studies in Amazonia and Pantanal', *Biological Conservation*, vol 147, no 1, pp213–221.

Naughton-Treves, L., Grossberg, R., and Treves, A. (2003) 'Paying for tolerance: Rural citizens' attitudes toward wolf depredation and compensation', *Conservation Biology*, vol 17, no 6, 1500–1511.

Nelson, M. P.,Vucetich, J. A., Paquet, P. C., and Bump, J. K. (2011) 'An inadequate construct?' *The Wildlife Professional*, vol 5, no 2, pp58–60.

Nie, M. (2003a) *Beyond Wolves: The Politics of Wolf Recovery and Management*, University of Minnesota Press, Minneapolis.

Nie, M. (2003b) 'Drivers of natural resource-based political conflict', *Policy Sciences*, vol 36, pp307–341.

Perreault, S., and Bourhis, R.Y. (1998) 'Social identificaiton, interdependence and discrimination', *Group Processes and Intergroup Relations*, vol 1, no 1, pp49–66.

Peterson, M. N., and Nelson, M. P. (2016) 'Why the North American model of wildlife conservation is problematic for modern wildlife management', *Human Dimensions of Wildlife*, vol 22, no 1, pp43–54.

Raik, D. B., and Decker, D. J. (2007) 'A multisector framework for assessing community-based forest management: Lessons from Madagascar', *Ecology and Society*, vol 12, no 1, p14.

Ramp, D., and Bekoff, M. (2015) 'Compassion as a practical and evolved ethic for conservation', *BioScience*, vol 65, no 3, pp323–327.

Shelley,V., Treves, A., and Naughton, L. (2011) 'Attitudes to wolves and wolf policy among Ojibwe tribal members and non-tribal residents of Wisconsin's wolf range', *Human Dimensions of Wildlife*, vol 16, no 6, pp397–413.

Sherif, M. (1967) *Group Conflict and Cooperation:Their Social Psychology*, Routledge and Kegan Paul Ltd, London.

Slagle, K., Zajac, R., Bruskotter, J.,Wilson, R., and Prange, S. (2013) 'Building tolerance for bears: A communications experiment', *The Journal of Wildlife Management*, vol 77, no 4, pp863–869.

Solomon, J. N., Gavin, M. C., and Gore, M. L. (2015) 'Detecting and understanding noncompliance with conservation rules', *Biological Conservation*, vol 189, pp1–4.

Sponarski, C. C., Semeniuk, C., Glikman, J. A., Bath, A. J., and Musiani, M. (2013) 'Heterogeneity among rural resident attitudes toward wolves', *Human Dimensions of Wildlife*, vol 18, no 4, pp239–248.

Tajfel, H. (1982) 'Social psychology of intergroup relations', *Annual Review of Psychology*, vol 33, pp1–39.

Tajfel, H., Billig, M. G., Bundy, R. P., and Flament, C. (1971) 'Social categorization and intergroup behaviour', *European Journal of Social Psychology*, vol 1, no 2, pp149–178.

Tajfel, H., and Turner, J. C. (1979) 'An integrative theory of intergroup conflict', in W. G. Austin and S. Worchel (eds) *The Social Psychology of Intergroup Relations*, Brooks/Cole, Monterey, CA.

Triezenberg, H. A. (2010) *Social Networks and Collective Actions Among Wildlife Management Stakeholders: Insights from Furbearer Trapping and Waterfowl Hunting Conflicts in New York State*, PhD thesis, Cornell University at Ithaca.

Triezenberg, H. A., Knuth, B. A., and Yuan, Y. C. (2011) 'Evolution of public issues in wildlife management: How social networks and issue framing change though time', *Human Dimensions of Wildlife*, vol 16, pp381–396.

Tyerman, A., and Spencer, C. (1983) 'A critical test of the Sherifs' Robber's cave experiments: Intergroup competition and cooperation between groups of well-acquainted individuals', *Small Group Research*, vol 14, no 4, pp515–531.

Webler, T., and Tuler, S. (2006) 'Four perspectives on public participation process in environmental assessment and decision making: Combined results from 10 case studies', *Policy Studies Journal*, vol 34, no 4, pp699–722.

Part III

Decision-making, stakeholder involvement, and policy in large carnivore conservation and management

12 Inappropriate consideration of animal interests in predator management

Towards a comprehensive moral code

Francisco J. Santiago-Ávila, William S. Lynn,
and Adrian Treves

Introduction

Wildlife managers frequently intervene in the lives of nonhuman animals ('animals' hereafter). For example, managers may attempt to condition, relocate or kill animals if these damage human property. Such interventions are usually averred to be guided strictly by the facts of science. However, facts and science, without values, are unable to decide how we ought to coexist with animals (Lynn, 2010; Nelson et al., 2011; Nelson and Vucetich, 2012). Whether our interventions in animals' lives are ethically appropriate is a value judgment and a question for ethics.

In this chapter, we provide a brief introduction to ethics and its role in establishing and fostering a moral community, which we define. We proceed to review the ethical and scientific case for including individual animals in the moral community (a.k.a., 'animal ethics', 'nature ethics', 'interspecies ethics'), which contends that dismissing individual animal interests is arbitrary and ethically inconsistent. With advances in environmental sciences highlighting our interdependence with other animals, and the harmful effects we have on them, have come advances in ethology confirming their commonly appreciated emotional and cognitive abilities. Individual animals have their own lives and interests that can be helped or harmed by human action. This is the root reason why carnivore management is unavoidably a matter of ethics.

We go on to explain why the ethical consideration of carnivores is crucial for ethical wildlife management. We examine why dismissal of animal ethics or ethical arguments of any kind inappropriately dismisses individual animals' moral standing, which can culminate in a version of 'might makes right' asserted by a minority of humans who claim paramount interests. We show how various institutions and actors at different levels of government are primarily responsible for deciding the scale and scope of lethal interventions into the lives of carnivores. Here, we make use of the formal and systematic analysis provided by an ethical framework (Horner, 2003) to perform an ethical and legal examination of the legal documents intended to guide and regulate decisions by the state of Wisconsin to kill grey wolves (*Canis lupus*) during the period in which federal protections for the species were removed (2012–2014).

We provide evidence that current laws and regulations lack appropriate consideration of animal ethics when intervening in the lives of grey wolves. Further, we discuss why wildlife management may be more prone than other applied fields to cater to powerful interest groups that fail to acknowledge the moral value of individual carnivores. We conclude by articulating a vision for wildlife management in the 21st century that explicitly injects ethics into carnivore policy and management.

Ethics, science, and predator management

Ethics is defined as 'the branch of knowledge that deals with moral principles' (OED, https://en.oxforddictionaries.com) that govern a person's conduct or behaviour. That is, ethics asks 'how we ought to live'. This question has epitomized ethics for millennia since its utterance by Socrates as recorded in Plato's *Republic*. Ethics evolves out of the human concern with what is right or wrong, good or bad, just or unjust, including what ends we should seek and what means are appropriate for pursuing them (Lynn, 1998a). In this evolving process, humans develop ethical arguments supported by reason and evidence into broadly accepted moral principles for analyzing and revising our conduct towards other living beings (Lynn, 1998a; Lynn, 2010).

Moral principles are indispensable for community living, serving as guidelines that foster not only our own personal good, but the good of the community as well. This provides a basis for moral cooperation and social living. One way of thinking about ethical principles is that they are truths considered ultimate (usually cannot be overridden), universal (apply to everyone within the community), impartial (treat everyone equitably), and other-regarding (the good of others is placed alongside self-interest) (Horner, 2003). Another way is to consider them rules of thumb that assist in revealing moral issues and how to address them. In this chapter, we assume the latter perspective, and view ethics as a set of situationally applied moral insights. We follow this ethical approach because moral conflict is inescapably rooted in specific situations. Which rules of thumb we use to guide us will differ based on the context and character of specific cases, providing for greater flexibility in considering the ethical principles and interests involved (Lynn, 1998a; Jonsen and Toulmin, 1988; Midgley, 1993).

When it comes to public policies like the management of carnivores, ethics and science are complementary. Science helps establish the empirical reality of the problems we face, and can provide options in addressing those problems. Ethics helps reveal the moral values at stake, and what options we are justified in choosing. Thus, lacking either science or ethics may result in a lack of relevant information or moral insight, respectively, culminating in the legal sanctioning of harmful yet unnecessary behaviour. In this sense, science and ethics are twin stars that can help guide carnivore policy and management into making better decisions; that is, decisions that are both ethically and scientifically sound (Lynn, 2006; Callahan and Jennings, 1983; Shrader-Frechette and McCoy, 1994). Science and ethics are thus powerful tools when combined and (if only used) for helping people, animals, and nature flourish.

Sentience and the moral community

Even though humans might be the only animals capable of engaging in ethics as a philosophical exercise (Lynn, 2007; Lewis et al., 2016), they are not the only beings to whom ethics apply. Peer-reviewed scientific studies from various disciplines – such as ethology, neuropsychology, and evolutionary biology – have gathered decades of evidence confirming that many animals are sentient and sapient, aware and self-aware beings with rich emotional and cognitive lives. Insofar as we know, qualities such as awareness may not apply to all animals equally (e.g., certain arthropods), but does characterize animals with more complex neuroanatomical structures, including all mammals (e.g., wolves) and birds (Low et al., 2012). And while there is great variability among species, and real differences between individuals of the same species, the idea that humans are unique as feeling or thinking creatures is wrong-headed (Bekoff and Goodall, 2004). The scientific evidence has invited scientists and ethicists to question and reject the inherent superiority of humans over other animals (Midgley, 1983; Peterson, 2013; Vucetich et al., 2015; Batavia and Nelson, 2016). This should not be taken to mean that substituting a rigid moral hierarchy focused on humans with one focused on other sentient animals is our point. It is rather that to understand the moral values at stake and that to make ethical distinctions, the particularities of individuals and species need to be considered.

It is for these reasons that ethicists increasingly argue that animals also value their lives, and have an interest in their own well-being. The *sentience* of animals (i.e., 'their ability to perceive or feel things' [OED]) is one of the main reasons – and, some argue, sufficient reason – to recognize the moral considerability of animals (Bentham, 1789; Singer, 1975; Feinberg, 1981; Midgley, 1983; Regan, 1987; Francione, 2009; Peterson, 2013). Sentient creatures have preferences and needs, and can experience subjective states such as stress, fear, and joy. In other words, they have interests – not the same, but akin to our own – in terms of freedom from avoidable or unjustified harm and death. Evaluating the avoidability or justification for harming another being would hinge on the weighing and consideration of all interests involved.

Philosopher Peter Singer argues that sentient beings deserve equal consideration of their interests when intervening in their lives (Midgley, 1983; Singer, 1993; Lynn, 1998a). This does not mean equal rights, a misunderstanding of terms by many observers. The moral principle of equal consideration refers to the equitable and explicit acknowledgement of all affected interests when determining the appropriateness of an action. Neither Singer nor we mean equal rights or equal treatment for every species or individual; "what this principle does require is for humans to give due consideration to the well-being of other creatures, and to do so without prejudice" (Lynn, 1998a, p. 291).

To reinforce the distinction between considering the interests of animals and not conflating this with animal rights theories or expectations of equal treatment among all species, we speak hereafter in terms of *equitable consideration*. This concept has both process-oriented and outcome-oriented dimensions. In terms of process, it argues for the fair consideration of animal interests in policy or management decisions that will impact their well-being in the world. Examples would include, but not be restricted to, population management through lethal control. In terms of outcomes, it argues that policy or management decisions must do more than consider animal interests as a pro forma matter of administrative process, only to subsequently dismiss them. Rather, the weight of reason and evidence for animal sentience is overwhelming (Bekoff et al., 2002; Bekoff and Goodall, 2004), and the outcomes of policy or management decisions should act upon this whenever appropriate.

Equitable consideration leads to similar treatment when interests are similar, but allows for differences in treatments when interests differ, or when the specific moral problems demand differential treatment. For example, both people and predators have a direct interest in avoiding harm. Only people have a direct interest in political participation. While predators might benefit from certain public policies and practices, they are not the kinds of beings for whom engagement in politics is applicable, because they do not have the capabilities to engage in it. So, while it is necessary to consider the interest of predators when human actions may harm or affect them, including in the political arena, only people can directly participate in the political deliberation necessary to set forth policies and management practices that do indeed consider these interests. Context plays a crucial role, and close attention to the type of beings and interests involved are part of the circumstances to which we apply moral principles (Lynn, 1998a).

As Mary Midgley (1983, p. 90) points out, the experience of other beings can be: "sufficiently like our own to bring into play the Golden Rule – 'treat others as you would wish them to treat you'". This punctuated continuity of interests between humans and carnivores grants them membership in what ethics calls a community of moral concern, or *moral community*. This has strong implications for predator management. For even if the sentience of carnivores is different from our own, we are obligated to consider their interests and well-being when it is impacted by human actions. Such actions include direct and indirect harms to their well-being, such as hunting, trapping, poisoning, habitat loss or degradation, and global warming. Unregulated and uncontrolled versions of these are contributing to the sixth great extinction of wildlife the world over (Kolbert, 2014) and underscore this point.

Although sentience by itself could be considered the ethically relevant trait for extending equal consideration to individual animals, our various relationships with animals add additional reasons for their moral consideration. Animals are an integral component of society and the environment. A purely 'human community' is a fiction; instead, humans live within a 'mixed-community' of species (Midgley, 1983). Human civilization has been built on the care and exploitation of individual animals (Peterson, 2013). Indeed, our instrumental use of animals has been made possible because of animal sentience, a prerequisite for the interspecies communication that facilitated domestication (Midgley, 1983). The food, fibre, labour, and companionship of animals are not free of moral weight.

Moreover, wild animals contribute to the health of the environment and the provision of ecosystem services indispensable for human well-being (Leopold, 1949; Favre, 1978; Callicott, 1980; Midgley, 1983; Lynn, 2007; Wallach et al., 2015). Our dependence on ecosystem processes such as pollination, seed dispersal, predation, scavenging, and water filtration, among others, are mediated through animals.

Altogether then, our species is neither ethically, socially, nor environmentally isolated. Instead, we have always lived our lives in deep relation to other species. This is nowhere truer than with respect to the only wild carnivore to have undergone domestication (wolves), and its domesticated descendants (dogs).

Ethical worldviews about people, animals, and nature

Thinking about animals as both sentient beings and contributors to ecological processes has resulted in questions of whether the interests of animals and the integrity of ecosystems can be aligned.

Carnivore policy and management is dominated by two worldviews of ethics. These worldviews are 'big-picture' approaches to thinking about humans and their place alongside the community of life on planet Earth.

The first worldview is *anthropocentrism*. Anthropocentrism asserts that only human beings have moral value and need to be considered from a moral point of view. Everything that is not human (e.g., the animals and ecosystems referred to as nature) is only of instrumental use for human beings. In their relation to the environment, humans are ends, while animals and natural objects are means to those ends. Humans, of course, can still have instrumental value to others, but their prevailing value is as ends in themselves. Anthropocentrism is the ideology behind the early conservation movement of Gifford Pinchot with his emphasis on the wise use of biotic and abiotic resources for the greater good of the nation and for future generations of citizens.

The second ideology is ecocentric holism, or *ecocentrism*. This is an ethical outlook that believes the needs of individual animals can be ignored or sacrificed if a population or species is protected. One touchstone for ecocentric holism is Aldo Leopold's 'land ethic' (1949; but see Millstein, 2018). The land ethic is mainly concerned with the ecosystem health of the 'land community'. It is based on two ideas: (1) nature has intrinsic value (meaning its existence has value for its own sake, irrespective of what it can provide for humans), and (2) this value lies in ecological aggregates or wholes (i.e.: species, ecosystems) rather than individual beings (Peterson, 2013).

Ecocentrism vies with anthropocentrism to be the dominant voice in carnivore management. To its credit, it has contributed immensely to global efforts to mitigate environmental degradation and the loss of biodiversity. Yet it has also justified ignoring the equitable consideration of interests that the subjective lives of animals makes mandatory (Peterson, 2013). For example, although unnecessary in most developed countries, ecocentrists may support scientifically managed subsistence hunting when it does not harm species or ecosystem integrity. Another example of ecocentrism is the prohibition on substantial impairment of environmental assets under the U.S. public trust doctrine, without a concomitant consideration of individual organisms that in part make up those biodiversity assets (Treves et al., 2015). Both anthropocentrism and ecocentrism may allow for relative consideration of certain animal interests through 'humane' treatment principles, but there is little attempt, if any, to question their killing if it serves a human interest.

The opposite to ecocentrism is frequently framed as biocentric individualism or *biocentrism* (Peterson, 2013). Biocentrists believe ecocentrism errs in its approach to parts and wholes, as only individual animals (human or otherwise) are thought morally considerable. Ecosystems, as wholistic entities, do not have moral value *per se*. Rather, they are the living context for morally valuable sentient lives. Our duty towards preservation of ecosystem integrity stems from ecosystems allowing wild organisms to flourish (Taylor, 2003). As a competing viewpoint to ecocentrism, biocentrism has become a dominant position of the animal rights movement. Conservationists following ecocentrism and animal rightists following biocentrism are often on polar opposite sides of management issues involving predators. In terms of their ideologies, the reasons for this are clear. For example, in Washington, USA, conservation organizations are evenly split on authorizing the killing of members of wolf packs involved in livestock depredations on public land. Meanwhile, animal advocates have publicly denounced some of these conservation organizations for sanctioning the killing of wolves as a form of subsidy to an unnecessary practice based on animal exploitation (breeding domestic ungulates).

The authors, however, do not believe this dichotomy to be helpful. Ecocentrism rightly recognizes ecosystems as ecological entities rooted in a network of interrelationships. Biocentrists rightly recognize individual animals as part of that ecology, many of whom are simultaneously sentient and sapient. Terrestrial predators exemplify this point. As individuals, all are subjective beings, members of our moral community, and deserving of ethical consideration. As predators, each of these individuals is also a functional unit of ecosystem processes and contributes to the ecological health of the biotic community.

To draw hard and fast lines between the parts and wholes, then, seems arbitrary to us, and creates a false dichotomy between *a priori* locations of moral value in aggregates or individuals. It is for that reason that we adopt a geocentrist approach. *Geocentrism* extends moral considerability to both individuals and ecological communities, recognizing that both have a well-being that needs to be explicitly considered at the same time. As a practical discipline, ethics should be rooted in context. Rather than arguing for a main location of moral value, geocentrism argues for a contextual accounting of the various overlapping sources of moral value (i.e., individual subjectivity as well as ecological and social relationships). It regards all animals as ends in themselves, yet acknowledges their intrinsic and extrinsic values (Lynn, 1998a).

An instructive example is that of predation. When wolves hunt deer, both predator and prey are manifestly sentient species. Individual wolves and individual deer matter from a moral point of view. Yet this does not mean it is wrong for the wolf to kill deer to survive or thrive. Predation is an ecological process necessary for life on earth. It would be irrational and unscientific to simply declare it immoral. Rather, the wolf killing the deer, and predation in general, exemplifies what is termed a 'sad good' (Lynn, 2012). The death of the individual deer is sad as a distinct individual has been extinguished. It is good, however, for the wolf or wolves that consume the deer, as well as for the ecological dynamics of trophic systems of which both the wolf and deer are a part. The case is similar when we talk about humans as the predator if this killing, as in the case of the wolf, is necessary for subsistence (Lynn, 2017). However, not all human motives for killing may override the vital claims of animals. Some kind of special urgency of the human claim in question (i.e., subsistence) should be established for the killing (as well as the treatment) to be ethically justified (Midgley, 1983).

We note that geocentrism should not be considered a superseding concept in a decision hierarchy that always or mostly justifies the lethal management of wildlife in pursuit of ecosystem health or function. Nor is it a typology providing categorical answers to contextual questions, or an imperialist theory seeking supremacy in self-serving academic debates (Lynn, 2002). It is rather a value-paradigm that seeks to untangle the complex ethical presuppositions and implications of varied worldviews (Lynn, 1998b).

Aligned with the pluralistic and interpretive ethics that gave it birth (Toulmin, 1950; Midgley, 1993; Weston, 2006), geocentrism does not claim to be uniquely true but rather helpfully insightful (Lynn, 2006). It is thus not concerned with being correct to the exclusion of insights from other paradigms about the intrinsic value of people, animals, or nature, respectively. Rather, it seeks to appreciate what each of these is right about, integrate their insights into a distinctive conceptual tool, and deploy this tool to better understand the nuances of ethical reasoning about predators and their management.

Law and ethics

Ethical judgments about the moral value and consideration of animals pervade policies about and the management of carnivores. Statutory laws, agency regulations, executive actions, and judicial decisions (collectively, the 'law') frequently focus on actions impacting individuals or groups in society, as well as various elements of the built and natural environment. These actions potentially have good or bad consequences for those they impact, and are thus legitimate topics of ethical scrutiny. Moreover, both the instruments and practice of the law are intrinsically bound up with ethics, as they encode a variety of value assumptions that recognize some (but perhaps not others) as part of the moral community served by legal and political institutions. In these senses, then, legal documents are moral documents – documents that matter in terms of the ethical positions they assume, convey, or impose (Caldwell and Shrader-Frechette, 1992; Beatley, 1994).

When ethical arguments have not been made explicit in them, one should not conclude that there are no ethical concerns. Nor should one conclude that those ethical concerns have been well considered but left unwritten because they were obvious. Rather, ethical reasoning should be made manifest and not left latent. This is done by seeking out the 'best' (e.g., accurate, comprehensive) accounts of ethical, legal, or scientific claims through reason and evidence.

Unfortunately, legal instruments relevant to individual carnivores are frequently not explicit about their ethics. Governments and individuals frequently resort to lethal 'management' methods for these individuals and populations when they are perceived to threaten human property or safety (Treves et al., 2016). This resort to lethal management, however, is most frequently predicated on an overt or implicit dismissal of their interests. For example, in the Global North, the costs of managing wild carnivores are relatively minor and attacks on people are vanishingly rare (say, compared to domestic dogs). Nevertheless, lethal management of carnivores is commonly the first intervention, and promoted when humans are not content with carnivore population numbers, the animals are considered a nuisance, hunters perceive competition for game, or based on the hypothesis that lethal management would promote the species' conservation (Treves, 2009).

To illustrate the absence of express ethical reasoning in carnivore management, we use a case study of grey wolf management in Wisconsin, USA. For our evidence, we rely on a close reading of the statutes and regulations that sanction lethal or harmful interventions into the lives of Wisconsin wolves.

Grey wolf extermination and management in the USA

Although grey wolves historically ranged throughout most of North America, the campaign to exterminate them and other large predators started soon after European settlers arrived (Lopez, 1978; Boitani, 2003). Predators were generally regarded by Europeans as pests that reduced game numbers and preyed on livestock. Persecution was widespread and government-sponsored (Lopez, 1978; Boitani, 2003). By the 1930s, the wolf had been eradicated from almost all 48 contiguous states, except for small pockets in Minnesota and Michigan (Bangs and Fritts, 1996; Boitani, 2003).

Recovery of predator populations was only possible after the enactment of the Endangered Species Act (ESA) in 1973. By 1974, the grey wolf was listed as an endangered species in the Great Lakes region (Minnesota, Wisconsin and Michigan), and by 1978 throughout the 48 contiguous states (Boitani, 2003). ESA listing placed the species under temporary federal authority, providing protection from 'take' ("to harass, harm, pursue, hunt, shoot, wound, kill, trap, capture, or collect, or to attempt to engage in any such conduct" according to the US Endangered Species Act [US, 1973]) and critical habitat protection (Treves et al., 2015). Full federal protections remain in place until the population is either 'down-listed' (reclassified recovery status removing certain protections) or 'delisted' (full removal of protections) federally, based on target recovery goals. ESA delisting also entails the transfer of management authority from the federal government to the states.

The Great Lakes wolf population was managed according to the US Fish & Wildlife Service's (USFWS) Eastern Timber Wolf Recovery Plan (Service, 1992). The plan classified the Minnesota population as 'threatened', which allowed for state removal of wolves through lethal management or translocation (16 USC §1531 Sec. 4d permits) but no public hunting or trapping season. The Wisconsin and Michigan populations were classified as 'endangered' (no 'take' except for imminent threats to human safety), because of their much smaller populations.

Likewise, the Wisconsin Department of Natural Resources (WDNR), developed a Wolf Management Plan (WMP) providing similar (to federal) protections from 'take' or lethal management until wolves reached their target recovery goals (DNR, 1999). Following the WMP, wolves were down-listed to state threatened status in 1999 and delisted in 2004, allowing the state to conduct public hunting and trapping seasons if delisted federally (Wydeven et al., 2009).

Simultaneously, the USFWS proceeded with plans to delist regional populations in preparation for delisting the whole species (Bruskotter et al., 2011; Bruskotter, 2013). In Wisconsin and Michigan, federal down-listing to threatened status first occurred in 2003 (Wydeven et al., 2009). Shortly after, in a series of agency decisions and legal battles between 2005 and 2014, the USFWS attempted to remove federal protections for wolves, while federal courts restored them, disagreeing with the USFWS determination that wolf populations were sufficiently recovered or protected to warrant delisting (Treves et al., 2015). Because of these lawsuits, as of December 2014 wolves in the Great Lakes were relisted as 'endangered' in Wisconsin and Michigan, and 'threatened' in Minnesota (HSUS, 2014). However, attempts at delisting the species via legislation are ongoing.

Grey wolf management in Wisconsin, 2012–2014

Despite their status as 'endangered' in Wisconsin as of writing, the species was federally delisted in the winter of 2012 for a period of approximately three years (2012–2014). By 2 April 2012, the state legislature had approved Act 169, authorizing the WDNR to plan for a public wolf hunting and trapping season, sanctioning wolf-killing for the first time since wolf bounties were terminated in 1957. The public hunting and trapping seasons, however, are regulated by the state to ensure the killing is sustainable (i.e., does not affect the viability of the population). The seasons were held from October through February. Hunters and trappers were allowed to kill wolves in hunting zones statewide (except inside Native American reservations) until the zone kill quotas were reached. Methods allowed included authorized firearms, bow and arrow or crossbow, cable restraints or steel-jawed foothold traps, subject to certain restrictions. Use of dogs to track or trail, predator calls, and some baiting were also allowed. Most likely, hounds harass or attack wolves, especially young wolves near dens and rendezvous sites during the summer hound-training period. There are no data on such incidents because hounds were left to run loose far from their owners for kilometres often tracked remotely by telemetry by owners on the nearest road. The evidence that hounds and wolves engaged in deadly confrontations is one-sided with reports of hounds injured or killed (see next paragraph); it is one-sided because injured or killed wolves were never reported. The WDNR went on to hold three wolf hunting and trapping seasons during which hunters and trappers killed 117, 257, and 154 wolves (WDNR, http://dnr.wi.gov/topic/hunt/wolf.html). State officials from the WDNR and supporters of the wolf hunt argued that the hunt would allow for maintaining the wolf population at target levels, bolstering political support for species' conservation and reducing conflicts over safety and property (DNR, 2013; Hogberg et al., 2015).

The state legislature also authorized the WDNR to remove wolves that were causing damage or nuisance (WI Stat §29.885[2]). Most complaints of this sort come from domestic animal breeders who perceive wolves as a threat or that have had domestic animals killed by wolves (depredation). Although depredation(s) can cost domestic animal breeders and the domestic animals certainly would not want it to happen, statistics also show that the amount of depredations is minuscule from an industry perspective, with wolves accounting for only 0.8% of cattle losses in Wisconsin in 2010 (NASS, 2011 http://usda.mannlib.cornell.edu/usda/current/CattDeath/CattDeath-05-12-2011.pdf). In 2016, with a wolf population of approximately 866–897, Wisconsin had a total of 52 depredations and six threats on domestic and farm animals (Wiedenhoeft et al., 2016). Complaints also come from people whose hounds or pets have been attacked by wolves. Between 2015–2016, wolves were involved in nine attacks on dogs outside of hunting situations, and killed 18 dogs while these were engaged in hunting activities (Wiedenhoeft et al., 2016). Besides these private concerns, there have also been public concerns about wolves threatening human safety, but at the time of writing no attacks have been confirmed in Wisconsin.

To carry out wolf removals, the WDNR reinstated a long-standing cooperative agreement with a federal agency within the Department of Agriculture named Wildlife Services (WS), charged with "providing federal leadership in managing conflicts with wildlife" (USDA-APHIS Wildlife Services, 2009, p. 1). WS would investigate complaints and determine whether to authorize the removal of individual wolves, following procedures outlined in the state WMP. The WMP calls for depredation control activities to "focus on preventive methods and mitigation" (DNR, 1999, p. 24), including non-lethal methods (Willging and Wydeven, 1997). In cases of confirmed and probable depredations (based on depredation verification procedures), the local WDNR wildlife biologist, the WDNR Regional Wildlife Expert, and WS staff determine the appropriate management activity (DNR, 1999) by analyzing the following criteria: (1) there are confirmed losses at the site, and (2) the producer signed a depredation management plan for the property and follows recommended abatement and husbandry recommendations. Other factors – such as location of depredation in relation to known wolves or wolf packs, severity of damage, and type and size of farm operation – seem to be considered (Willging and Wydeven, 1997), but no measurement criteria for any of these are included in the WMP. Thus, if the previous two criteria are met, the WMP provides the WS Depredation Specialist with discretion to recommend and implement 'euthanasia', contingent on WDNR approval (DNR, 1999). That approval was not specified in any regulation or policy we could find and therefore represents another individual's discretion, we surmise. Landowner lethal management was also allowed "by WDNR permit after Federal delisting has occurred" (DNR, 1999, p. 20). Once the population reached target levels, "proactive depredation control can be authorized" (DNR, 2007, p. 6). Proactive control involves the implementation of interventions (lethal or non-lethal) prior to the occurrence of any incident with the objective of mitigating or preventing them.

Ethical considerations in grey wolf management

We reviewed the text of laws, regulations, and related documents relevant to wolf management in Wisconsin to evaluate the appropriateness and thoroughness of the clearest and most ethical arguments explicit in them, if any. We evaluated if these documents accurately acknowledge individual animals as members of the moral community by appropriately considering their interests, in addition to the interests of humans and ecological wholes. We conducted an ethical examination of statements expressing the main concerns of each agency regarding wildlife management, as well as specific statutes and regulations relevant to grey wolf management. The passages included (Annex 12.1) were identified by reviewing the texts looking for sections revealing the types of interests considered behind certain interventions or views relating to wildlife or their treatment, be these human, ecological wholes, or individual animals. These passages contain all the statements that suggest even remotely that the authors considered the interests of other individual animals, particularly wolves.

Our analysis focuses on providing evidence of what interests (human, ecological wholes, and individual animals) are being considered in laws and regulations, and to what extent (partially or equitably considered). By accounting for these interests, we address anthropocentrism, ecocentrism, biocentrism and, if all were equitably considered, geocentrism.

Annex 12.1 presents the ethical concerns stated in all nine federal and state documents governing wolf management in Wisconsin, as established by federal and state agencies involved in management, along with empirical observations detailing what interests each document made explicit in their ethical justifications (corresponding passages in italics). We classified these interests as focusing on instrumental (human) concerns, ecological wholes or individual animals, allowing for overlap based on the interests made explicit in the document. For each interest category, we awarded a rank of '2' if those interests categories were explicitly and appropriately considered, as stated in each document. We awarded a rank of '1' to an interest category if the document contained explicit yet limited or inappropriate consideration of the interests represented by that category (for example, if a statute considers an organism's desire in freedom from harm, but not desire to continue living, as an interest; see discussion of individual animal interests in the next paragraphs). We awarded a rank of '0' if we found no explicit consideration of a particular category. For comparison, we provide rank summaries per interest category (total rank value and median rank, Annex 12.1). After identifying the relevant ethical passages and agreeing on the coding scheme, two co-authors (FSA & WL) separately coded each statement, with perfect agreement on the coding for each statement-interest combination.[1]

In our coding, we exercised the principle of charitable interpretation when accounting for consideration of nonhuman interests (wholes or individuals) within each text. Thus, seemingly ambiguous statements such as calls for environmental stewardship or respect and humaneness towards wildlife or species, when lacking an anthropocentric statement, were taken to denote at least limited consideration for nonhuman interests.

We find evidence that the documents in Annex 12.1 do not adequately consider the most basic needs and interests of individual animals (Annex 12.1, median rank = 0). In contrast, the texts suggest that wolf management is almost exclusively concerned with instrumental (i.e., human) interests (total rank value = 18; median rank = 2), and specifically, human enjoyment of wildlife. Multiple documents reference an agency mission to engage in environmental stewardship "for the continuing benefit of the American people" (USFWS, 1998), species' values "to the Nation and its people" (United States, 1973) or with the objective of "increasing or maintaining populations to provide hunting opportunities" (WI NR Stat Ch 1). It is no surprise that these documents mention instrumental interests; what is more striking is the almost complete lack of mention of other non-instrumental and non-human interests.

The interests of ecological wholes (total rank value = 7; median rank = 1) seem implicit in calls for "environmental stewardship" (USFWS, 1998), respect and humaneness towards wildlife (USDA-WS documents), provision of healthy life systems (WI NR Stat Ch 1) or healthy populations (WMP), but animals' intrinsic value or individual interests are hardly mentioned (Annex 12.1). Concerns related to ecological wholes are restricted to "respect" or "ecological diversity and health", which are ambiguous if they lack an explicit mention of the intrinsic value of ecological wholes and their respective interests. Despite the apparent overlap in consideration of interests of humans and ecological wholes, if conflicting the focus on the former limits consideration of the latter. For example, a "healthy viable population of grey wolves in the state" (WMP) may still have an interest in freedom from unnecessary harm and social stability (just as healthy groups of humans surely would), yet this interest is not addressed further, particularly when addressing hunting. The human interest in hunting an animal for recreation trumps these interests of the wolf population. Thus, more urgent interests of ecological wholes beyond those relevant for provisioning ecosystem services for humans are not equitably considered. The lack of consideration is even starker for individual animal interests.

Individual animal interests are not mentioned in most of the legal documents we examined (total rank value = 4; median rank = 0). Some documents addressing lethal control do contain some mention of the welfare of animals and striving for respect and humaneness, but this apparent consideration of animal interests also raises ethical concerns, and so is awarded a '1' for each document where we identified these types of concerns. A legal document expressing an animal welfare concern implicitly acknowledges individual animals as sentient; otherwise, their welfare would be irrelevant. However, this concern for their welfare may also concede to relatively trivial human interests or may be implemented arbitrarily because the documents do not safeguard any animal interests against infringement. For example, USDA-WS directive 1.301(USDA-APHIS, 2010) and their Supplement to the Environmental Assessment (USDA-WS SEA) illustrate concerns for the suffering of individual animals when implementing lethal methods using "the most selective and humane methods available" or "minimizing harmful effects of damage management measures on . . . wolves", respectively. However, use of non-lethal methods is limited to cases where it is "practical and effective" (USDA-WS SEA), without any guidance for weighing these criteria against wolves' vital interests. Hence, practicality and effectiveness would also seem to hinge on purely human interests, which relegates concern for the vital interests of individual wolves to cases where it is convenient or does not conflict with instrumental ones.

Another example, WI Stat Ch 951, illustrates concern for animals' interests in freedom from unnecessary or unjustifiable harm or death. But, again, consideration may stop where arguably trivial human interests (i.e.: recreation) are negatively affected by it. For example, although §951.02 prohibits cruel treatment of individual animals, §951.015(1) and Wisconsin v Kuenzi (2011) clarified that this prohibition only applies to game animals if the behaviour in question is not normally considered 'hunting' (see ★★ in Annex 12.1 for clarification on this point). Thus, in the case of a wolf hunt, concern for the welfare of individual wolves is reduced to minimizing their suffering (through undefined codes such as 'clean kill' or 'fair chase'), and is left begging the question of how ethical or legitimate is the killing in the first place. Gary Francione (2009, p. 7) critiques this 'welfarist' position:

> Although the animal welfare position supposedly prohibits the infliction of 'unnecessary' suffering on animals, we do not ask whether particular institutional uses are themselves necessary because we assume that these

uses are acceptable and because our only concern is treatment. It is clear, however, that most of our animal uses are transparently frivolous and cannot be described as involving any 'necessity'.

Once moral consideration is recognized, additional steps would be required to examine the appropriateness of the behaviour in question. Acknowledging the moral standing of wolves would demand an examination of how ethically appropriate would be their killing prior to sanctioning it, equitably weighing the vital interests of wolves against human interests, be these protection, recreation, or convenience. We cannot just assume their killing is appropriate because it conflicts with any human interest, and proceed to examine only the killing technique.

When we consider the state's wolf hunt, the supposed concern for wolf welfare is anthropocentric, given that, despite there being no clear urgent claim to a wolf's life, the vital interest of wolves in living is subordinated to the unnecessary and trivial human interest in recreation (Vucetich and Nelson, 2014). Efforts to justify Wisconsin's wolf-killing in other ways that appear less trivial have not found strong evidence. For example, authorities claim social or ecological chaos without wolf killing (e.g., wolf populations are out of control), threats to human subsistence (e.g., livestock producers and deer hunters cannot compete), or a need for political support of wolf conservation (e.g., social tolerance for wolves depends on state lethal management). Years of scientific testing have come up empty for each claim (Treves, 2009; Treves et al., 2013; Browne-Nuñez et al., 2015; Hogberg et al., 2015; Treves et al., 2015; Chapron and Treves, 2016a, 2016b; Callan et al., 2013; Storm et al., 2015). In Wisconsin, wolf presence has been linked to an increase in ecosystem diversity, while there is no evidence of them driving down the state's deer population. Moreover, there is no evidence that lethal management or liberalized wolf-killing is an effective conflict mitigation strategy, or that these policies increase tolerance for the species.

In sum, as written, and despite our conservative approach, the examined statutes and regulations governing wolf management in Wisconsin lack important ethical principles safeguarding the interests of nonhuman members of the mixed community. The legal documents are more than twice as concerned with human instrumental interests than concerns for ecological wholes and more than four times more than concerned with individual animals' interests (Annex 12.1). We find no evidence that these documents provide an adequate account of the scientifically backed sentience and sapience of wolves or that individual animal interests are being appropriately considered (median rank = 0), especially when weighed against trivial human interests such as recreation or trophies. Nods to animal ethics through welfare concerns are inadequate because the documents fail to justify the foreseeable harm to animals against the standard of necessity. Individual animals may be granted magnanimity when convenient, but we find no evidence that the documents acknowledge individual animals as members of the moral community, let alone evidence of application of the principle of equitable consideration. Moreover, the documents seem to limit the relevance of moral consideration to instances where they would not conflict with human interests. The documents also fail to state explicitly their moral presuppositions so that the law, managers and public are adequately informed about their ethical implications.

The lack of appropriate and equitable consideration of animal interests present in the regulations precludes geocentrism, which would demand equitable consideration at all scales. Rather, the lack of consideration suggests that the prevailing paradigms within these regulations are anthropocentric and ecocentric. By dismissing the interests of individual animals, both paradigms fail to appropriately consider all loci of moral value and moral perspectives in nature, suggesting that these are ethically incomplete and inappropriate tools for regulating interactions with nature and individual animals that would allow all to flourish equitably.

Given this lack of consideration in explicit regulations, the level of discretion and guidance afforded to government agents is worth examining. As described for Wisconsin, management documents often provide government staff or private citizens with wide discretion for implementing harmful interventions against animals. Based on our results, we hypothesize that the lack of explicit mention of animal interests in these documents would result in their dismissal or inadequate guidance for considering them. Although that assessment is beyond the scope of this chapter, we believe our recommendations might contribute to correcting these ethical flaws, when present.

Moving towards equitable consideration

Equitable consideration entails the equitable and explicit acknowledgement of all affected interests when deciding on the appropriateness of an action. The current disregard for individual animal interests is not inevitable. The interdisciplinary work of ethicists, ethologists, and environmental scientists, among others, sheds light on what ethical coexistence with wildlife might look like. Ethical concerns go hand in hand with the best available science in these fields. We cannot simply dismiss these scientific advances because the ethical implications would prove inconvenient. Nor can we claim a lower standard of consideration for animals, simply because we have not read the latest science. Although there are no simple answers to the complex ethical dilemmas, we propose that an indispensable component of ethical coexistence is the promotion and codification of equitable consideration for individual animals.

We propose that humans have duties to the more vulnerable members of our moral community (marginality ... compassion, respect, tolerance, kindness), and therefore a responsibility for the ethical handling of conflicts. Without adherence to that ethical principle, regulations cannot be said to imbue any of the moral principles that allow for prosperous community living. The codifying of equitable consideration of animal interests is a powerful way to safeguard against unjustified infringement. As our analysis has shown, leaving codes of conduct unstated and ignoring the interests of individuals allows cruelty, sadism, and illiberal actions that can affect humanity adversely – as well as the direct victims, the animals.

Codifying moral consideration of animals should be complemented by ethical education. Government sponsored ethical education is mentioned in the regulations, but only when related to hunting and trapping (WI Stat §29.591[1b], WI NR §1.11) through codes of "fair chase" and "clean kill", and the promotion of "wildlife as a renewable natural resource" (WI NR §1.11). Such efforts treat wildlife as a resource instead of sentient beings, with the underlying assumption that certain wildlife interests (such as living) can be trumped by trivial human interests.

As the institution responsible for policies regulating the environment and human-nonhuman interactions, legislatures and wildlife agencies should provide proper ethical guidance incorporating the scientific and ethical advances in understanding and respecting individual animals. Accomplishing this will require ethics education for appropriate legislators, agency personnel, and interested constituency groups. How this is to be accomplished is not the subject of this chapter, but we envision it as minimally involving some combination of mandatory and voluntary training, and partnerships with ethical specialists in animal and environmental ethics. It is imperative that professionals receive the most complete ethical training if wildlife management aspires to manage animals ethically and conform to society's evolving moral codes of conduct.

Our recommendations are not a panacea. Instrumental interests will continue to dominate dialogue and perhaps practice. However, our recommendations provide a starting point for explicitly considering and retaining ethics in our intrusions into the lives of all animals. Their implementation would aid in allowing humans and our entire mixed moral community to flourish.

Annex 12.1 Statements expressing ethical standpoints in statutes, regulations and agency documents relevant to grey wolf management in Wisconsin, USA

Statute, regulation or agency document	Statements related to ethics and wildlife management (emphasis added)	Interests explicit in ethical justification[1]		
		Human	Ecological wholes	Individual animals
Federal level				
022 FW 1, Creation, Authority and Functions (USFWS)	D. Objectives. The U.S. Fish and Wildlife Service has three basic objectives: (1) to assist in the development and application of an *environmental stewardship ethic* for our society, based on ecological principles, scientific knowledge of fish and wildlife, and *a sense of moral responsibility;* (2) to guide the conservation, development, and management of *the Nation's fish and wildlife resources;* and (3) to administer a national program to provide the public opportunities to *understand, appreciate, and wisely use fish and wildlife resources.* These objectives support the Service mission of conserving, protecting, and enhancing fish and wildlife and their habitats *for the continuing benefit of the American people.*	2	1	0
Endangered Species Act	Findings (Sec. 2(a)(3)): "… these species of fish, wildlife and plants are of *aesthetic, ecological, educational, historical, recreational, and scientific value to the Nation and its people;* (4) the United States has pledged itself as a sovereign state in the international community to conserve to the extent practicable the various species of fish or wildlife and plants facing extinction …"	2	0	0

(Continued)

Annex 12.1 (Continued)

Stature, regulation or agency document	Statements related to ethics and wildlife management (emphasis added)	Interests explicit in ethical justification[1]		
		Human	Ecological wholes	Individual animals
USDA-WS Directive 1.201 Mission and Philosophy of the WS Program	4. WS Management Philosophy. (a) General Philosophy: "In the United States, *wildlife is a publicly-owned resource held in trust and managed by State and Federal agencies.* Government agencies, including WS, are required by law and regulation to *conserve and mange wildlife resources while being responsive to public desires, views and attitudes.* In so doing, agencies must also respond to requests for resolution of damage and other problems caused by wildlife. . . . Actions considered and employed should be *biologically sound, environmentally safe, scientifically valid, and socially acceptable.* . . . WS' vision is to improve the coexistence of people and wildlife, *while considering a wide range of public interests that can conflict with one another.* These interests include *wildlife conservation, biological diversity, and the welfare of animals, as well as the management of wildlife for purposes of enjoyment, recreation, and livelihood.*"	2	1	1
USDA-WS Directive 1.301 Code of Ethics of the WS Program	4. Policy. WS Code of Ethics: "d. Will show exceptionally high levels of *respect for people, property and wildlife*; g. Will utilize the WS Decision Model to resolve wildlife damage problems and strive to use *the most selective and humane methods available,* with preference given to nonlethal methods *when practical and effective.*"	2	1	1
USDA-WS Supplement to the Environmental Assessment: Management of Wolf Conflicts and Depredating Wolves in Wisconsin (2013)	"WS uses and/or recommends the full range of legal, practical and effective nonlethal and lethal methods for preventing or *reducing wolf damage while minimizing harmful effects of damage management measures on humans, wolves, other species, and the environment* in accordance with the WDNR guidelines for wolf depredation control." (p. 2)	2	1	1

State level

	2	1	0
WI Statutes Chapter NR 1 – Natural Resource Board Policies	Management of wildlife, Preamble (NR 1.015(2): "The primary goal of wildlife management is to provide *healthy life systems* necessary to sustain Wisconsin's *wildlife populations* for their *biological, recreational, cultural and economic values*. Wildlife management is the application of knowledge in the protection, enhancement and regulation of wildlife resources *for their contribution toward maintaining the integrity of the environment and for the human benefits they provide*." Wildlife Management (NR 1.11): "(2) Recognizes the need to strengthen the educational efforts of the department relating to hunter competence, *standards of ethical hunting behaviour* (see 29.591 below) and respect for landowners rights; educational efforts must also be directed toward nonhunters to improve *their knowledge and understanding of wildlife as a renewable natural resource* and of *hunting as both a method of controlling wildlife populations and as a form of outdoor recreation*.* (7) Supports *the maintenance of ecological diversity and health*, and will do everything in its power to protect and maintain free-living populations of all species of wildlife currently existing in Wisconsin; extirpated species will be reintroduced whenever feasible ecologically, economically and socially. (8) Supports the management of game species and habitat with the *objective of increasing or maintaining populations to provide hunting opportunities*. (9) Supports the *regulated use of wildlife for human benefits*, including hunting and trapping where legal harvests do not reduce *subsequent population levels of these renewable wildlife resources or where population reduction of certain species is a deliberate objective*."		

Annex 12.1 (Continued)

Statute, regulation or agency document	Statements related to ethics and wildlife management (emphasis added)	Interests explicit in ethical justification[1]		
		Human	Ecological wholes	Individual animals
WI Statutes Chapter 29 – Wild Animals and Plants	Title to Wild Animals (29.011(1)): "The legal title to, and the custody and protection of, all wild animals within this state is vested in the state *for the purposes of regulating the enjoyment, use, disposition, and conservation of these wild animals*." Rule-making of this chapter (29.014(1)): "The department shall establish and maintain open and closed seasons for fish and game and any bag limits, size limits, rest days and conditions governing the taking of fish and game that will *conserve the fish and game supply and ensure the citizens of this state continued opportunities for good fishing, hunting and trapping*." Removal of wild animals (29.885(2)): "The department may remove or authorize the removal of all of the following: (a) *a wild animal that is causing damage or that is causing a nuisance*; (b) *a structure of a wild animal that is causing damage or that is causing a nuisance*." Trapper Education program (29.591(1b)): "The courses of instruction under these programs shall provide instruction to students in the responsibilities of *hunters to wildlife, environment, landowners and others, how to recognize threatened and endangered species that cannot be hunted and the principles of wildlife management and conservation*."	2	1	0
WI Statutes Chapter 951 – Crimes Against Animals	Definitions (951.01(2)): "'Cruel' means causing *unnecessary and excessive pain or suffering or unjustifiable injury or death*." Construction and Application (951.015(1)): "*This chapter may not be interpreted as controverting any law regulating wild animals that are subject to regulation under ch. 169 [hunting], the taking of wild animals, as defined in s. 29.001 (90), or the slaughter of animals by persons acting under state or federal law*."	2	0	1

Mistreating animals (951.02): "No person may treat any animal, whether belonging to the person or another, in a cruel manner. This section does not prohibit normal and accepted veterinary practices."			0
WI Wolf Management Plan (WMP) I. Introduction: "These guidelines provide a conservation strategy for *maintaining a healthy viable population of grey wolves in the state*, and contribute toward national recovery, while addressing problems that may occur with wolf depredation on livestock or pets." (p. 8)	2	1	
B. Population Monitoring and Management: "Harvest by private citizens is controversial, but *the taking of wolves in a recovered population is consistent with the management of other furbearers in the state of Wisconsin*." (p. 21)			
E. Wolf Depredation Management: "*WDNR is charged with protecting and maintaining a viable population of wolves in the state, but also must protect the interests of people who suffer losses due to wolf depredations*." (p. 23) 1. Depredation Management Plan. "The objective of the wolf depredation program is to *minimize depredations and compensate people for their losses*." (p. 24)			
Total rank value (max.18)	18	7	4
Median rank	2	1	0

1 '2' – interests in category were explicitly and appropriately considered; '1' – interests in category explicit yet limited or inappropriately considered; '0' – no explicit consideration of a particular interest category.

* Wisconsin's Hunter Education Course Manual (Kalkomey Enterprises, Inc., 2012) limits ethical behaviour towards individual nonhuman animals to codes of "fair chase", "clean kill" and use of all usable parts (p. 66). Although it also advocates for treating "both game and non-game animals ethically", it does not expand on this to explain what that would mean (how the human [subsistence, recreation] and non-human [life, flourishing] interests involved should be weighted) and how it would impact the practice of hunting and the field of conservation.

** However, this chapter only "controverts" those statutes if the behaviour in question is not normally considered 'hunting', following Wisconsin v Kuenzi (2011).

Note

1 Despite highlighting our high level or agreement, we offer no quantitative measure of inter-observer reliability (IRRI) for our coded analysis of the texts. Such quantitative measures are usually provided in qualitative studies as an indicator of rigour ('the quality of being extremely thorough and careful', https://en.oxforddictionaries.com/definition/rigour). We disagree with such an interpretation, and highlight the difference between measures of agreement and measures of rigour.

Coder reliability measures can indeed display a good faith effort in teasing out all the meanings of a text, but they are not a measure of validity ('the quality of being logically or factually sound; soundness or cogency', https://en.oxforddictionaries.com/definition/validity) in qualitative or interpretive research. Such an index may provide a false sense of rigour based on shared value judgements between observers, rather than reason and the evidence presented. One example helpful to illustrate this point is the landmark *Dred Scott v. John F.A. Sandford* US Supreme Court decision (1857). The court overwhelmingly (7–2) agreed that slaves were not entitled to their freedom despite residing in a free state; thus, African Americans could never be US citizens. An IRRI-like index for such a decision would have validated the decision, reflecting high coder (the judges) reliability and 'rigour'. But we acknowledge today, as was argued then, that the argumentation and evidence was flawed.

Moreover, these kinds of methodological misunderstandings inhibit ethical and interpretive contributions to science-based research. Different theories and paradigms of science are amenable to different kinds of data and methods. Rigour in qualitative and interpretive research is based on reason and evidence. Well-intentioned attempts to extend the methods of quantitative science to qualitative and interpretive research, despite their inadequacy, may provide a false sense of objectivity, bolster false empiricist claims of a dichotomy between the subjective and objective, limit interpretation, and value empiricism over reason.

Therefore, we present our interpretations and the original text side by side so the reader can evaluate our interpretations for themselves, without making a claim of independent objectivity implied by an IRRI. Challenges to our interpretation of the language should be based on a review of the same sources (which are provided and endlessly replicable), with evidence for why our interpretation may be incorrect or incomplete.

References

Bangs, E. E., and Fritts, S. H. (1996) 'Reintroducing the gray wolf to central Idaho and Yellowstone National Park', *Wildlife Society Bulletin*, vol 24, no 3, pp402–413.

Batavia, C., and Nelson, M. P. (2016) 'Heroes or thieves? The ethical grounds for lingering concerns about new conservation', *Journal of Environmental Studies and Sciences*, vol 7, no 3, pp394–402.

Beatley, T. (1994) *Ethical Land Use: Principles of Policy and Planning*, Johns Hopkins University Press Baltimore, MD.

Bekoff, M., Allen, C., and Burghardt, G. M. (2002) *The Cognitive Animal: Empirical and Theoretical Perspectives on Animal Cognition*, MIT Press, Cambridge, MA.

Bekoff, M., and Goodall, J. (eds) (2004) *Encyclopedia of Animal Behavior*, Greenwood Press, Westport, CT.

Bentham, J. (1789) *An Introduction to the Principles of Morals and Legislation. Printed in the Year 1780, and Now First Published. By Jeremy Bentham*, T. Payne, and Son, London.

Boitani, L. (2003) 'Wolf conservation and recovery', in D. Mech and L. Boitani (eds) *Wolves: Behavior, Ecology, and Conservation*, The University of Chicago Press, Chicago.

Browne-Nuñez, C., Treves, A., Macfarland, D., Voyles, Z., and Turng, C. (2015) 'Tolerance of wolves in Wisconsin: A mixed-methods examination of policy effects on attitudes and behavioral inclinations', *Biological Conservation*, vol 189, pp59–71.

Bruskotter, J. T. (2013) 'The predator pendulum revisited: Social conflict over wolves and their management in the western United States', *Wildlife Society Bulletin*, vol 37, no 3, pp374–379.

Bruskotter, J. T., Enzler, S. A., and Treves, A. (2011) 'Science and law: Rescuing wolves from politics: Wildlife as a public trust resource', *Science*, vol 333, no 6051, pp1828–1829.

Caldwell, L. K., and Shrader-Frechette, K. S. (1992) *Policy for Land: Law and Ethics*, University Press of America, Lanham, MD.

Callahan, S., and Jennings, B. (1983) *Ethics, the Social Sciences, and Policy Analysis*, Plenum, New York.

Callan, R., Nibbelink, N. P., Rooney, T. P., Wiedenhoeft, J. E., and Wydeven, A. P. (2013) 'Recolonizing wolves trigger a trophic cascade in Wisconsin (USA)', *Journal of Ecology*, vol 101, no 4, pp837–845.

Callicott, J. B. (1980) 'Animal liberation: A triangular affair', *Environmental ethics*, vol 2, no 4, pp311–338.

Chapron, G., and Treves, A. (2016a) 'Blood does not buy goodwill: Allowing culling increases poaching of a large carnivore', *Proceedings of the Royal Society of London B: Biological Sciences*, vol 283, no 1830, pp20152939.

Chapron, G., and Treves, A. (2016b) 'Correction to "Blood does not buy goodwill: Allowing culling increases poaching of a large carnivore"', *Proceedings of the Royal Society B*, vol 283, no 1845, pp20162577.

DNR, WI. (1999) *Wisconsin Wolf Management Plan*, Resources, W. D. O. N., Madison, WI.

DNR, WI. (2007) *Wisconsin Wolf Management Plan Addendum 2006 & 2007*, W. D. O. N., Madison, WI.

DNR, WI. (2013) 'DNR Secretary Cathy Stepp statement on lawsuit to re-list Western Great Lakes Wolves', *News Release*, 12 February 2013, http://dnr.wi.gov/news/Breaking News_Print.asp?id=2659, accessed November 2016.

Favre, D. S. (1978) 'Wildlife rights: The ever-widening circle', *Environmental Law*, vol 9, p241.

Feinberg, J. (1981) 'The rights of animals and unborn generations', in Landau, R.-S. (ed) *Ethical Theory: An Anthology*, Blackwell Publishing Ltd, West Sussex.

Francione, G. L. (2009) *Animals as Persons: Essays on the Abolition of Animal Exploitation*, Columbia University Press, New York.

Hogberg, J., Treves, A., Shaw, B., and Naughton-Treves, L. (2015) 'Changes in attitudes toward wolves before and after an inaugural public hunting and trapping season: Early evidence from Wisconsin's wolf range', *Environmental Conservation*, vol 43, no 1, pp45–55.

Horner, J. (2003) 'Morality, ethics, and law: Introductory concepts', *Seminars in Speech and Language*, vol 24, no 4, pp263–274.

HSUS. (2014) Humane Society of the United States, et al. v. Sally Jewell, Secretary of the Interior, et al. v. State of Wisconsin, et al. 1:13-cv-00186-BAH. United States District Court for the District of Columbia.

Jonsen, A. R., and Toulmin, S. E. (1988) *The Abuse of Casuistry: A History of Moral Reasoning*, University of California Press, Berkeley.

Kalkomey Enterprises, Inc. (2012) *Today's Hunter: A Guide to Hunting Responsibly and Safely*, Kalkomey Enterprises, Dallas, TX.

Kolbert, E. (2014) *The Sixth Extinction: An Unnatural History*, Picador, New York.

Leopold, A. (1949) *A Sand County Almanac: And Sketches Here and There*, Oxford University Press, New York.

Lewis, P.-M., Burns, G. L., and Jones, D. (2016) 'Response and responsibility: Humans as apex predators and ethical actors in a changing societal environment', *Food Webs*, vol 12, pp49–55.

Lopez, B. (1978) *Of Wolves and Men*, Scribner, New York.

Low, P., Panksepp, J., Reiss, D., Edelman, D., Van Swinderen, B., Low, P., and Koch, C. (2012) *Cambridge Declaration on Consciousness in Non-Human Animals*, University of Cambridge, Churchill College.

Lynn, W. S. (1998a) 'Animals, ethics and geography', in J. Wolch and J. Emel (eds) *Animal Geographies: Place, Politics, and Identity in the Nature-Culture Borderlands*, Verso, London.

Lynn, W. S. (1998b) 'Contested moralities: Animals and moral value in the Dear/Symanski debate', *Ethics, Place & Environment*, vol 1, no 2, pp223–242.

Lynn, W. S. (2002) 'Canis lupus cosmopolis: Wolves in a cosmopolitan worldview', *Worldviews: Global Religions, Culture, and Ecology*, vol 6, no 3, pp300–327.

Lynn, W. S. (2006) 'Between science and ethics: What science and the scientific method can and cannot contribute to conservation and sustainability', in D. M. Lavigne and S. Fink (eds) *Gaining Ground: In Pursuit of Ecological Sustainability*, International Fund for Animal Welfare, Guelph, ON.

Lynn, W. S. (2007) 'Practical ethics and human-animal relations', in Bekoff, M. (ed) *Encyclopedia of Human-Animal Relationships*, ABC-CLIO, Inc., Greenwood Press, Westport.

Lynn, W. S. (2010) 'Discourse and wolves: Science, society, and ethics', *Society & Animals*, vol 18, no 1, pp75–92.

Lynn, W. S. (2012) *Barred Owls in the Pacific Northwest: An Ethics Brief*. George P. Marsh Institute, Clark University, Worcester, MA.

Lynn, W. S. (2017) 'Ethics', in J. Urbanik and C. Johnston (eds) *Humans and Animals: A Geography of Coexistence*, ABC-Clio, Santa Barbara, CA.

Midgley, M. (1983) *Animals and Why They Matter*, Cambridge University Press, Cambridge.

Midgley, M. (1993) *Can't We Make Moral Judgements?*, St. Martin's Press, New York.

Millstein, R. L. (2018) 'Debunking myths about Aldo Leopold's land ethic', *Biological Conservation*, vol 217, pp391–396.

NASS, National Agricultural Statistics Service (2011) 'Cattle Death Loss', Released May 12, 2011, http://usda.mannlib.cornell.edu/usda/current/CattDeath/CattDeath-05-12-2011.pdf

Nelson, M. P., and Vucetich, J. A. (2012) 'Environmental ethics and wildlife management', in D. D. Decker, S. J. Riley and W. F. Siemer (eds) *Human Dimensions of Wildlife Management*, Johns Hopkins University Press, Baltimore, MD.

Nelson, M. P., Vucetich, J. A., Paquet, P. C., and Bump, J. K. (2011) 'An inadequate construct? North American model: What's flawed, what's missing, what's needed', *The Wildlife Professional*, vol 5, pp58–60.

Peterson, A. L. (2013) *Being Animal: Beasts and Boundaries in Nature Ethics*, Columbia University Press, New York.

Regan, T. (1987) 'The case for animal rights', in *Advances in Animal Welfare Science 1986/87*, Springer, Netherlands.

Service, USFW (1992) *Recovery Plan for the Eastern Timber Wolf*, Service, U. S. F. W., Twin Cities, MN.

Shrader-Frechette, K. S., and Mccoy, E. D. (1994) 'How the tail wags the dog: How value judgments determine ecological science', *Environmental Values*, vol 3, no 2, pp107–120.

Singer, P. (1975) *Animal Liberation*, Avon, New York.

Singer, P. (1993) *Practical Ethics*, Cambridge University Press, Cambridge, MA.

Storm, D., Walrath, R., Norton, A., Peterson, B., Watt, M., Van Deelen, T., Martin, K., and Rolley, R. 2015. *Wisconsin Deer Research Studies: Annual Report, 2013–2014*. Wisconsin Department of Natural Resources.

Taylor, P. W. (2003) 'The ethics of respect for nature', *Environmentalism: Critical Concepts*, vol 1, no 3, p61.

Toulmin, S. E. (1950) *An Examination of the Place of Reason in Ethics*, Cambridge University Press, Cambridge.

Treves, A. (2009) 'Hunting for large carnivore conservation', *Journal of Applied Ecology*, vol 46, no 6, pp1350–1356.

Treves, A., Chapron, G., López-Bao, J.V., Shoemaker, C., Goeckner, A. R., and Bruskotter, J.T. (2015) 'Predators and the public trust', *Biological Reviews*, vol 92, no 1, pp248–270.

Treves, A., Krofel, M., and Mcmanus, J. (2016) 'Predator control should not be a shot in the dark', *Frontiers in Ecology and the Environment*, vol 14, no 7, pp380–388.

Treves, A., Naughton-Treves, L., and Shelley, V. (2013) 'Longitudinal analysis of attitudes toward wolves', *Conservation Biology*, vol 27, no 2, pp315–323.

United States (1973) 'Endangered Species Act', 16 U.S.C. § 1531 et seq

USDA-APHIS Wildlife Services (2013) 'Supplement to the Environmental Assessment: Management of Wolf Conflicts and Depredating Wolves in Wisconsin', Released February 2013, https://www.aphis.usda.gov/regulations/pdfs/nepa/2013_WI_Wolf_EA_Supplement_Final.pdf

USDA-APHIS Wildlife Services (2009) 'WS Directive 1.201, Mission and Philosophy of the WS Program', Released July 20, 2009, https://www.aphis.usda.gov/wildlife_damage/directives/1.201_mission_and_philosphy.pdf

USDA-APHIS Wildlife Services (2010) 'WS Directive 1.301, Code of Ethics', Released August 31, 2010, https://www.aphis.usda.gov/wildlife_damage/directives/1.301_code_of_ethics.pdf

USFWS (1998) '022 FW 1, Creation, Authority and Functions', Released March 6, 1998, https://www.fws.gov/policy/022fw1.html

Vucetich, J. A., Bruskotter, J. T., and Nelson, M. P. (2015) 'Evaluating whether nature's intrinsic value is an axiom of or anathema to conservation', *Conservation Biology*, vol 29, no 2, pp321–332.

Vucetich, J., and Nelson, M. P. (2014) 'Wolf hunting and the ethics of predator control', in L. Kalof (ed) *The Oxford Handbook of Animal Studies*, Oxford University Press, Oxford.

Wallach, A. D., Bekoff, M., Nelson, M. P., and Ramp, D. (2015) 'Promoting predators and compassionate conservation', *Conservation Biology*, vol 29, no 5, pp1481–1484.

Weston, A. (2006) *A Practical Companion to Ethics*, Oxford University Press, Oxford.

Wiedenhoeft, J. E., Macfarland, D., Libal, N. S., and Bruner, J. (2016) *Wisconsin Gray Wolf Monitoring Report, 15 April 2015 Through 14 April 2016*, Resources, W. D. O. N., Madison.

Willging, R. C., and Wydeven, A. P. (1997) 'Cooperative wolf depredation management in Wisconsin', in C. D. Lee and S. E. Hygnstrom (eds) *Thirteenth Great Plains Wildlife Damage Control Workshop Proceedings*, Kansas State University Agricultural Experiment Station and Cooperative Extension Service, Lincoln, NE.

Wisconsin State Legislature 'WI Statutes Chapter NR 1 - Natural Resources Board Policies', https://docs.legis.wisconsin.gov/code/admin_code/nr/001/1

Wisconsin State Legislature 'WI Statutes Chapter 29, Wild Animals and Plants', https://docs.legis.wisconsin.gov/statutes/statutes/29

Wisconsin State Legislature 'WI Statutes Ch 951, Crime Against Animals', https://docs.legis.wisconsin.gov/statutes/statutes/951/Title

Wisconsin v Kuenzi (2011) WI APP 30 (Court of Appeals of Wisconsin)

Wydeven, A. P., Wiedenhoeft, J. E., Schultz, R. N., Thiel, R. P., Jurewicz, R. L., Kohn, B. E., and Van Deelen, T. R. (2009) 'History, population growth, and management of wolves in Wisconsin', in A. P. Wydeven, T. R. Van Deelan and E. Heske (eds) *Recovery of Gray Wolves in the Great Lakes Region of the United States*, Springer, New York.

13 Science, society, and snow leopards

Bridging the divides through collaborations and best practice convergence

Charudutt Mishra, Justine Shanti Alexander, Yash Veer Bhatnagar, Örjan Johansson, Koustubh Sharma, Kulbhushansingh R. Suryawanshi, Muhammad Ali Nawaz, and Gustaf Samelius

Introduction

The existence of multiple societal priorities, often competing against each other, represents a major challenge to nature conservation and large carnivore management. Competing priorities have led to gaps and divides that continue to widen, and manifest in various forms of ideological and land use conflicts and power struggles. At the most fundamental level, economic development and nature conservation are often in direct conflict with each other. Large scale economic development today comes at significant costs of environmental degradation and fragmentation of habitats. Our role as conservationists often requires opposing development-related projects that are seen by many other sections of the society as being critical for economic growth or poverty alleviation, and we rarely engage with the industry except for accessing funds from corporate foundations. While we recognize the critical need for governmental support for conservation, we often find ourselves in conflict with governments as well, as the pressures and priorities of economic development in the government order often trump the concerns for conservation. This problem is intensified by the poor recognition and lack of financial accounting of the value of ecosystem services that wild or semi-wild landscapes provide to humanity, alongside their providing the last refuges for large carnivores.

Similarly, conservation conflicts arise because nature, while providing immense and varied benefits and services (e.g., Murali et al., 2017a), also comes with disservices such as wildlife causing damage to human life and property, and potential competition between wildlife and humans for natural resources. These can lead to a divide between conservation objectives on the one hand and livelihoods and security of local people on the other. Large carnivores, where they persist, kill livestock, and are a threat to livelihoods and even human life (Madhusudan and Mishra, 2003). This often leads to conflicts between the need for conservation and peoples' legitimate desires to secure their livelihoods and safety.

A large gap similarly exists between conservation science and practice. This is a problem that has been recognized for a while (Pfeffer and Sutton, 1999), but not much has improved over the years (Arlettaz et al., 2010). Despite a continuous increase in the number of peer-reviewed scientific publications in conservation science, the use of this knowledge in on-the-ground conservation has been rare (Sutherland et al., 2004). Insufficient access and ability to use research-based knowledge due to funding constraints and language barriers have been significant limitations for conservation practitioners in many countries (Arlettaz et al., 2010). Use of already limited conservation funding to access conservation research articles that are locked behind expensive paywalls is seldom justified. Although there is a growing trend in the use of open access journals to publish conservation research, progress on this has been far from adequate. Use of this option is also limited by the ability of conservation scientists to pay the journals to publish their findings as open access. Websites promoting evidence-based conservation have tried to bridge this gap by making scientific knowledge available to practitioners for free and in a language of their preference. However, such websites are few and limited in their ability to translate the large body of scientific literature being generated (Sutherland et al., 2004).

The lack of political support for evidence-based conservation is further an important reason for the gap between conservation science and practice (Jacobson and Duff, 1998). Even when all these barriers are overcome, on-the-ground involvement of the scientific community is hampered by an academic system that rewards scientific publications and not conservation impact (Lach et al., 2003; Arlettaz et al., 2010). Professional priorities of researchers therefore often focus on publication outcomes rather than conservation outcomes. This also means that research that is produced is focused on high academic impact publications, which need not necessarily imply high conservation value. Thus, the research information being generated is often biased towards issues that are marginal to conservation (McNie, 2007).

In this chapter, we underscore the challenges posed by these widening gaps between conservation needs, approaches and scientific pursuits. With the snow leopard (*Panthera uncia*) as our flagship and focal taxon, we suggest that these major gaps within society must be bridged for large carnivores and their habitats to be conserved effectively. We describe our efforts – started nearly 20 years ago but in many ways still nascent – to bridge these divides. We suggest that as scientists, we will achieve little in conservation, which is increasingly a political and socio-economic pursuit. However, we are hopeful that if scientists can step out and engage respectfully with other sections of the society and the government, together we still have a chance to secure the future of the planet's surviving large carnivores and their habitats.

The vexing problem of large carnivore conservation

Efforts to conserve large carnivores and associated biodiversity often suffer from major gaps between theory and practice. Many carnivore species today face a serious risk of extinction; of the 31 large carnivore species, 77% are thought to be declining, and 61% are listed as threatened (vulnerable, endangered, or critically endangered) in the IUCN's Red List (Ripple et al., 2014). Species such as the tiger (*Panthera tigris*) have faced widespread killing and vast tracts of habitat have been lost (tigers have seen a reduction of 90% of their original range; Dinerstein et al., 2007), they face continued threats and are victims of conflict in the regions where they still survive.

Many large carnivore species are specialized for ungulate predation and readily kill domesticated ungulates when opportunities arise (Treves and Karanth, 2003). Diverse livestock are affected, including cattle, sheep, horses, goats, and yak, depending on the location and predators present (Karanth, 2002; Mishra et al., 2016). In many areas where large carnivores and livestock still occur together, uptake by carnivores is thought to be increasing due to increase in livestock populations, reduction in wild ungulate populations, inadequate livestock management practices, and, sometimes, even an increase in wild ungulate populations that lead to higher carnivore populations (Meriggi et al., 1996; Mishra, 1997; Treves and Karanth, 2003; Suryawanshi et al., 2017). Because livestock can represent a major source of livelihood for many communities in the affected areas (Mishra et al., 2016), large carnivores pose a direct threat to human activities and interests (Treves and Karanth, 2003). Large carnivores are also potentially dangerous to humans themselves.

This has meant, historically, that lethal control has been a common approach to managing the problems of livestock depredation, competition for game animals, and threat to human life (Woodroffe et al., 2005). Large carnivores have therefore been victims of exterminations and continued persecution (Woodroffe et al., 2005). Some have also been assigned with special meanings of power and danger, which has led to a demand for their body parts in clothing or traditional medicines and ceremonies, which, in turn, leads to poaching being a major threat for species such as the tiger (Sharma et al., 2014). The illicit demand for products such as bone and pelt add to the pressure of poaching (Graham-Rowe, 2011). The modern carnivore conservation paradigm has come to be dominated by a separationist or land sparing approach that aims at segregation of carnivores and people through emphases on protected areas, human relocations, fences, and armed guards, while the alternate coexistence model has rarely been given appropriate consideration (Chapron et al., 2014).

Yet, research widely shows and ecologists widely agree that large carnivores have relatively large spatial needs and often occur at very low densities, which makes the land sparing approach particularly challenging (Gittleman, 2001; Chapron et al., 2014). The average sized protected area in Asia, for example, is often too small to conserve even small populations of large carnivores (Figure 13.1). Meaningful conservation therefore must look beyond the separationist

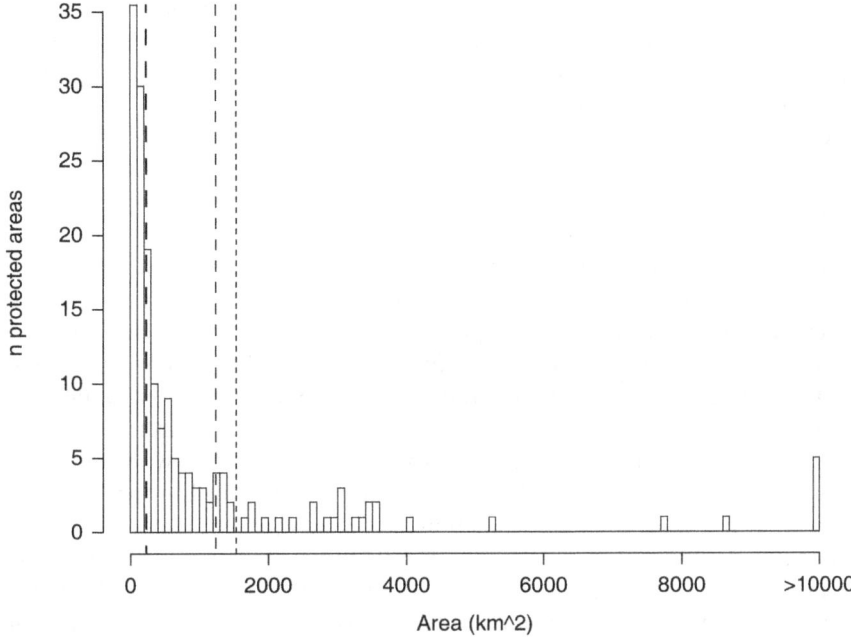

Figure 13.1 The size distribution of all protected areas within the snow leopard range in Asia shows how small the protected areas are in relation to the space needs of the species. Vertical hashed lines from left to right show the mean home range size for adult males (— — —) and the habitat area required, on average, by 15 adult females (– – –) with no overlap between neighbours; both estimated using 95% Local Convex Hulls. The third vertical line shows the area required by 15 adult females (- - -) based on 95% Convex Polygons with overlap between neighbours (see Johansson et al., 2016).

model of trying to segregate large carnivores and humans (e.g., Packer et al., 2013). It requires exploring ways of promoting coexistence by creating conditions that increase ecological and social carrying capacities for carnivores across large, multiple-use-landscapes extending well beyond protected area boundaries, or, as we note later, making the idea of protected areas more nuanced. Many large carnivores can persist even in high human densities when the management regimes are more favourable (Linnell et al., 2001).

The land sharing conservation paradigm aims at multiple-use landscapes with conservation taking place alongside human activities, while land sparing aims at separating conservation and human activities, mainly through protected areas (Crespin and Garcia-Villalta, 2014). Both land sharing and land sparing paradigms face great challenges for large carnivore conservation due to the predatory nature of carnivores and their extensive spatial needs (Linnell et al.,

2001). For example, land sharing almost inevitably takes place along with conflicts over livestock killing, threat to human life, and competition for game animals. On the other hand, land sparing is increasingly difficult given the large spatial needs of large carnivores and the human pressures on the land, and usually comes at a substantial cost on local people, their societies, cultures, and livelihoods (Linnell et al., 2001).

What is possible

Achieving sustainable large carnivore conservation across vast multiple-use landscapes is possible, as exemplified by the recovery of large carnivore populations in substantial parts of Europe. The European continent currently hosts twice as many wolves (*Canis lupus*) (more than 11,000) as the United States despite being half the size (4.3 million km^2 versus 8 million km^2) and more than twice as densely populated (97 inhabitants/km^2 versus 40 inhabitants/km^2). Significant populations of large carnivores including the wolf and brown bear (*Ursus arctos*) occur largely outside protected areas in Europe, a development that has been enabled by supportive public opinion, legislation and law enforcement, political and economic stability, human migration from rural areas to urban centres and the associated abandonment of agricultural land, compensation programs, and maintenance and revival of traditional livestock protection measures (e.g., livestock guarding dogs, night corrals, and shepherds), as well as investments in newer techniques (e.g., electric fences) (Chapron et al., 2014).

Land sharing and coexistence between large carnivores and people is, however, particularly challenging in low income economies where the human communities involved often represent the relatively poor sections of the society (Mishra et al., 2003). Economic losses caused by carnivores are hard to tolerate or mitigate, and poaching can represent a means to control the predator population as well as be an additional source of income.

In large parts of Asia and Africa, protected areas have been established at a huge economic and social cost to local people who have lost traditional access to natural resources and places of religious or cultural importance, societies have been fragmented due to relocations, and people have been progressively disempowered and marginalized (MacKay and Caruso, 2004; Mishra, 2016). Conservation in the face of poverty is therefore largely viewed as a pursuit of the privileged (Mishra, 2016). We therefore strongly believe that in the high mountains of South and Central Asia, where snow leopards occur alongside livestock-based economies, enabling tolerance and land sharing between people and large carnivores is essential for carnivore conservation, rather than trying to conserve the carnivores at the cost of alienating people and causing social injustices. We believe that carnivore conservation can be achieved through sustainable economic development of local people and an inclusive conservation approach based on community participation and empowerment.

Snow leopard conservation with local communities

The Snow Leopard Trust and our partners pioneered and have remained involved in community-based efforts to conserve snow leopards and their habitats for more than 20 years now, with conservation programs with local communities in five of the most important snow leopard range countries: China, Mongolia, India, Pakistan, and Kyrgyzstan, which together cover over three-fourths of the global snow leopard range.

Like other large carnivores, snow leopards kill livestock. This is particularly of concern as livestock rearing is one of the predominant forms of livelihoods across the snow leopard's range (Mishra et al., 2003). The species is a common victim of retaliatory killing, and such local persecution also forms linkages with illegal trade in snow leopard body parts. We have worked with local communities to build long-term partnerships to improve social and biodiversity outcomes, and to facilitate coexistence of people and snow leopards. Our work with communities is typically long-term (some of our community partnerships are nearly two decades old) and includes initiatives for enhancing conservation and livelihood opportunities (e.g., handicrafts, sustainable cashmere production practices), as well as collaboratively setting up and co-managing wildlife reserves on community land. We also set up conflict management programs such as collaborative corral improvement (predator-proofing), community-managed livestock insurance schemes, and creating greater conservation awareness (see Mishra, 2016). Our ongoing efforts with nearly 150 communities in snow leopard habitats of five countries are helping protect the species over an estimated 110,700 km^2 of community land (Mishra et al., 2017). Our work has also included joining forces with local communities to persuade the government to create a large (c. 7,500 km^2) state nature reserve in Mongolia's South Gobi, where mining threatened to destroy important snow leopard habitats (Mishra, 2016). Today, the Tost Nature Reserve is being managed by an alliance of local people, conservationists, and the government. Further, we have assisted the Indian government to formulate a landscape-level, participatory national strategy for snow leopard conservation, and have also assisted in management planning and implementation in snow leopard landscapes of India, Pakistan, Mongolia, and Kyrgyzstan (e.g., USL, 2011; Bhatnagar and Mishra, 2014).

While considerable literature has emerged on 'community-based natural resource management', especially from Africa, there has been little available for practitioners in the form of principles or guidelines for community-based conservation (Mishra et al., 2017). We have therefore formalized our approach to community-based conservation into a set of eight principles described by the acronym PARTNERS (Figure 13.2). The principles underscore the importance of relationship and trust-building with communities through the sustained and long-term *Presence* of conservationists amidst the local community; the *Aptness* of community-based interventions with respect to addressing the main threats to biodiversity, the underlying science, the local culture, socio-economics, the available or potential social capital, and the value of multi-faceted programs;

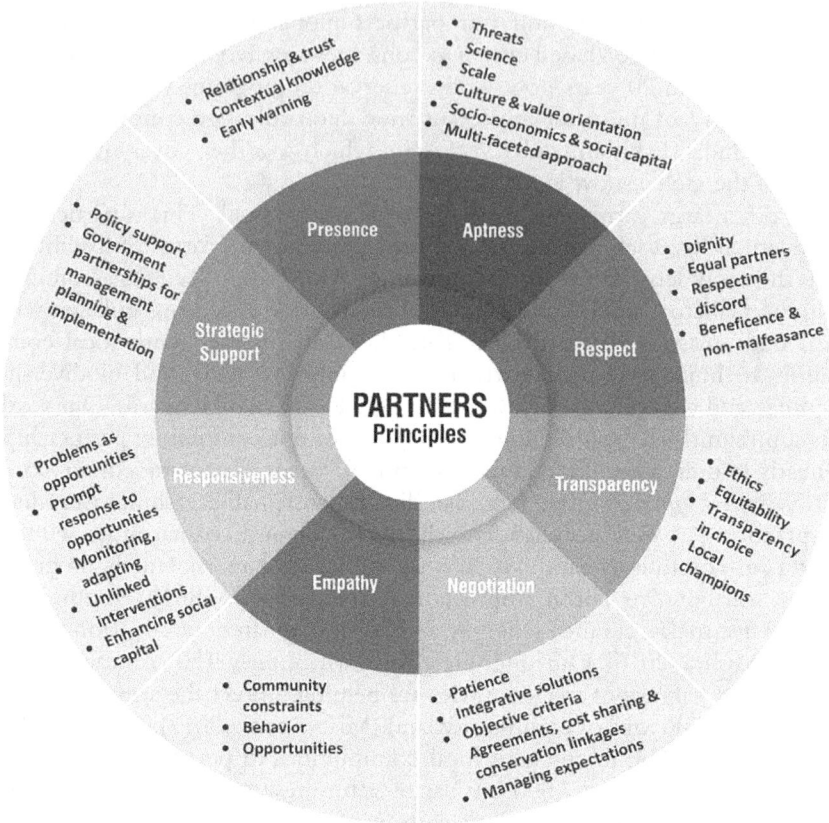

Figure 13.2 Principles for effective and ethical community-based conservation, developed from 20 years of working with local communities in Asia's mountains to enhance coexistence with snow leopards (*Panthera uncia*) (Mishra, 2016).

a relationship that views the community with dignity and *Respect*, and interactions based on beneficence and non-maleficence; high *Transparency* in interactions with local communities with open communication regarding each other's interests, and visible equitability in program benefits to community members; integrative *Negotiations* with local communities and interventions based on formal agreements and conservation linkages; the ability to view problems, constraints and opportunities from the community's perspective with a high level of *Empathy*; the ability to adaptively improve the programs and address emerging problems and opportunities with a high level of *Responsiveness* and creativity; and *Strategic support* to increase the resilience and reach of community-based conservation efforts through partnerships with governments in management planning and implementation, and policy and legal support.

These PARTNERS Principles have emerged from our field experiences and research, but have also been informed by ideas developed in fields of ecology, conservation and natural resource management, community health, social psychology, rural development, negotiation theory, and ethics (Mishra, 2016). Through a collaboration with social scientists, a training program has been founded on this approach and close to 100 conservationists have been trained so far in community-based conservation of not just snow leopards, but also large carnivores including the tiger and leopard (*Panthera pardus*).

The PARTNERS Principles approach has been effective in helping us create strong and long-term conservation partnerships with local communities across cultures (Mishra et al., 2017). Our interventions founded on these principles have been effective in getting people involved in snow leopard conservation, and in addressing especially the traditional threats such as poaching and retaliatory killing. Yet, today, there are other intensifying and emerging threats to snow leopards and their habitats that cannot be addressed adequately by local communities and conservationists, however strong their partnerships may be.

Landscape level efforts

With average home range size ranging from more than 124 km^2 for females and more than 200 km^2 for males (Johansson et al., 2016), snow leopard is a landscape species, and as mentioned earlier, most of Asia's protected areas are too small to support viable populations of this carnivore (Figure 13.1). Conservation of snow leopards must look beyond protected areas and focus on the larger landscape, including community lands as well as government lands outside the protected area network (Mishra et al., 2010; Bhatnagar and Mishra, 2014).

While on the one hand, there are huge opportunities for large carnivore conservation in Asia's high mountains since wildlife populations still persist over much of the region, on the other, conservation in these large landscapes is immensely challenging. These once remote mountainous habitats used by local communities and wildlife face immense pressures from competing priorities: mining, road construction and other linear intrusions, and large scale industrial and infrastructure development (Snow Leopard Network, 2014). The high mountains and snow leopard habitats are also particularly vulnerable to climate change (Forrest et al., 2012; Li et al., 2016).

Conservation in such multiple use landscapes with competing priorities requires a different paradigm. Unlike managing protected areas, such landscape-level conservation efforts need to engage with a much greater diversity of stakeholders, and transcend administrative, tenurial, land use, and, sometimes, international boundaries. For example, our stakeholder analysis in the snow leopard landscape of Spiti Valley in India generated a list of nearly 50 different stakeholder groups including various departments of the government, local community institutions, and the private sector that must be engaged with for meaningful snow leopard conservation (USL, 2011).

Conservation efforts in these landscapes require finding a balance and compromise between the needs of economic, infrastructural, and industrial development, and those of wildlife and biodiversity. They require promoting relatively less environmentally damaging green industry, and a more explicit recognition and valuation of ecosystem services, especially those that are used by local communities. To achieve snow leopard conservation at meaningful scales, therefore, it is not enough to engage with the traditional sectors of wildlife management and law enforcement. Engaging with local communities is critical. The other two important groups that conservationists must engage with are the government and the industry.

Engaging with governments

As mentioned earlier, for several years, we had been working with national governments in some of the snow leopard range countries to assist with national-level policies and action plans (e.g., India and Mongolia) and with relevant provincial governments to help create and implement management plans (e.g., India, Pakistan, and Mongolia). These efforts have been useful and have highlighted the importance of engaging with multiple stakeholders. The scale of these efforts, however, has not been enough to help address the challenges posed by the larger forces of development and associated habitat destruction across the snow leopard range. For example, mining started in the Tost Mountains in the South Gobi province of Mongolia despite our team having worked with local communities and the government for many years. A change in mining laws towards the turn of the century in Mongolia created a very liberal investment climate, and important wildlife habitats have since come under the purview of mining companies. It took our Mongolia team, our partner local communities, and the local government six years to convince the Mongolian Parliament to elevate the protection status of Tost Mountains in 2016 to a State Nature Reserve, which now prevents mining from expanding into snow leopard habitats. This was an important lesson for us that reiterated how critical it is to engage with national governments proactively for conservation.

In 2013, we started actively working with various partners to start a new initiative whereby we jointly aimed to raise awareness and resources for snow leopards and high mountain conservation at the highest levels of the range country governments. This effort has involved striking new and unusual alliances for us, as we started collaborating not just with other conservation organizations or enforcement agencies, but also with senior international bureaucrats and diplomats from multilateral agencies such as the United Nations, the USAID, and the World Bank, and with political leaders of the range countries.

The idea of this intergovernmental initiative, which has come to be known as the Global Snow Leopard and Ecosystem Protection Program (GSLEP), was born during a chance meeting and discussions between one of us (CM) and colleagues from Germany's Nature and Biodiversity Conservation Union (NABU) in Bishkek, Kyrgyzstan, in 2011. Following this meeting, NABU brought the idea to Rosa Otunbayeva, then president of the Kyrgyz Republic. In February 2012, the next president, Almazbek Atambayev, provided momentum to the initiative by writing to the World Bank, which in turn requested the Snow Leopard Trust to act as the technical partner. Together with the Global Tiger Initiative of the World Bank, we created a broad alliance of all major international conservation organizations and UN agencies to support this initiative. One of us (CM) along with other scientists and conservationists drafted various sections of the GSLEP document (Snow Leopard Working Secretariat, 2013), while scientists and conservationists based in the range countries (including YVB and MAN) assisted their respective governments in creating National Snow Leopard Ecosystem Priorities (NSLEP) as part this initiative. Each of the 12 national governments was requested to designate a senior official as the National Focal Point (NFP) for the initiative. A core group including the NFPs and representatives from the main conservation partners met a few times to finalize GSLEP during 2012–2013, often along the sidelines of other international meetings such as the Conference of Parties. On October 23, 2013, at a forum presided over by President Atambayev, the Bishkek Declaration for Snow Leopard Conservation was endorsed by all snow leopard range countries (www.globalsnowleopard.org/who-we-are/bishkek-declaration/). This Declaration prioritizes conservation of snow leopards and high mountain ecosystems with great emphasis on local community welfare and involvement, stronger protection, engagement of the industry, and multi-stakeholder participation. October 23 was designated as the International Snow Leopard Day. A steering committee comprising the environment ministers of the range countries was established to review progress annually and provide direction to the program. A Secretariat was created in Bishkek to support and coordinate the activities of GSLEP, and is currently managed by one of us (KS) along with a team of Kyrgyz staff and international advisors.

The GSLEP program aims to secure approximately 500,000 km^2 of habitat through community-based conservation, sustainable development and anti-poaching efforts in more than 20 large landscapes, each capable of harbouring at least 100 breeding snow leopards (Snow Leopard Working Secretariat, 2013). The foundation of the GSLEP program is formed by the individual NSLEPs developed by each range country's government with the help of their scientists and conservationists, supported by five cross-cutting global support themes including combating poaching and illegal trade, knowledge sharing and capacity enhancement, trans-boundary cooperation, addressing the conservation challenges of large scale infrastructural development, and research and monitoring. At various stages, current or former ministers and senior bureaucrats, including diplomats, have played an important role in providing support and momentum to the initiative.

This effort has created an important and influential international conservation alliance and unprecedented interest and visibility for the issues of snow leopard and high mountain conservation across Asia. As a result, range country governments have been demonstrating high political will to enable snow leopard conservation and sustainable development of local communities in snow leopard landscapes. Funds to the tune of US $50 million are already being channelled into snow leopard conservation by the range countries and funding partners such as the Global Environment Facility. The last steering committee meeting of GSLEP in Kathmandu in January 2017 was chaired by the prime minister of Nepal. At the Snow Leopard Forum in August 2017 in Bishkek, the participants included the president of the Kyrgyz Republic, the vice president of Afghanistan, deputy speaker of the parliament of Uzbekistan, and ministers and senior officials of all snow leopard range country governments. Countries are at various stages of preparation, endorsement, and implementation of management plans under the GSLEP program for the large landscapes that have been identified for focused conservation efforts.

Under this program, the range country governments also endorsed a new initiative for assessing snow leopard population across the range, described by the acronym PAWS (Population Assessment of the World's Snow Leopards). Currently, less than 2% of the entire estimated snow leopard habitat of Asia has been ever sampled for snow leopard abundance using scientifically valid techniques (camera trapping and genetics), with study areas being biased towards high density populations. All available global estimates of the snow leopard population are guesses. Yet, in what we think of as a highly ill-advised move (Ale and Mishra, 2018), IUCN recently changed the Red List status of snow leopards from Endangered to the lower threat category of Vulnerable, based on a notion that there are more snow leopards in Asia's mountains than previously believed. While categorically rejecting IUCN's move to change the listing of snow leopards through a joint statement (www.globalsnowleopard.org/blog/2017/09/26/statement-of-concern-regarding-the-status-of-the-snow-leopard-on-the-iucn-red-list/), range country governments have expressed the need and support for PAWS, which will aim to arrive at a global estimate of the snow leopard population using standardized techniques and based on stratified random sampling of up to 20% of the snow leopard habitat.

From a situation of relative ignorance at national levels where the existence of snow leopards was barely acknowledged, GSLEP has brought a high level of awareness and visibility to issues surrounding snow leopard and high mountain ecosystem conservation. For the first time, national political leaders of the range countries are showing interest and support in snow leopard conservation, and are participating and collaborating at the common platform provided by GSLEP. The potential influence of this high-level alliance is reflected in the new Secretary General of the United Nations Antonio Guterres issuing a video statement recently in support of snow leopard conservation (www.youtube.com/watch?v=sh9ecRzDyN8).

We have continued to play the role of key technical advisor to the GSLEP program since its inception, and have supported the functioning of the Snow Leopard Secretariat in Bishkek together with other partners such as WWF and NABU. However, we fully recognize that as scientists and conservationists, we could never have created such a program on our own, which has largely been an exercise in international diplomacy rather than in science, though backed by key pieces of information on snow leopard and prey ecology, threats and tested conservation models. While the on-the-ground results of this new conservation effort are still to be seen, this experience has shown us how much can be potentially achieved for large carnivore conservation through broad-minded partnerships with various sectors. The GSLEP Program is also helping conservationists to start engaging with the industry – the main agent of economic growth, and the main vehicle, if not the cause, of biodiversity loss (Mishra, 2016).

Engaging with the industry

Asia supports 55% of the world's human population, and approximately two-thirds of the world's poor (Asian Development Bank, 2014). There is high government emphasis on development and poverty eradication currently across Asia and in the snow leopard range countries that include some of the largest regional and even global economies such as China and India. Snow leopard habitats across Asia are increasingly being impacted by intensified emphasis on mining, road and rail construction, and consequent opening up of remote habitats, construction of dams, greater military infrastructure, and general industrial growth and investment (Snow Leopard Network, 2014). Climate change and increased frequency of natural disasters in Asian mountains due to habitat destruction and infrastructure construction (e.g., Kala, 2014) are making it essential that ecologically sensitive development alternatives and industries are considered for these landscapes urgently.

At the recent Snow Leopard Forum in Bishkek, apart from conservationists, industry leaders from sectors such as construction and infrastructure came together to discuss the possibilities of less damaging infrastructure designs, and investments into relatively green industrial development in snow leopard landscapes that could also assist with sustainable development of local communities. The initiation of this dialogue represents the first positive step to be undertaken towards the involvement of business and industry constructively in the conservation of snow leopards, though there is a long way to go. Interest has been expressed in setting up national level wildlife business councils within the existing national associations of commerce and industry. Such business councils could be expected to provide a mechanism to share conservation concerns within the private sector, a platform to facilitate green investments in high mountain landscapes, and to provide directions for conservation investments by the industry through their corporate social responsibility programs.

At a more local scale, we have initiated a new program to bring together the industry and the farmers in promoting sustainable livestock grazing in snow leopard habitats. This program, called Snow Leopard Friendly Cashmere, addresses the challenge of intensifying and unsustainable goat grazing pressures on snow leopard habitats due to the global demand for cashmere, which is a high value fibre (Berger et al., 2013). The program, being piloted in Ladakh, strengthens the ecological and cultural sustainability of cashmere production.

We assist farmers in adopting various predator-friendly and sustainable grazing practices that include, amongst others, rotational grazing, setting aside part of the grazing land for recovery of rangelands and wild prey populations, collaborative corral improvement, and community-based livestock insurance programs. They commit to protection of carnivores and their prey, and to follow their traditional festal calendars, paying special attention to reviving livestock-related ceremonies and events. The farmers are able to earn a modest premium amount, while we expect the industry to be able to create a niche market for the sustainably grown cashmere.

Through our research on ecosystem services, we also hope, over time, to create greater awareness of the contribution of nature to economic security and development. Our recent studies show that the estimated economic value of ecosystem services going into the production of 1 kg of cashmere in South Gobi of Mongolia was at least 18 times the average price that the farmer earned per kilogram (Murali et al., 2017b). More specifically, the estimated economic value of fodder and water that farmers derived from snow leopard habitats to produce 1 kg of cashmere was US $704, compared to US $39 that they earned per kilogram of cashmere. Elsewhere, we have also found a high level of economic dependence of local communities on ecosystem services that involve subsistence goods that they depend on (those that are not sold) and not related directly to their livelihoods. In the Spiti Valley of the Indian Himalayas, the economic value of subsistence goods people derived from the ecosystem was estimated to be 2.8 times higher than those directly contributing to their livelihood (Murali et al., 2017a). We believe that the general lack of such quantitative knowledge and recognition of the value of ecosystem services used by local people is one of the factors that often swings land use decisions against relatively sustainable practices such as livestock grazing in favour of damaging ones such as mining.

Concluding thoughts

In this chapter, we have suggested that competing societal priorities create conflicts over large carnivore conservation, and that there is a strong need for conservationists to work with competing sectors and stakeholders in order to conserve the world's large carnivores and wild spaces, while ensuring the welfare of local people and their sustainable economic development.

Efforts to promote coexistence between humans and snow leopards, as indeed other carnivores, require an effective combination of ecological and social-science research, increasing conservation capacity, promoting conservation awareness, involving local communities in conservation and conflict management programs, and strategic engagement with governments to facilitate policies that promote knowledge sharing and community-based conservation. It requires enabling sustainable development in snow leopard landscapes through partnerships with the industry. These can only come through effective multi-stakeholder collaborations.

We have discussed the inadequacy of Asia's protected areas in being able to conserve landscape species such as the snow leopard. From a policy perspective, we suggest that the management paradigm of protected areas must be debated and made more nuanced. Protected areas currently cover at least 14% of the earth's land and inland water areas (Juffe-Bignoli et al., 2014), and the Convention on Biological Diversity aims to increase this to 17% by 2020 (Target 11 under the Aichi Biodiversity Targets, https://www.cbd.int/sp/targets/default. shtml). It is important to recognize that the establishment of protected areas in large parts of the world has imposed immeasurable social and cultural injustices on local communities, in addition to curtailing their traditional access to natural resources (MacKay and Caruso, 2004; West et al., 2006; Mishra, 2016). In the context of local community dependence on ecosystem services, therefore, we need to ask what must a protected area protect from. From everything except tourism and recreation for the elite that the current protected area model encourages and facilitates? Based on our experience mentioned earlier in the Tost Mountains of Mongolia, where the local communities are now co-managing the newly created state nature reserve, we suggest that protected areas must be focused on affording legal protection to lands and ecosystems from ecologically damaging land uses and industries. At the same time, they must stop marginalizing and criminalizing local communities as has been historically the case in large parts of the world (West et al., 2006; Mishra, 2016). If the idea of protected areas can provide more space and tolerance for less damaging, relatively sustainable activities and livelihoods of local people who are dependent on ecosystem services, and can involve them in protected area management, we may even have a chance to expand the planet's terrestrial protected area coverage considerably, well beyond the 17% that the Aichi Biodiversity Targets aim for.

Whether or not such multi-stakeholder, landscape level grassroots to government and industry level efforts as our own described in this chapter will succeed or not remains to be seen. However, it is clear that effective and ethical large carnivore conservation in the future will require multi-disciplinary knowledge, and collaborations and respectful negotiations among affected communities, conservation practitioners, political representatives, public administrators, and the industry.

Acknowledgement

We are thankful to the Whitley Fund for Nature and the Acacia Conservation Fund for supporting many of our research and conservation programs.

References

Ale, S. B., and Mishra, C. (2018) 'The snow leopard's questionable comeback', *Science* vol 359, p1110.

Arlettaz, R., Schaub, M., Fournier, J., Reichlin, T. S., Sierro, A., Watson, J. E., and Braunisch, V. (2010) 'From publications to public actions: When conservation biologists bridge the gap between research and implementation', *BioScience*, vol 60, pp835–842.

Asian Development Bank (2014) *Poverty in Asia: A Deeper Look*, Asian Development Bank, Philippines.

Berger, J., Buuveibaatar, B., and Mishra, C. (2013) 'Globalization of the cashmere market and the decline of large mammals in Central Asia', *Conservation Biology*, vol 27, pp679–689.

Bhatnagar, Y. V., and Mishra, C. (2014) 'Conserving without fences: Project snow leopard', in M. Rangarajan, M. D. Madhusudhan and G. Shahabuddin (eds) *Nature Without Borders* (pp. 157–177), Orient Blackswan, New Delhi.

Chapron, G., Kaczensky, P., Linnell, J. D., Von Arx, M., Huber, D., Andrén, H., López-Bao, J. V., Adamec, M., Álvares, F., Anders, O., Balčiauskas, L., et al. (2014) 'Recovery of large carnivores in Europe's modern human-dominated landscapes', *Science*, vol 346, pp1517–1519.

Crespin, S. J., and Garcia-Villalta, J. E. (2014) Integration of land-sharing and land-sparing conservation strategies through regional networking: The Mesoamerican Biological Corridor as a lifeline for carnivores in El Salvador. *Ambio*, vol 43, pp820–824.

Dinerstein, E., Loucks, C., Wikramanayake, E., Ginsberg, J., Sanderson, E., Seidensticker, J., Forrest, J., Bryja, G., Heydlauff, A., Klenzendorf, S., and Leimgruber, P. (2007) 'The fate of wild tigers', *AIBS Bulletin*, vol 57, pp508–514.

Forrest, J. L., Wikramanayake, E., Shrestha, R., Areendran, G., Gyeltshen, K., Maheshwari, A., Mazumdar, S., Naidoo, R., Thapa, G. J., and Thapa, K. (2012) 'Conservation and climate change: Assessing the vulnerability of snow leopard habitat to treeline shift in the Himalaya', *Biolgical Conservation*, vol 150, pp129–135.

Gittleman, J. L. (2001) *Carnivore Conservation*, Vol. 5, Cambridge University Press, Cambridge.

Graham-Rowe, D. (2011) 'Biodiversity: Endangered and in demand', *Nature*, vol 480, no 7378, ppS101–S103.

Jacobson, S. K., and Duff, M. D. (1998) 'Training idiot savants: The lack of human dimensions in conservation biology', *Conservation Biology*, vol 12, no 2, pp263–267.

Johansson, Ö., Rauset, G. R., Samelius, G., McCarthy, T., Andrén, H., Tumursukh, L., and Mishra, C. (2016) 'Land sharing is essential for snow leopard conservation', *Biological Conservation*, vol 203, pp1–7.

Juffe-Bignoli, D., Burgess, N. D., Bingham, H., Belle, E. M. S., de Lima, M. G., Deguignet, M., Bertzky, B., Milam, A. N., Martinez-Lopez, J., Lewis, E., Eassom, A., Wicander, S., Geldmann, J., van Soesbergen, A., Arnell, A. P., O'Connor, B., Park, S., Shi, Y. N., Danks, F. S., MacSharry, B., and Kingston, N. (2014) *Protected Planet Report 2014*, UNEP-WCMC, Cambridge.

Kala, C. P. (2014) 'Deluge, disaster and development in Uttarakhand Himalayan region of India: Challenges and lessons for disaster management', *International Journal of Disaster Risk Reduction*, vol 8, pp143–152.

Karanth, K. U. (2002) 'Mitigating human-wildlife conflicts in southern Asia', In *Making Parks Work: Strategies for Preserving Tropical Nature*. Island Press, Washington, DC.

Lach, D., List, P., Steel, B., and Shindler, B. (2003) 'Advocacy and credibility of ecological scientists in resource decision making: A regional study', *BioScience*, vol 53, no 2, pp170–178.

Li, J., McCarthy, T. M., Wang, H., Weckworth, B. V., Schaller, G. B., Mishra, C., Lu, Z., and Beissinger, S. R. (2016) 'Climate refugia of snow leopards in High Asia', *Biological Conservation*, vol 203, pp188–196.

Linnell, J. D. C., Swenson, J. E., Andersen, R. (2001) 'Predators and people : Conservation of large carnivores is possible at high human densities if management policy is favourable', *Animal Conservation*, vol 4, pp345–349.

MacKay, F., and Caruso, E. (2004) 'Indigenous lands or national parks?', *Cultural Survival Quarterly*, vol 28, no 1, p14.

Madhusudan, M. D., and Mishra, C. (2003) 'Why big, fierce animals are threatened: Conserving large mammals in densely populated landscapes', in V. K. Saberwal and M. Rangarajan (eds) *Battles Over Nature: Science and the Politics of Wildlife Conservation* (pp. 31–55), Permanent Black, Delhi.

McNie, E. C. (2007) 'Reconciling the supply of scientific information with user demands: An analysis of the problem and review of the literature', *Environmental Science and Policy*, vol 10, no 1), pp17–38.

Meriggi, A., Brangi, A., Matteucci, C., and Sacchi, O. (1996) 'The feeding habits of wolves in relation to large prey availability in northern Italy', *Ecography*, vol 19, no 3, pp287–295.

Mishra, C. (1997) 'Livestock depredation by large carnivores in the Indian trans-Himalaya: Conflict perceptions and conservation prospects', *Environmental Conservation*, vol 24, pp338–343.

Mishra, C. (2016) *The PARTNERS Principles for Community-Based Conservation*, Snow Leopard Trust, Seattle. 180 pp. ISBN: 978-0-9773753-1-8; ISBN: 978-0-9973753-0-1 (e-book).

Mishra, C., Allen, P., McCarthy, T., Madhusudan, M. D., Bayarjargal, A., and Prins, H. H. T. (2003) 'The role of incentive programs in conserving the snow leopard', *Conservation Biology*, vol 17, no 6, pp1512–1520.

Mishra, C., Bagchi, S., Namgail, T., and Bhatnagar, Y. V. (2010) 'Multiple use of Trans-Himalayan rangelands: Reconciling human livelihoods with wildlife conservation', in J. Du Toit, R. Kock and J. Deutsch (eds) *Wild Rangelands: Conserving Wildlife While Maintaining Livestock in Semi-Arid Ecosystems* (1st ed.), Blackwell Publishing, West Sussex.

Mishra, C., Redpath, S. R., and Suryawanshi, K. S. (2016) 'Livestock predation by snow leopards: Conflicts and the search for solutions', in T. McCarthy and D. Mallon (eds) *Snow Leopards of the World* (pp. 59–67), London, Elsevier.

Mishra, C., Young, J. C., Fiechter, M., Rutherford, B., and Redpath, S. M. (2017) 'Building partnerships with communities for biodiversity conservation: Lessons from Asian mountains', *Journal of Applied Ecology* vol 54, pp1583–1591.

Murali, R., Lkhagvajav, P., Saeed, U., Kizi, V. A., Zhumbai-Uulu, K., Nawaz, M. A., Bhatnagar, Y. V., Sharma, K., and Mishra, C. (2017b) *Valuation of Ecosystem Services in Snow Leopard Landscapes of Asia*. Snow Leopard Trust, Nature Conservation Foundation, Snow Leopard Conservation Foundation, Snow Leopard Foundation Kyrgyzstan, and Snow Leopard Foundation Pakistan. Report Submitted to the Global Environment Facility (GEF) funded United Nations Development Program (UNDP) project on Transboundary Cooperation for Snow Leopard and Ecosystem Conservation.

Murali, R., Redpath, S. R., and Mishra, C. (2017a) 'The value of ecosystem services in the high altitude spiti valley, Indian trans-himalaya', *Ecosystem Services*, vol 28, pp115–123.

Packer, C., Loveridge, A., Canney, S., Caro, T., Garnett, S. T., Pfeifer, M., Zander, K. K., Swanson, A., Macnulty, D., Balme, G., and Bauer, H. (2013) 'Conserving large carnivores: Dollars and fence', *Ecology Letters*, vol 16, no 5, pp635–641.

Pfeffer, J., and Sutton, R. I. (1999) 'Knowing "what" to do is not enough: Turning knowledge into action', *California Management Review*, vol 42, no 1, pp83–108.

Ripple, W. J., Estes, J. A., Beschta, R. L., Wilmers, C. C., Ritchie, E. G., et al. (2014) 'Status and ecological effects of the world's largest carnivores', Science vol 343, 1241484.

Sharma, K., Bayrakcismith, R., Tumursukh, L., Johansson, O., Sevger, P., et al. (2014) 'Vigorous dynamics underlie a stable population of the endangered snow leopard Panthera uncia in Tost Mountains, South Gobi, Mongolia', PLoS One, https://doi.org/10.1371/journal.pone.0101319.

Snow Leopard Network (2014) *Snow Leopard Survival Strategy*, 1 Snow Leopard Network, www.snowleopardsurvival.org.

Snow Leopard Working Secretariat (2013) *Global Snow Leopard and Ecosystem Protection Program*, Bishkek, Kyrgyz Republic.

Suryawanshi, K. R., Redpath, S. M., Bhatnagar, Y. V., Ramakrishnan, U., Chaturvedi, V., Smout, S. C., and Mishra, C. (2017) 'Impact of wild prey availability on livestock predation by snow leopards', *Royal Society Open Science*, http://dx.doi.org/10.1098/rsos.170026.

Sutherland, W. J., Pullin, A. S., Dolman, P. M., and Knight, T. M. (2004) 'The need for evidence-based conservation', *Trends in Ecology & Evolution*, vol 19, no 6, pp305–308.

Treves, A., and Karanth, K. U. (2003) Human-carnivore conflict and perspectives on carnivore management worldwide', *Conservation Biology*, vol 17, pp1491–1499.

USL (2011) *Management plan for Upper Spiti Landscape including the Kibber Wildlife Sanctuary*, Himachal Pradesh Forest Department, Shimla and Nature Conservation Foundation, Mysore, India.

West, P., Igoe, J., and Brockington, D. (2006) Parks and peoples: The social impact of protected areas. *Annual Review of Anthropology*, vol 35, pp251–277.

Woodroffe, R., Thirgood, S., and Rabinowitz, A. (2005) *People and Wildlife, Conflict or Co-Existence?* (No. 9), Cambridge University Press, Cambridge.

14 Between politics and management

Governing large carnivores in Fennoscandia

*Camilla Sandström, Annelie Sjölander-Lindqvist,
Jani Pellikka, Juha Hiedanpää, Olve Krange, and
Ketil Skogen*

Introduction

The governance and management of complex conflicts over the presence of large carnivores at national, regional, or local level often comes into opposition with concerns for saving threatened species (Sjölander-Lindqvist et al., 2015; Redpath et al., 2017). Environmental collaborative governance or decentralization are increasingly promoted as useful means to manage conflicting goals, balancing different interests, and reconciling local concerns without compromising wildlife population viability (Council of Europe, 1979; UNEP, 1992; UNCED, 1992; UNECE, 1998; Sabatier et al., 2005; Newig et al., 2009; Emerson et al., 2012; Lange et al., 2013; Decker et al., 2016; Redpath et al., 2017). The three countries in Fennoscandia – Finland, Norway, and Sweden – are no exception to this development. In all three countries, new approaches to large carnivore governance and management have emerged since 2000, each including some elements of collaborative governance or decentralization of authority (Sandström et al., 2009; Sjölander-Lindqvist et al., 2015).

The approaches can be seen as a response to policy changes made in the 1960s and 1970s, when the three countries, after centuries of government-sponsored persecution of large carnivores, decided to protect the bear (*Ursus arctos*), the lynx, (*Lynx lynx*), the wolf (*Canis lupus*), and the wolverine (*Gulu gulo*) and thereby ending the bounty-based system that had been the norm since the 1600 (Pohja-Mykrä et al., 2005; Eriksson, 2017). The policy changes, which have strong support among the public in the three countries, have had a rather profound effect on the population numbers of the different species, although with a certain time lag of 20–30 years (Ratamäki, 2008 Dressel et al., 2015; Krange et al., 2017; Mykrä et al., 2017). However, partly due to the existence of varying ecocentric to anthropocentric values among the general public, different interests, and positive and negative experiences gained from increasing large carnivore presence, human–animal conflicts, and human–human conflicts are prevailing in the three countries (Eriksson, 2017; Krange et al., 2017).

In this chapter, we will compare the recent responses in terms of institutional changes that have been initiated to promote large carnivore recovery amidst human land uses. First, we will synthesize existing country-specific literature on policies and institutional design to advance the knowledge on the challenges, incentives for, and constraints on attempts to mitigate these types of conflicts. Thereafter, we will compare how different stakeholders and communities of interest and practice perceive, conceptualize, and understand their participation in decentralized large carnivore management. Finally, we discuss the findings of our comparison.

Policies and institutional design

Finland

Large carnivores, and wolves in particular, have been the subject of conflict in Finland for decades. The debate intensified after 1995, when Finland joined the European Union and committed itself to the Habitats Directive (European Commission, 1992) and its conservation goals (e.g., Bisi and Kurki, 2008; Ratamäki, 2008). As a consequence, tensions increased among some stakeholders, but also in general between residents of urban and rural areas and between ordinary citizens and authorities/researchers (Pohja-Mykrä, 2014). In light of this development, wolf hunting became one of the core policy issues (Bisi et al., 2010). Six years after Finland joined the European Union, the Commission initiated an infringement procedure against Finland. While the Finnish Hunting Act (1993) was harmonized to match with the Habitat Directive, the Finnish officials had allegedly issued too loosely derogations to the strict protection, required by the Habitat Directive Annex IV (only Finnish Lapland belongs to Annex V). The hunting had played a too dominant role in wolf management. The procedure, the calling of Finland to the European Court of Justice (ECJ) in 2005, and finally, the rendered judgement in June 2007, initiated many adjustments to better fit the aims and methods of government, the routines of administration, the practice of wildlife science, and the lifestyles of civil society (Hiedanpää and Bromley, 2010; Hiedanpää, 2013). To solve this, wolf management plans were prepared and enforced (MAF, 2005). Very similar guidelines were soon formulated for the lynx (MAF, 2007a) and the brown bear (MAF, 2007b). However, the latter two species give reason to notably smaller societal tension than wolves. The management plan for wolverine was published much later (MAF, 2014).

The problems in wolf management initiated a new era in Finnish wildlife policy. While the wolf population had increased since the beginning of the 2000s and spread to western Finland, it started to decrease few years after the implementation of management plans, and after the ECJ found Finland guilty of the unselective hunting of wolves. The monitoring of the population trend indicated an increase in illegal killing of wolves – possibly as reaction to the initial population increase and disability of killing them legally. After the ECJ judgment, the administrative rules became stricter, i.e., wolf hunting licenses were issued by the wildlife agency only regarding "problem individuals", defined as individual wolves causing demonstrably harm to private property. At that time, the Ministry of Agriculture and Forestry (MAF) issued licenses outside of the hunting season, which slowed down the process. Consequently, the Finnish Hunting Act was revised in 2008 to speed it up. It led the year-round issuing of the wolf, lynx, and brown bear licenses being decentralized to the regional administrative level and the regional wildlife agencies. However, the MAF held power to decide the hunting-season-specific maximum quota for the killing of wolves, lynx, and brown bear to ensure the sustainability of large carnivore populations and to decide solutions to human-carnivore conflicts. To further enforce the protection of large carnivores, the MAF also multiplied the nominal value of illegally killed large carnivores (MAF Decree 214/2010). A year later, a new penalty category, i.e., aggravated hunting offence, was added in the Criminal Code of Finland (39/1889).

In parallel to the formal policy planning, there were regional attempts to increase collaboration between stakeholders and mitigate tensions. This resulted in the establishment of Regional Large Carnivore Committees (RLCCs), with the first being established prior to the planning process in the Province of North-Karelia in 1999. The initiative to set up the RLCCs came from the members of the non-governmental and governmental bodies for the sake of increasing the face-to-face-communication and cooperation among representatives of stakeholders with conflicting interests (Sandström et al., 2009; Pellikka and Sandström, 2011). This arrangement initiated a missed dialogue between stakeholders and encouraged to establish similar committees, first in Kainuu in 2001, and then in Northern Savo in 2004. At the same time, some other regions established forums that focused not only on large carnivores but also moose and other ungulates, rabbit (metropolitan area), great cormorant, and grey seal management. One could characterize the first decade of the 2000s as the era of informal regional wildlife forums in Finland.

The consultative RLCCs and other forums did not have any formal tasks. The informal role allowed them to rather freely define their specific purposes and activities (Pellikka and Sandström, 2011). According to the representatives, the main role of RLCCs was to exchange knowledge between stakeholders and authorities. When inquired, nearly 70% of the respondents reported being moderately to highly satisfied with the RLCCs' overall activity in 2008 (Sandström and Pellikka, 2008). The representatives were largely satisfied with the internal communication, and reported that stakeholder relations were more trustworthy. However, the opinions about the usefulness of the RLCCs and overall satisfaction to activity varied among the stakeholders. The stakeholders more concerned about the human interests were more positive to the RLCCs compared with the ecocentric-minded stakeholders, such as the nature conservation NGOs.

The reform of the wildlife administration in 2011 affected both the role of the RLCCs (see next section) and the procedure of issuing wolf, lynx and brown bear hunting licenses. The issuing of the licenses was re-centralized (back) to the national level (Wildlife and Game Administration Act 158/2011). This adjustment made it possible to unify the specific conditions for issuing licenses that had varied to some extent from region to region and to still take advantage of regional expertise (the regional officials prepare the decision). The digitalization of the process ensured that the process remained fast enough. The reform thus contributed to the separation of wildlife administration from interest politics by removing the decision-making powers from the boards (directors) of regional wildlife agencies (locally elected officials, most of them representing hunters) to the director for public administration functions in the Finnish Wildlife Agency.

The MAF commissioned an evaluation of the Finnish large carnivore policy in 2013 to explore how the prevailing policy and its implementation had succeeded in accomplishing ecological, economic, and social sustainability in population management. The evaluators (Pohja-Mykrä and Kurki, 2014) concluded that the management measures regarding brown bear and lynx were found to be quite feasible but lacked place-based policy, i.e., the policy where local and regional policy objectives influence decisions in the desired manner. According to evaluators, "in the case of wolf outside the reindeer-herding area, and wolverine in the reindeer-herding area, severe conflicts have arisen" due to in particular loss of livestock but also, in the case of the wolf, negative impact on hunting. By conflicts, they referred to multitude of issues, rooted in the low legitimacy of top-down conservation. The evaluation report concluded (Pohja-Mykrä and Kurki, 2014, p. 12, see also Ratamäki, 2008) that future large carnivore policy should include a clear societal strategy and give equal weight to social and ecological concerns.

The evaluation was the starting point for the updating processes of the management plans of wolf, lynx, and brown bear. The main aim of the updated wolf management plan was to co-create concrete actions and thereby better facilitate human-wolf coexistence in rural areas of Finland (Hiedanpää et al., 2016). The management unit of the updated plan was not anymore based on the six management regions with different population trend goals, defined by the Working Group for Large Terrestrial Carnivores in 1996 (MAF, 1996), or three management regions defined in the original wolf management plan (MAF, 2005), but on each wolf territory separately.

The implementation of the updated management plan (MAF, 2015) started in 2015 with the managerial hunting of the wolf, the establishment of the territory-level cooperation between groups representing local stakeholders, and the intervention threshold pilot with the police, plus a dozen of other measures. A survey from October 2016 showed that about one-third of adult citizens living in the wolf regions actually knew that the wolf policy had changed in 2015 (Taloustutkimus, 2016). Among the measures initiated prior to the survey, the ones focusing on fencing the livestock or sheep pastured were generally regarded as having had the highest positive impact on both wolves and local citizens, while killing or flushing wolves further from house yards were regarded as having had the highest positive impact on local citizens.

Norway

During the late 1980s, large carnivore populations started to increase in Norway, and conflicts over them have been quite intense ever since. Several stakeholder groups, like sheep farmers, reindeer herders, hunters, dog owners, and ordinary rural dwellers claim that the growing number of large carnivores has a negative impact on their livelihoods, lifestyles and well-being. Among these groups, there is no doubt that those involved in the livestock grazing industry, and especially the sheep farmers, have had the strongest voice. The situation spurred the first parliamentary white paper on large carnivore management in 1992 (Ministry of the Environment, 1992), outlining a dual goal for Norwegian large carnivore management. The primary goal was to preserve viable large carnivore populations. At the same time, the conflicts they caused should be limited as much as possible. This is not an easy goal to reach and the environmental management agencies have not been able to reduce conflicts. Among researchers, it has been claimed that the large carnivore management system is unsustainable, since viability of the Norwegian populations is dependent on influx from the larger Swedish populations of large carnivores (Krange et al., 2016). From early on, large carnivore management was facing legitimacy problems, and the system Norway has today is designed as a response to this challenge.

In 2005, the Norwegian parliament decided to divide the country into eight management regions for large carnivores. In an effort to mitigate conflicts by bringing large carnivore management closer to the "local level", they transferred authority to regional large carnivore management boards. Population goals are still decided on the national level, but the boards can set hunting and culling quotas as long as population goals are met. The boards must produce management plans for their region, and they have fairly generous budgets for damage prevention and conflict mitigation measures. Board members are indirectly elected politicians from the county assemblies and from the Sámi parliament (in regions with Sámi reindeer herding). Hence, board members are nominated by the assemblies, but formally appointed by the Ministry of Climate and Environment. Although politicians in the boards are democratically elected to the county assemblies, which nominate them as their representative in the carnivore committees, they are accountable to a government bureaucracy and not their constituents. If not unique, this is an atypical system within Norwegian public administration. The manager role is obviously unfamiliar to most politicians. They are, however, expected to have intimate knowledge of the situation in their regions, and it was hoped that decisions informed by such insight would give large carnivore management increased legitimacy (St.meld. nr. 15 [2003–2004]).

The county governor's environmental department acts as secretariat for each board, preparing proposals for decisions and providing scientific knowledge on the different species. The national Environment Agency has no power to instruct the boards, unless regional population goals are not met, but the boards are obliged to seek the agency's advice. In the breeding period from February 16 to May 31, the Agency has the management authority over wolves, wolverines, and lynx, and from October 16 to May 31 for the brown bear.

The boards are intended to be the central management bodies in Norwegian large carnivore management. However, the system is still fundamentally controlled at the national level, e.g., through exact population goals set by the parliament, and hence the boards are faced with strict limitations to their powers. For instance, when they make decisions on hunting quotas, they make them within very detailed frames that the national parliament has determined. The regional population goals are precise, meaning that they are both a maximum and a minimum, and the boards can only allow hunting if the population size is spot on or exceeds the target. If the populations decrease below the target, management is handed over to the national level and decisions on quotas, etc. are made by the Environment Agency. This creates a mechanism whereby the boards can manage themselves out of power, and they have done so on several occasions. Another example of the detailed frames and the related constraints that the national level has set up is that all boards are required to make a management plan where the core management tool is zoning, separating grazing areas for livestock (and breeding areas for reindeer) from breeding areas for large carnivores. The topography, area size, large carnivore biology, and even sheep and reindeer behavior make this a very difficult, if not impossible, task. In general, it is safe to say that the boards have been less than successful in their efforts to reduce livestock depredation (Krange et al, 2016). Some of the boards have found zoning to be contra-productive and have adopted practices that go against what the ministry demands. This is, for instance, evident in the management plan for the southernmost large carnivore management region, where only a very small area in the southwestern part of the region is prioritized for grazing livestock.

A special management zone for wolves has been established outside this system, due to particular management challenges, including the virtual impossibility of combining territorial wolf packs and free-ranging sheep and reindeer husbandry. This zone involves four counties, covers two management regions partially, and is managed by the two regional boards in collaboration. This is an extremely complicated system, and even more so because the two regions are different in terms of economic structure and demographics (e.g., one of them contains Oslo, the capital and by far the largest city in the country).

Considerable resources are used on compensation for predation and measures to prevent livestock loss also within the wolf zone, although livestock production is a limited economic activity there. Norwegian regional policy is a factor that underpins the centrality of the farming perspective, which also explains where the wolf zone boundaries were drawn (i.e., around an area with few sheep). Maintaining rural settlement has been a stable political goal in Norway, and stimulating the agricultural sector has been a cornerstone of this policy. During the long period when large carnivores were almost absent, husbandry methods entailing free-ranging sheep with limited supervision were developed. When brown bears, lynx, and wolverines increased in numbers and expanded their ranges, effects were serious, and conflicts have been flourishing ever since. Thus, the livestock/farming focus was well established when the wolves returned. It has led to a single-sided implementation of mitigation efforts (including the wolf zone and its intricate management system) that are misguided when it comes to wolves, as sheep farming is very limited in Norwegian wolf areas. Wolf conflicts in Norway are based on predominantly other sources than livestock predation (Skogen et al., 2017).

The regional management system was evaluated in 2016 (Krange et al., 2016), and one main conclusion was that it does not contribute to the reduction of conflicts. Management of the four Norwegian large carnivore species has almost exclusively focused on livestock owners. Policies are quite blind to conflicts other than those involving sheep farmers and reindeer herders (Skogen et al., 2017).

Sweden

In a similar vein as in the neighboring countries, large carnivores have returned to the Swedish landscape. Since the return of the brown bear, lynx, wolf, and wolverine, large carnivore management has been conflict ridden, and debates over wolf protective strategies, in particular, have been centered around a contrast between urban and rural interests (Ericsson and Heberlein, 2003; Cinque, 2008; Heberlein and Ericsson, 2008; Eriksson, 2017). People living in rural areas have perceived themselves as not having sufficient control over wolf management, which has led segments of the rural population to consider Swedish wolf management as illegitimate (Sjölander-Lindqvist, 2009; Sjölander-Lindqvist, 2011). Furthermore, the increasing numbers of wolves were jeopardized by inbreeding and illegal hunting, with the latter reflecting current social conflict (Committee of Environment and Agriculture, 2009; Liberg et al., 2005, 2011; Sand et al., 2010; Government Official Report SOU, 2007).

Due to increasingly intense conflicts between stakeholder groups and reduced trust in the authorities, the government decided to introduce the first coherent large carnivore policy in Sweden in 2001. The policy, an adaptation of the EU Habitats Directive (European Commission, 1992), defines ecological criteria for the four species based on population targets. Already at this time, both the brown bear and lynx had met the population target and were assigned minimum levels for annual regeneration, opening up for license hunting above these targets. Wolf and wolverine species were assigned temporary population targets which would be re-evaluated once achieved. The fact that the long-term survival of large carnivores was to be taken as a starting point was, however, not exclusive: According to the governmental investigation of 1999, hunting interests and the prospects for continued livestock husbandry should also be taken into account since "the presence of large carnivores may affect the daily lives of people who reside in areas of large carnivores" (Government Official Report SOU, 1999, p. 185). The policy opened up for concerned stakeholders and agencies to be included in the management process within Regional Large Carnivore Committees (RLCCs) in order to support a socially sustainable implementation of the predator policy (Cinque, 2008; Sandström et al., 2009). The RLCCs were built according to ideas of interest representation, and consisted of a mix of governmental actors such as representatives of the County Administrative Boards (CAB), municipalities, the police, and nongovernmental actors such as hunters' organizations, landowners, nature conservancy, and reindeer herding communities (RHC). RLCCs were established in every county with residential large carnivore populations (Duit et al., 2009; Sandström et al., 2009).

A decade later, the RLCCs were deemed to have failed to address the perceived lack of legitimacy, and were abolished (Government Bill, 2009); Lundmark et al., 2014). Instead, a new policy, adopted in 2010, replaced the RLCCs with Wildlife Management Delegations (WMDs), to further increase regional and local influence over large carnivore management. The WMDs are led by the county governors and includes political party representatives and representatives of the following interests: forestry, local business, outdoor recreation, hunting, nature conservation, agriculture, reindeer herding, fishery, mountain farming, and the Sámi Parliament (SFS 2009:1474). In contrast to the RLCCs, the WMDs have been delegated formal decision-making powers. This means that the WMDs have been given the authority to suggest minimum and interim levels of county carnivore populations. The internal structure of the WMDs decision making is planned to be consensus-driven, but the delegations are allowed – when consensus cannot be achieved – to employ a majority vote, with the county governor having the casting vote, in the case of a tie (SFS, 2014:1242). Due to critique, both from research and specific interest groups (e.g., environmental advocacy groups) about the representativeness of the WMDs, the delegations were evaluated by the government in 2016. The evaluation concluded that the representation was unbalanced and as a result, environmental groups as well as representatives of eco-tourist organizations have been granted one seat extra in every WMD by the government. This has caused strong protests among organizations representing hunters, farmers, reindeer herders, and landowners, further aggravating the conflicts instead of reducing it, as intended.

Perceptions of participation in large carnivore management

Finland

The management plans (MAF, 2005, 2007a, 2007b) were the first nationwide attempts to bring structure and shared baseline understanding to plural and often conflicting issues over large carnivores. While the process provided new channels for collecting the ideas, views, experiences, and goals of the people, the regional public hearings were criticized, in particular, on the basis that participants did not widely represent the demographics or general citizens' views in the area (Bisi and Kurki, 2008). While this was a true challenge to any participation process, such a critique may also have arisen if the outcomes of the process were not regarded as pleasing to one's own interest (Hiedanpää and Pellikka, 2017).

Some of the RLCCs, including the first one established in 1999, were still operating in 2017. However, most of them finished their activity in the beginning of the 2010s. The reason behind this turn was mainly the reorganization of the wildlife administration in 2011, which has included, amongst other things, the establishment of 15 regional wildlife councils and one national, appointed by the MAF. The new National Wildlife Council (NWF) had 24 members, while the new Regional Wildlife Councils (RWF) had 10 members each. Most members in the latter were the same as in the RLCCs. However, a considerable number of the animal husbandry, hunting, (hunting) dog breeding, and environmental NGOs participating in the RLCCs were excluded from regional councils – but not from the national council. According to Wildlife and Game Administration Act (2011), regional wildlife council membership was granted to the representatives from the Regional Council (the public authority responsible for the general development of the region); the regional Centre for Economic Development, Transport and the Environment; to the regional Forestry Centre; and to a landowner organization (NGOs).

Both the national and the regional wildlife councils have a formal consultative role regarding large carnivore management. They participate in strategic planning, meet 3–4 times a year, and organize annually wildlife forums for regional stakeholders to inquire their perceptions as additional input into the planning processes. The annual feedback collected by the Finnish Wildlife Agency indicates that three-fourths of the members view their influence on strategic planning from moderate to very high (Finnish Wildlife Agency, 2017, unpublished). Over the years, the proportions of members who regarded their influence as low has gradually increased, however. The major strengths of wildlife councils, according to feedback given by participants, lay in exchange of expert knowledge, while the main challenges were the mitigation of large carnivore conflicts, and a perceived lack of influence on large carnivore policy.

A recent development in the participation of citizens and stakeholders was the updating of the wolf management plan in 2014. The participatory process started with facilitated online forum discussions, offering an opportunity for citizens to share their wolf concerns nationwide (Salo et al., 2017). The forum had five sections for different regions, each including several wolf territories, and a sixth section for people living in areas with no known wolf territories. The purpose was to identify participant beliefs about what they understood as problematic behavior of the wolf, why these problems have emerged, and how to alleviate the problems. In total, 100 participants took part in the discussions by writing nearly 600 comments, suggesting different solutions to the related problems.

The process continued with a survey made among a random sample of adult citizens, stratified within and outside wolf regions. The survey inquired the ordinary citizens' preferences for large carnivore management methods, alternative damage compensation schemes and their stated willingness to personally participate in various management actions, or to serve as private donors in support of Finnish wolf management (Hiedanpää et al., 2016). The results revealed, for example, that a major proportion of the adult population was willing to participate in wolf monitoring, at least to a minor extent. Nearly half of the respondents said they were willing to assist in the building of wolf-proof fences around pastures. The willingness to voluntarily donate varied from 15–20% of the population, depending on the specific target, showing public participation had much potential to move beyond to commenting on management or participating in the planning. The process ended with a series of eleven workshops held in selected wolf territory areas (e.g., Hiedanpää et al., 2016). From 20–30 participants were invited to each of them, based on their residence or employment in a particular wolf territory. The idea was to identify, co-create, and come to an agreement on territory-level management actions and potential projects that could be pursued. The reception of the workshops was positive, based on the verbal feedback given by participants, and raised hopes among participants that many problems can be mitigated by new measures. As a concrete output of the workshops, individual workshop participants identified 375 new measures. These were condensed in the group discussions to 100 which could be further elaborated. Each workshop also suggested some 5–10 draft development projects. The whole process, the key findings of the material collected, and the proposed management actions were introduced to the RWFs and stakeholders, and their comments and suggestions were inquired. Before publication, the draft of the management plan was again commented by some 30 organizations and NGOs nationwide.

All the participatory methods applied in the updating of the wolf management plan, except for the workshops, were also applied later in the updating processes of lynx and brown bear management that took place in 2016.

Norway

As mentioned previously, the Norwegian carnivore management system seeks to balance multiple interests that come into conflict with each other. One important objective for the regional management boards is to reduce conflicts between interest groups, and while grazing (sheep and reindeer) is given special focus in the mandate of regional management, the system requires boards to consider other stakeholders and interest groups as well. Research carried out in areas with large carnivores has shown that interviewees have scant knowledge about and, indeed, little interest in talking about details in the formal structure of large carnivore management or the decision-making procedures that determine the current regime (see Skogen et al., 2017). Whatever opinions they might have about carnivores, people are more concerned about the general political level (to which they often refer in vague terms but with keen interest) and highly specific local issues, like observations or incidents near their own doorstep. The regional level is, at best, favored with general comments. The interviews show that both the politicians and the processes by which they are appointed in the grand scheme of thing is unknown to the public (Skogen et al., 2017). The fact that boards are made up of politicians only came as a surprise to some otherwise well-informed interviewees (Bekk Norstad and Skogen, 2017).

People often express a desire for more local power in decision-making (especially in connection with culling permits), but very few know much about the role of the local authorities or have any practical ideas about how it should be done. However, the power structures in this field seem unclear or even unknown to many people. Moreover, few care how the management is organized, as long as they are not adversely affected by decisions made by the board. Regardless of people's attitudes toward large carnivores and their management, most with a clear view on the situation link these consequences to elevated political levels – that is, to the general policy on large carnivores – not what are perceived to be organizational details that make no difference in the bigger picture. Few see the regional boards or other institutional arrangements as relevant factors in this context.

The regional management system was evaluated in 2016 (Krange et al., 2016). One of the main conclusions was that it had relatively little influence on the large carnivore conflicts. Central actors in the field did not view the boards as being especially important. Some interest groups (farming, hunting) defended the boards on principled grounds, e.g., because they could have more relevance if they were given greater authority, or because the regional model represents a "normalization" of carnivore management (i.e., more similar to management for other wildlife species). Few believe that the boards have any practical consequences for management today.

The choice of management model reflects which aspects of the conflict are given priority. A continuation of the present model will clearly have the support of farming interests and, probably, hunting interests. It is clearly possible to choose to prioritize these interests, and especially grazing, if it is concluded that they are the most important. However, this would only reflect a part of the overall conflict, and there are interest groups that do not think their interests are best served by the present system. Both conservation groups and the reindeer herder's organization see problems with the boards, which they believe systematically favor sheep grazing interests. These groups would be more open to a system where all interest groups would have a seat, where no point of view would be privileged or ignored, and where it would be possible to lead open discussions.

Lack of trust among actors is another big problem in large carnivore management, partially because carnivore conflicts are embedded in wider societal conflicts, entailing urban-rural power relations as well as socio-economic and cultural change. Despite these tensions, dialogue and information exchange should be core objectives. Even if it will not solve conflicts, it will certainly give management authorities a more complete understanding of how the large carnivore issue is viewed from different perspectives. The mandate of the regional boards clearly requires that such dialogue is to be established.

However, there is a great deal of variation in the way that different boards manage to keep a dialogue with interest groups and the general public. Some clearly bias their contact towards farming interests, which they see as their special responsibility to accommodate. Although meetings are open to everybody (to attend, but not partake the floor) and there is an opening for prepared talks from NGOs, etc., daytime meetings and huge travel distances in some regions clearly favor resourceful actors like the farmers' trade organizations.

Everybody expresses a desire for including more "knowledge" into management, but the different interest groups are not talking about the same type of knowledge, or knowledge from the same sources. Some would prefer more scientific knowledge from biology, others point to economics and livestock research, while still others refer to experience-based local knowledge. The exchange of diverse knowledge forms could be enabled by dialogue forums, but these do not exist. We can conclude that the regional management system does not contain many elements of collaborative governance. It consists largely of interactions between the county governor's office, large carnivore management boards, the national Environment Agency, and the Ministry of Climate and Environment.

Sweden

The purpose of the first measure introduced, namely the establishment of the RLCCs, was to generate trust in, credibility for, and commitment to predator policy implementation, and to reduce conflict (Cinque, 2008; Sandström et al., 2009; Cinque and Sjölander-Lindqvist, 2011; Hallgren and Westberg, 2015, Lundmark and Matti, 2015). In 2005, the Swedish government evaluated the operation of RLCCs. The evaluation concluded that the institutional structure had been insufficiently designed, and the authorities have not been successful in establishing efficient, trustworthy policy implementation and meaningful participation standards (Faugert et al., 2005). Farmers and hunters continued to feel excluded from management and decision making (Sjölander-Lindqvist, 2008; Sandström et al., 2009). This inspired the establishment of Large Carnivore Emergency Groups in the counties of Dalarna and Värmland – an institutional structure intended to encourage dialogue between County Administrative Boards (CABs), hunters, and farmers who had suffered economic damages due to large carnivore attacks on dogs or livestock. The Large Carnivore Emergency Groups were also expected to eventually lead to a reconstructed interface between the state and the public that would more effectively address the challenges introduced by large carnivore presence (Cinque and Sjölander-Lindqvist, 2011).

Despite these intentions, local leverage and locally-approved decisions have failed. Continuously, some perceive the reappearance of large carnivores in the forest fringe and mountainous areas as jeopardizing the prospects for the survival of the rural community. By establishing RLCCs, the authorities could learn more about the local contexts and consequences of large carnivore presence. It was also assumed that this would facilitate the dissemination of information to the local level. RLCC members were encouraged to share information and knowledge with their respective organizations. In this sense, the RLCCs were planned to be sensitive to process, context, and time, and function as connective and transformative arenas in which appropriate understandings of the management effort and related actions would evolve (Sandström et al., 2009). Despite these intentions, polarized understandings have remained. As indicated, the RLCCs did not succeed in finding ways to deal with contradictory perspectives. Instead of promoting dialogue and the exchange of views on large carnivore presence, the RLLCs had allowed social conflict to continue (Government Official Report SOU, 2007), primarily due to different interpretations of mandate. The officials understood the RLCCs as a measure to facilitate policy implementation and increase public acceptance, while carnivore skeptical parties interpreted the RLCCs as a channel to give voice to local concerns, and as a means for contributing to a rearrangement of politics and administration (Sjölander-Lindqvist, 2008, 2009).

If we move to the WMDs, we see similar patterns. Despite substantive decentralization of decision-making, broadened base of competence and the inclusion of additional interests, studies suggest that the WMD has failed to achieve legitimacy (Duit and Löf, 2015). The methods used have not been able to mitigate between value conflicts within the WMDs, forming the base for two distinct coalitions built on anti-carnivore/pro-WMD and pro-carnivore/anti-WMD beliefs (Lundmark and Matti, 2015). Interestingly, this attitudinal divide can also be found among the general public. When asked about what actors should be included in large carnivore management, the respondents in urban areas preferred to be represented by conservation interest groups, while respondents in rural areas preferred to be represented by utilitarian interest groups. The WMDs thus tend to represent the attitudes found among the general public. However, studies show that the actors within the WMDs, probably spurred by the presidency of the WMDs, tend to focus on coalition building and strategic voting, relating to one or the other value, instead of trying to bridge the gap through deliberative processes as intended by the legislator (Duit et al., 2009; Lundmark and Matti, 2015; Sjölander-Lindqvist et al., 2015).

Problems related to this gap are reflected in an ongoing study (Sjölander-Lindqvist and Sandström, forthcoming) investigating in depth the ideas and conceptualizations of participation in two WMDs. According to the results, the current structure of the WMDs seems even to perpetuate the understanding that there is an imbalance in representation, having to do with the design of the delegation. In particular, we find that some members understand their interests to be underrepresented, rendering a difficult position in the delegation. One dimension refers to members who represent a concerned interest who find that they should have more to say on issues that potentially may affect their livelihoods: "I believe that we are underrepresented. We who bear the costs of having large carnivores in the landscape. It is the farming and the reindeer interests that should have more to say". The view of the delegation being improperly designed when it comes to representation is shared by the nature conservation interests who also refer to the representational imbalance as problematic, as more or less the only member, apart from some political party representatives, proposing a careful protection of large carnivores. The issue of underrepresentation also holds a gender dimension: "Sometimes I think I should stand on her side [because they are both women] to show some sisterhood". However, deviation from the line of representation on the grounds of gender imbalance is not understood as a proper action but, allying with the issue of certain interests being underrepresented that ""at least, there should be more than one [representing the nature conservation interest] in the delegation". In contrast, hunting interests seem to be more optimistic about their presence: "We have two politicians who always stand up for us. I guess it has to do with them being hunters".

Another dimension affecting the outcome of regionalized delegation of power concerns interpretation of mandate and representation. Among Swedish delegates, uncertainty seems to prevail regarding their actual role, and the role of the CABs in the work of the WMDs. As held by the CABs:

> [Y]ou have been nominated as a delegate because you are familiar with these questions, but when you attend the meetings of the delegations, you are in effect, a civil servant. Once a delegate, you are no longer a representative of an organization, you are here as your own with your knowledge.

Many delegates find this order of things confusing, referring that they are in fact a representative for their organization, such as: "I must represent those who have put me there, I must point out the shortcomings and stand up for the views my organization has when it comes to the predator policy" (Sjölander-Lindqvist and Sandström, forthcoming).

A third theme which recurred in the interviews was the lack of support from many of the delegates' respective organizations. While the hunting interest representatives felt strongly supported by their organizations and had networks of knowledge to consult if necessary, many delegates of the other organizations indicated the opposite. They described lack of interest, that their organization had no clue about the role and mandate of the WMDs, and that they wished to have more support so they could hold a more knowledgeable position in the delegation. The issue of support also referred to the regional authority. The delegates asked for improved preparedness before meetings, including material and meeting agenda in advance to allow for discussing the tasks of the upcoming meeting with their organization. They also asked for more training, both to raise their competence on the different aspects of large carnivore management but also as an opportunity to better familiarize themselves with fellow delegates. In sum, several issues relating to how the actors perceive the procedure, their role and mandate, as well as representation and accountability, were highlighted as problematic in relation to the current operation of the WMDs.

Discussion

When comparing the policies and institutional design, all three countries share attempts to decentralize power over large carnivore management through various forms of regional forums (Table 14.1). The setup of these bodies is motivated by the need to establish relationship-building processes among the involved actors being close to the problems experienced so that they can follow on the presence of large carnivores. In all three cases, the processes are assumed – or at least hoped – to result in the sharing of information and knowledge, joint agreements, dialogues, and conflict mitigation measures. These endeavors are in line with international conventions, such as the Convention on Biodiversity and the Bern Convention, which the three countries have ratified. It also resonates well with research findings on identified benefits with decentralization and collaborative governance (Ansell and Gash, 2007; Emerson et al., 2012). However, despite the ambitions to mitigate large carnivore related conflicts, the results from different studies and evaluations show a rather mixed outcome of the processes.

When it comes to the institutional arrangements, such as carnivore committees, Norway clearly differs from the two other countries. Instead of involving stakeholders, Norway has chosen to involve indirectly elected politicians at the regional level. This can partly be explained by the more decentralized structure of the Norwegian society. It is also in line with recent developments in Norwegian nature conservation policy, where the management of some protected areas and national parks has been decentralized to lower level authorities (Sandström et al., 2009; Hongslo et al., 2016). Despite the decentralized character of Norwegian politics and administration, this form of decentralization in which politicians more or less assume the role of wildlife managers is rather foreign to the Norwegian system and has not been very successful in delivering the desired output in terms of conflict management, according to a recent evaluation (Krange et al., 2016).

Sweden and Finland have chosen another path for stakeholder involvement in the 1990s and 2000s. While the Swedish parliament at first decided to set up formal RLCCs based on stakeholder representation, the Finnish RLCCs were based on informal, bottom-up processes, involving a wide array of interest groups (NGOs) and authorities at the regional level. In both Finland and Sweden, the RLCCs provided arenas for face-to-face communication, building of trust, and exchange of knowledge and ideas. While consensus on difficult large carnivore-related issues was not always reached, the need for such committees as instruments for mitigating tensions was largely recognized both by stakeholders and state authorities.

Based on evaluations and academic studies, both Finland and Sweden decided to substitute the RLCCs in the 2010s. In Sweden, the RLCCs were replaced by

Table 14.1 Institutional design for large carnivore management in Finland, Norway, and Sweden

	Finland	Norway	Sweden
International	Bern Convention, CBD	Bern Convention, CBD	Bern Convention, CBD
European	EU Habitats Directive		EU Habitats Directive
National	The parliament of Finland	The parliament of Norway	The parliament of Sweden
Ministry	Ministry of Agriculture and Forestry	Ministry of Environment and Climate	Ministry of Environment and Energy
Agency	Finnish Wildlife Agency	Norwegian Environmental Agency	Swedish Environmental Protection Agency
Council	National Wildlife Council (consultative)		Three Cooperation councils between the County Administrative Boards and the Sámi Parliament (north, middle, south Sweden)
Regional Large Carnivore Committees (RLCCs)	• 15 regional wildlife councils (formal, consultative) • Two RLCCs (informal arenas, consultative)	• The county governors • Eight RLCCs (indirectly elected politicians with a decisive role)	• The county administrative boards • 21 wildlife management delegations (indirectly elected politicians and stakeholder groups with a decisive role)
Local	• 295 game management associations (formal, consultative) • 21 wolf territory based co-operation groups (informal arenas, consultative)		
Responsibilities at regional and local level	• Supporting, steering, and assisting in the activity of the regional office of the Finnish Wildlife Agency • Participation in strategic planning of large carnivore management • Arranging of stakeholder participation	• Approval of regional management plan including zonation • Preventive and conflict-mitigation measures • Decide LC hunting quotas	Approval of management plan for the county, including proposals for minimum levels for the county's large carnivore populations, zoning, license hunting and protective hunting, compensation schemes, and distribution

21 WMDs. This change was somewhat influenced by the Norwegian RLCCs, with political representation. However, the stakeholders from the various interest organizations represented in the RLCCs did not want to lose their influence over the process and this is why, ultimately, a combination of politicians and stakeholder groups became the solution (Sjölander-Lindqvist et al., 2015). In Finland, 15 regional wildlife councils were established with the purpose of exchanging information between regional experts and fostering the participation of involved actors in strategic planning (http://mmm.fi/en/wildlife-and-game/wildlife-and-game-administration). Later on, wolf territory level cooperation groups were established as informal arenas to facilitate face-to-face communication and more concrete bottom-up action initiated at the local level. Although the changes in Finland implied a re-centralization of power, Finland has a substantially higher formal local presence through the local game districts and territory-level cooperation compared to the other countries. An important difference between the countries is that the Finnish regional wildlife councils, unlike the Swedish WMDs and the Norwegian RLCCs, have no authority in deciding regional large carnivore quotas. However, the devolved power in Norway and Sweden is also limited, primarily focusing on the implementation of national policy objectives through zoning, compensatory schemes, and other mitigation measures. This approach has contributed to confusion and to some extent conflict regarding roles and responsibilities of the WMDs and RLCCs in Sweden and Norway, respectively. As elaborated previously, the actors involved at the regional level – in particular, stakeholder groups or politicians representing utilitarian interests – rather see themselves as primarily accountable downwards towards their constituencies or stakeholder organizations, instead of acting as if they were accountable upwards, which would better correspond with the current situation, since they are formally appointed at the national level, i.e., by the ministries (Sandström et al., 2009; Krange et al., 2016).

Since large carnivore governance is embedded in wider societal conflicts, entailing urban-rural power relations, and socioeconomic and cultural change, the level of trust – both from within but also from the wider public – towards the institutional bodies tends to vary quite substantially, with the risk of undermining their legitimacy (Krange et al., 2016; Eriksson, 2017). This is among other things reflected in perceptions of how various interests were considered to be included in the process. In Norway, the conservation interests but also reindeer husbandry were perceived to be underrepresented compared to farming interests (Krange et al., 2016). A similar pattern can be found in Sweden, where in particular conservation interests perceive themselves to be underrepresented compared to hunting interests. Although it is possible to discern similar attitudinal patterns in Finland as well, it is interesting to note that the setting up of management plans in combination with participatory processes seem to have played a substantial role in the wider involvement of both interest organizations and the public (for the process and involvement, see Hiedanpää et al., 2016). These processes may, at least to some extent, have contributed to closing the gap between different interests. Although similar management plans have been set up in both Norway and Sweden, they do not seem to have played a similar role as a participatory planning tool to mitigate conflicts and develop new and shared knowledge on how to manage conflicts, opening up opportunities for social learning between actors and policy learning among the three countries.

Conclusion

The attempts to govern and manage complex socio-ecological conflicts such as the presence of large carnivores in Finland, Norway, and Sweden through decentralized or collaborative approaches has proven to be a demanding task. Given the challenges identified in this chapter, the three countries are still searching for modes of governance capable for accommodating multiple objectives and activities.

In relation to this search, one of the problems that the three countries share is the difficulty of delegating or decentralizing authority from the national to regional levels. Despite the ambitions to introduce collaborative or decentralized measures, often adapted to fit into already established institutional structures, our review showed that the competent authority primarily remains at the central level. Any decentralized tasks mainly focus on conflict management, and thereby, regional or local levels are responsible for the implementation of already defined policies. There are limited opportunities to adapt these policies to local needs and desires. As a consequence, the conflicts related to large carnivore governance do not only include different values, e.g., anthropocentric vs. ecocentric values. They also relate to multi-level governance and divergent opinions regarding the most appropriate level for decision making (see, e.g., the Malawi principles which were developed already in 1998; UNEP, 2017).

Another problem is that there is no panacea, quick fix, or blueprint for a single type of governance system that has the capacity to accommodate multiple objectives and activities. The three different modes of governance that have been chosen in the three countries have both strengths and weaknesses. As our comparison showed, the countries have tried to successively adjust their governance systems to handle the weaknesses. However, given that the three countries more or less share large carnivore populations, what seems to be missing is a more designated learning process not only within the respective countries, but also among the three countries. As our review demonstrated, considerable efforts should be undertaken to share experiences and best practices between the countries.

References

Ansell, C., and Gash, A. (2008) 'Collaborative governance in theory and practice', *Journal of Public Administration Research and Theory*, vol 18, pp543–571.

Bekk Norstad, A., and Skogen, K. (2017) *Jervejakt og jervejegere, NINA-rapport*, Oslo.

Bisi, J., Liukkonen, T., Mykrä, S., Pohja-Mykrä, M., and Kurki, S. (2010) 'The good bad wolf – Wolf evaluation reveals the roots of the Finnish wolf conflict', *European Journal of Wildlife Research*, vol 56, no 5, pp771–779.

Bisi, J., and Kurki, S. (2008) 'The wolf debate in Finland: Expectations and objectives for the management of the wolf population at regional and national level', in *Julkaisuja 3. Ruralia Institute*, University of Helsinki, Helsinki.

Cinque, S. (2008) *I vargens spår. Myndigheters handlingsutrymme i förvaltningen av varg* [In the Wolf Track: Administrative Discretion in Wolf Management], PhD thesis, University of Gothenburg, Gothenburg.

Cinque, S., and Sjölander-Lindqvist, A. (2011) 'L'evoluzione degi esperimenti partecipativi in Svezia', in A.Valastro (ed) *La Regole della Democrazia Partecipativa: Itinerari per la Costruzione di un Metodo di Governo*, Joveno Editore, XX.

Committee of Environment and Agriculture (2009/10:MJU8) (2009) 'A new large carnivore management' (En ny rovdjursförvaltning) – Swedish Parliament, 21 October 2009.

Council of Europe (1979) *Convention on the Conservation of European Wildlife and Natural Heritage*, Bern, Switzerland, http://conventions.coe.int/Treaty/EN/Treaties/Html/104. html.

Criminal Code of Finland (39/1889) Unofficial translation, www.finlex.fi/fi/laki/kaannok set/1889/en18890039.pdf.

Decker, D., Smith, C., Forstchen, A., Hare, D., Pomeranz, E., Doyle-Capitman, C., Schuler, K., and Organ, J. (2016) 'Governance principles for wildlife conservation in the 21st century', *Conservation Letters*, vol 9, pp290–295.

Dressel, S., Sandström, C., and Ericsson, G. (2015) 'A meta-analysis of studies on attitudes toward bears and wolves across Europe 1976–2012', *Conservation Biology*, vol 29, no 2, pp565–574.

Duit, A., Galaz, V., and Löf, A. (2009) 'Fragmenterad förvirring eller kreativ arena? Från hierarkisk till förhandlad styrning i svensk naturvårdspolitik', in J. Pierre and G. Sundström (eds) *Samhällsstyrning i förändring*, Liber, Malmö.

Duit, A. and Löf. A. (2015) 'Dealing with a wicked problem? A dark tale of carnivore management in Sweden 2007 – 2011'. Administration & Society, DOI: 10.1177/ 0095399715595668.

Emerson, K., Nabatchi, T., and Balogh, S. (2012) 'An integrative framework for collaborative governance', *Journal of Public Administration Research and Theory*, vol 22, no 1, pp1–29.

Ericsson, G., and Heberlein, T. A. (2003) 'Attitudes of hunters, locals, and the general public in Sweden now that the wolves are back', *Biological Conservation*, vol 111, no 2, pp149–159.

Eriksson, M. (2017) *Changing Attitudes to Swedish Wolf Policy Wolf Return, Rural Areas, and Political Alienation*, PhD thesis, Department of Political Science 2016:4, Umeå University, Umeå.

European Commission (1992) 'The Habitats Directive (Council Directive 92/43/EEC)', http://ec.europa.eu/environment/nature/legislation/habitatsdirective/index_en.htm, accessed 11 April 2018.

Faugert, S., Sandberg, B., and Blomfeldt, D. (2005) 'Rovdjursrådet. En utvärdering av Rovdjursrådets funktion för att främja arbetet med rovdjursfrågor' Faugert and Co. Utvärdering, Stockholm, Sweden.

Finnish Hunting Act (1993) Unofficial translation www.finlex.fi/fi/laki/kaannokset/1993/ en19930615.pdf, accessed 21 April 2018.

Finnish Wildlife Agency (2017) 'Palautekysely sidosryhmille ja riistaneuvostoille 2016' by Antti Impola, unpublished document.

Government Bill (2009) 'En ny rovdjursforvaltning', https://data.riksdagen.se/fil/0A22 C7CA-D5C8-48D3-9855-E150760B3DBF, accessed 21 April 2018.

Government Official Report SOU (1999) *Rovdjursutredningen-Slutbetänkande om en sammanhållen rovdjurspolitik*, Miljö-och samhällsbyggnadsdepartementet (Ministry of Environment and Social Structure).

Government Official Report SOU (2007)'Rovdjurens och deras förvaltning', http://www.regeringen.se/49bbac/contentassets/4ccf9b41c310420a9e1030fc0459ab05/rovdjuren-och-deras-forvaltning.-hela-utredningen.-sou-200789, accessed 21 April 2018.

Hallgren, L., and Westberg, L. (2015) 'Adaptive management? Observations of knowledge coordination in the communication practice of Swedish game management' *Wildlife Biology*, vol 21, no 3, pp165–174.

Heberlein, T. A., and Ericsson, G. (2008) 'Public attitudes and the future of wolves, canis lupus, in Sweden', *Wildlife Biology*, vol 14, no 3, pp391–394.

Hiedanpää, J. (2013) 'Institutional misfits: Law and habits in Finnish wolf policy', *Ecology and Society*, vol 18, no 1.

Hiedanpää, J., and Bromley, D. W. (2010) 'The harmonization game: Reasons and rules in European biodiversity policy', *Environmental Policy and Governance*, vol 21, no 2, pp99–111.

Hiedanpää, J., Kalliolevo, Salo, M., Pellikka, J., and Luoma, M. (2016) 'Payments for improved ecostructure (PIE): Funding for the coexistence of humans and wolves in Finland', *Environmental Management*, vol 58, no 3, pp518–533.

Hiedanpää, J., and Pellikka, J. (2017) 'Preadaptative transactions and institutional change: Wolf-critical activism in southwestern Finland', *Environmental Policy and Governance*, vol 27, no 3, pp270–281.

Hongslo, E., Hovik, S., Zachrisson, A., and Aasen Lundberg, A. K. (2016) 'Decentralization of conservation management in Norway and Sweden – Different translations of an international trend', *Society and Natural Resources*, vol 29, no 8, pp998–1014.

Krange, O., Odden, J., Skogen, K., Linnell, J. D. C., Stokland, H. B., Vang, S., and Mattisson, J. (2016) *Evaluering av regional rovviltforvaltning*, Oslo: NINA Rapport 1268, 190.

Krange, O., Sandström, C., Tangeland, T., and Ericsson, G. (2017) 'Approval of wolves in Scandinavia: A comparison between Norway and Sweden', *Society and Natural Resources*, vol 30, no 9, pp1127–1140.

Lange, P., Driessen, P. P. J., Sauer, A., Bornemann, B., and Burger, P. (2013) 'Governing towards sustainability: Conceptualizing modes of governance', *Journal of Environmental Policy and Planning*, vol 15, no 3, pp403–425.

Liberg, O., Andrén, H., Pedersen, H-C., Sand, H., Sejberg, D., Wabakken, P., Åkesson, M., and Bensch, S. (2005) 'Severe inbreeding depression in a wild wolf (canis lupus) population', *Biology Letters*, vol 1, no 1, pp17–20.

Liberg, O., Chapron, G., Wabakken, P., Pedersen, H.C., Hobbs, N.T. and Sand, H. (2011) 'Shoot, shovel and shut up: cryptic illegal killing slows restoration of a large carnivore in Europe', Proceedings of the Royal Society B: Biological Sciences, vol 279, pp910-915.

Lundmark, C., and Matti, S. (2015) 'Exploring the prospects for deliberative practices as a conflict-reducing and legitimacy-enhancing tool: The case of Swedish carnivore management', *Wildlife Biology*, vol 21, no 3, pp147–156.

Lundmark, C., Matti, S., and Sandström, A. (2014) 'Adaptive co-management: How social networks, deliberation and learning affect legitimacy in carnivore management', *European Journal of Wildlife Research*, vol 60, no 4, pp637–644.

MAF (1996) 'Suomen maasuurpetokannat ja niiden hoito. Suurpetotyöryhmän raportti', *Maa-ja metsätalousministeriön julkaisuja 6/1996*.

MAF (2005) 'Management plan for the wolf population in Finland', *Ministry of Agriculture and Forestry 11b/2005*.

MAF (2007a) 'Management plan for the lynx population in Finland', *Ministry of Agriculture and Forestry 1b/2007*.

MAF (2007b) 'Management plan for the bear population in Finland', *Ministry of Agriculture and Forestry 2/2007*.

MAF (2014) 'Management plan for the wolverine population in Finland', *Ministry of Agriculture and Forestry 9/2014*.

MAF (2015) '*Management Plan for the Wolf Population in Finland*', Ministry of Agriculture and Forestry, Helsinki, Finland.

MAF Decree (241/2010) 'Maa- ja metsätalousministeriön asetus elävän riistaeläimen ohjeellisista arvoista'.

Ministry of the Environment (1992) 'Om forvaltning av bjørn, jerv, ulv og gaupe', St.meld. nr.27(1991-92).

Mykrä, S., Pohja-Mykrä, M., and Vuorisalo, T. (2017) 'Hunters' attitudes matter: diverging bear and wolf population trajectories in Finland in the late nineteenth century and today', European Journal of Wildlife Research, vol 63, DOI 10.1007/s10344-017-1134-1.

Newig, J., and Fritsch, O. (2009) 'Environmental governance: Participatory, multi-level – And effective?', *Environmental Policy and Governance*, vol 19, pp197–214.

Pellikka, J., and Sandström, C. (2011) 'The role of large carnivore committees in legitimising large carnivore management in Finland and Sweden', *Environmental Management*, vol 48, no 1, pp212–228.

Pohja-Mykrä, M. (2014) 'Vahinkoeläinsodasta psykologiseen omistajuuteen. Petokonfliktien historiallinen tausta ja nykypäivän hallinta', *Julkaisuja 33* Ruralia-instituutti, Helsingin yliopisto.

Pohja-Mykrä, M., and Kurki, S. (2014) 'Kansallisen suurpetopolitiikan kehittämisarviointi', *Raportteja 114*, Ruralia-Instituutti, Helsingin yliopisto.

Pohja-Mykrä, M., Vuorisalo, T., and Mykrä, S. (2005) 'Hunting bounties as a key measure of historical wildlife management and game conservation: Finnish bounty schemes 1647–1975', *Oryx: The Journal of the Fauna Preservation Society*, vol 39, no 3, pp284-291

Ratamäki, O. (2008) 'Finland's wolf policy and new governance', The Journal of Environment & Development, vol 17, no 3, pp316–339.

Redpath, S. M., Linnell, J. D. C., Festa-Bianchet, M., Boitani, L., Bunnefeld, N., Dickman, A., Gutiérrez, R. J., Irvine, R. J., Johansson, M., Majić, A., McMahon, B. J., Pooley, S., Sandström, C., Sjölander-Lindqvist, A., Skogen, K., Swenson, J. E., Trouwborst, A., Young, J., and Milner-Gulland, E. J. (2017) 'Don't forget to look down – Collaborative approaches to predator conservation', *Biological Reviews*, vol 92, no 4, pp2157–2163.

Sabatier, P., Focht, W., Lubell, M., Trachtenberg, Z., Vedlitz, A., and Matlock, M. (2005) *Swimming Upstream: Collaborative Approaches to Watershed Management*, MIT Press, Cambridge, MA and London.

Salo, M., Hiedanpää, J., Luoma, M., and Pellikka, J. (2017) 'Nudging the impasse? Lessons from the nationwide online wolf management forum in Finland', *Society and Natural Resources*, vol 30, no 9, pp1141–1157.

Sand, H., Wabakken, P., and Liberg, O. (2010) 'Vargens biologi: karaktärer och konsekvenser för små populationer', in H. Sand et al. (eds) *Den skandinaviska vargen. En sammanställning av kunskapsläget 1998–2010 från det skandinaviska vargforskningsprojektet SKANDULV*, Grimsö forskningsstation, SLU, Rapport till Direktoratet for Naturforvaltning, Trondheim, Norge.

Sandström, C., and Pellikka, J. (2008) 'Rovdjursgrupper i Finland och Sverige – form, funktion och framtid', *Technical Report to the Swedish Environmental Protection Agency*.

Sandström, C., Pellikka, J., Ratamäki, O., and Sande, A. (2009) 'Management of large carnivores in Fennoscandia: New patterns of regional participation', *Human Dimensions of Wildlife*, vol 14, no 1, pp37–50.

SFS (Svensk författningssamling) (2009) 'Förordning om viltförvaltningsdelegationer', https://www.riksdagen.se/sv/dokument-lagar/dokument/svensk-forfattningssamling/forordning-20091474-om_sfs-2009-1474, accessed 21 April 2018.

SFS (Svensk författningssamling) (2014: 1242) 'Förordning om ändring i förordningen (2009:1474) om viltförvaltningsdelegationer https://www.notisum.se/rnp/sls/sfs/2014 1242.pdf, accessed 21 April 2018.

Sjölander-Lindqvist, A. (2008) 'Local identity, science and politics indivisible: the Swedish wolf controversy deconstructed', Journal of Environmental Policy and Planning, vol 10, pp71–94.

Sjölander-Lindqvist, A. (2009) 'Social-natural landscape reorganised: Swedish forest-edge farmers and wolf recuperation', *Conservation and Society*, vol 7, no 2, pp130–140.

Sjölander-Lindqvist, A. (2011) 'Predators in "agri-environmental" Sweden: Rural heritage and resistance against wolf propagation', in H. Gökçekus, U. Türker, and J. W. LaMoreaux (eds) *Survival and Sustainability: Environmental Concerns in the 21st Century*, Springer Verlag, Berlin.

Sjölander-Lindqvist, A., Johansson, M., and Sandström, C. (2015) 'Individual and collective responses to large carnivore management: The roles of trust, representation, knowledge spheres, communication and leadership', *Wildlife Biology*, vol 21, no 3, pp175–185.

Sjölander-Lindqvist, A., and Sandström, C. (forthcoming) 'Collaborative governance of large carnivores: A pathway to consensus or continued dispute?' Paper to be presented at the conference 'Democracy and Public Administration' 14–15 March 2018. Institute for Futures Studies, Stockholm.

Skogen, K., Krange, O., and Figari, H. (2017) *Wolf Conflicts: A Sociological Study* (Interspecies encounters; volume 1), Berghahn Books, New York.

Taloustutkimus (2016) 'Susitutkimus 2016', unpublished report of the survey commissioned by the Finnish Wildlife Agency.

United Nations Conference on Environment and Development (UNCED) (1992) *Agenda 21*, United Nations, New York.

United Nations Economic Commission for Europe (UNECE) (1998) *Aarhus Convention*, www.unece.org/env/pp/treatytext.html, accessed 21 April 2018.

United Nations Environment Program (UNEP) (1992) *Convention on Biodiversity*, www.cbd. int, accessed 21 April 2018.

United Nations Environment Program (UNEP) (2017) Ecosystem approach, www.cbd.int/ecosystem/principles.shtml, accessed 21 April 2018.

Wildlife and Game Administration Act (158/2011) (2011) Unofficial translation, www.fin lex.fi/fi/laki/kaannokset/2011/en20110158.pdf, accessed 21 April 2018.

15 Trans-boundary and trans-regional management of a large carnivore

Managing brown bears across national and regional borders in Europe

Vincenzo Penteriani, Djuro Huber, Klemen Jerina, Miha Krofel, José Vicente López-Bao, Andrés Ordiz, Alejandra Zarzo-Arias, and Fredrik Dalerum

Introduction

Because of uneven spatial distribution of resources, population processes are often not homogenously distributed in space. This frequently leads to spatial variation in features such as density, mortality rate, phenology, movement, and space use (e.g., Turgeon and Kramer, 2012; Gervasi et al., 2015). However, this spatial variation is rarely aligned with political or administrative borders, which may result in mismatches between the scale of population processes and the level at which management and conservation actions are implemented (Trouwborst, 2010). Such incongruities can be deeply problematic, partly because management strategies may not be well designed for local conditions, and partly because demographically cohesive units may be subject to contrasting management strategies and interventions. Indeed, it has been long recognized that different management regimes within different portions of the same population can alter vital processes and, thus, negatively influence local management goals (Andreasen et al., 2012; Gervasi et al., 2015).

The wide-ranging nature of large carnivores makes them difficult to manage within national or regional borders (Linnell et al., 2005, 2016). For instance, 29 out of 33 large carnivore populations in Europe are trans-boundary (Linnell et al., 2008). Populations are often subject to different or even potentially contrasting management regimes, so that a trans-boundary population can be simultaneously fully protected or heavily harvested on the two sides of an administrative unit (Gervasi et al., 2015). Therefore, the sustainability of management actions executed in a given area also depends on adopted policies in neighbouring areas (Bischof and Swenson, 2012). Subsequently, integrated trans-boundary decision-making is paramount for effective management of many large carnivore populations (Linnell and Boitani, 2011; Blanco, 2012; Chapron et al., 2014).

The European populations of brown bear (*Ursus arctos*) are particularly prone to conflicting management regimes. Many bear populations inhabit large areas (Dahle and Swenson, 2003; Mertzanis et al., 2005; Krofel et al., 2010, Ćirović et al., 2015; Gavrilov et al., 2015), and single populations often extend across multiple administrative borders at multiple levels, from borders between neighbouring countries to different jurisdictions within the same country. Hence, the same brown bear populations are often subject to different monitoring programmes, management, and conservation policies (see Annexes 15.1 and 15.2). Under European Union (EU) legislation, the brown bear is strictly protected (Swenson et al., 2000). Brown bear populations are listed in Annexes II and IV of the Habitats Directive (European Commission, 1992). However, EU legislation does allow for relatively flexible management through local derogations (e.g., hunting/culling), which consequently open up the possibility for conflicting management strategies among countries. For example, the Dinaric Mountains in southeast Europe contain one of the largest and most important brown bear meta-populations in Europe, stretching from Greece to the Alps (Linnell et al., 2016). The species is protected in Greece, Serbia, and the Former Yugoslav Republic of Macedonia, protected but hunted through derogations in Slovenia and Croatia, and managed as game species in Bosnia and Herzegovina and Montenegro (Knott et al., 2014; see also Annex 15.2). Actually, although the management or conservation status is always the same within each bear population, even if it is trans-boundary, heterogeneity arises in conflict management policies and hunting practices (Annex 15.2). In some cases, such heterogeneity in bear management between neighboring countries may make it more challenging for individual countries to achieve their national management goals. For example, the increasing bear population in Croatia and resulting emigration to neighboring Slovenia augmented the need for bear culling in the latter country, where the management goal was to keep the population stable to prevent an increase of human-bear conflicts (Jerina and Adamič, 2008). High culling rates in turn caused public opposition towards bear management in the country (Kryštufek et al., 2003).

In the present chapter, we present an overview on brown bear conservation and management in Europe, including population monitoring, hunting regulations, damage compensation schemes, management of problematic bears, and viewing tourism concentrated on the bear. This chapter will thus address the interplay among management, policies, population sizes, and the economics of the management of brown bears, which can help improve our understanding of how to efficiently manage large carnivores under conflicting management goals caused by their trans-boundary distribution.

Bear conservation and management in Europe

After a long period of intense persecution, the brown bear is currently the most common large carnivore species in Europe and is present in 22 European countries (excluding Belarus, Ukraine, and Russia; Chapron et al., 2014).

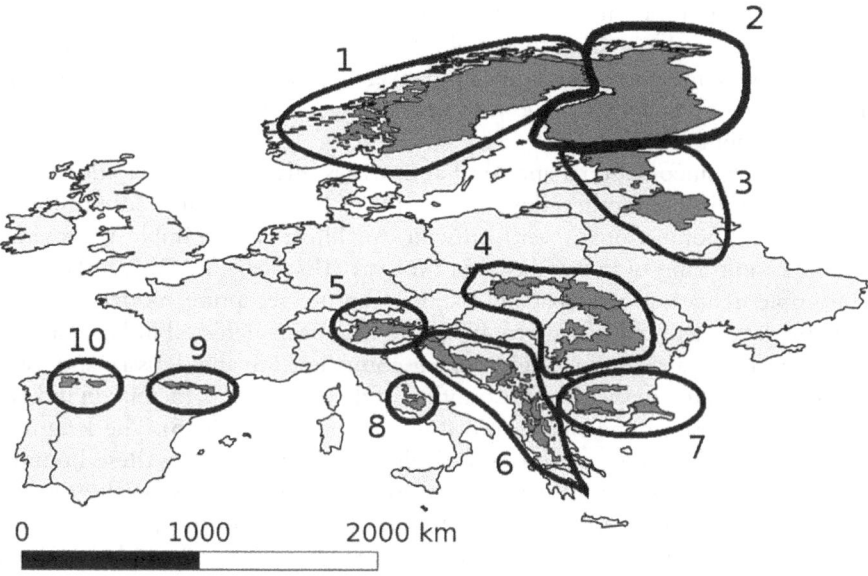

Figure 15.1 Geographic ranges of the 10 different sub-populations of brown bears in Europe (dark shaded areas). 1: Scandinavian; 2: Karelian; 3: Baltic; 4: Carpathian; 5: Alpine; 6: Dinaric–Pindos; 7: Eastern Balkan; 8: Central Apennine; 9: Pyrenean; 10: Cantabrian. All but two (the Central Apennine and the Cantabrian populations) are trans-boundary at an international level. The distribution map is taken from IUCN's official range data (www.iucnredlist.org/technical-documents/spatial-data) and the delineation of sub-populations from Chapron et al. (2014).

The estimated 17,000 bears in Europe can be clustered into 10 populations (Figure 15.1). All of these populations are relatively stable or slightly increasing, although a few remain critically small (Chapron et al., 2014). Eight out of the 10 brown bear populations in Europe have distributions that overlap several countries (three countries on average, range 1–9; Annex 15.1). Only the Central Apennines (Italy) and Cantabrian Mountains (Spain) brown bear populations occur within a single country. The management of bears occurs from national down to regional levels, and in several countries management is carried out at several levels simultaneously. Hunting, which we here define as the deliberate killing of bears for either recreational or conflict prevention and resolution purposes, occurs in 12 of the 22 European countries that are home to brown bears (Annex 15.1).

Population monitoring

Despite the heavy reliance on accurate estimates of population size for most bear management strategies, there is a general absence of trans-boundary and trans-regional approaches towards homogeneous and coordinated monitoring. Instead, multiple methods are being used throughout Europe to estimate brown bear numbers (Chapron et al., 2014). These methods include genetic capture–mark–recapture, hunting data, damage reports, direct observations (e.g., monitoring of females with cubs), snow tracking, telemetry, questionnaires, and camera trapping.

Moreover, since national and local authorities often conduct monitoring in isolation from one another, the comparison of bear data across administrative borders and populations is very difficult. An illustrative example is provided by the monitoring of brown bears in Norway (Bischof et al., 2015). Recently, the implementation of a spatially explicit capture-recapture model based on genetic data suggested that up to 49% of female bears detected in Norway had their centres of activity in neighbouring countries (Finland, Russia, and Sweden). This resulted in the double counting of these bears and, thereby, in inflated estimates of bear abundance in Norway (Bischof et al., 2015). The length of the borders with other countries, as well as the terrain along these boundaries, might dramatically influence these estimates, which may be the case for both Norway with Sweden and the Dinaric-Pindos population, if we take into account all borders between countries that share the latter population. Such biased estimates can have strong implications for policy making, such as population status assessments, establishment of sport hunting quotas, and the adoption of population control actions. Nevertheless, it is worth noting that some trans-boundary initiatives have been put into practice. For example, considerable effort is currently being expended in the Pyrenees, which are inhabited by bears on both the Spanish and French sides of the mountain range (Piédallu et al., 2016) and in the Northern Dinaric Mountains and the south-eastern Alps, where trans-boundary non-invasive genetic monitoring of bears has been implemented (http://dinalpbear.eu).

Conflict resolution

Human-bear conflicts represent a major challenge for brown bear management across neighbouring countries (Majić and Krofel, 2015). Trans-boundary management of human-bear conflicts are complex because conflicts involving bears are particularly diverse, including damage to crops and property, damage to beehives, livestock depredation, and attacks on humans. Given these numerous types of human-bear conflicts, there is a requirement for a variety of conflict resolution measures. These policies need to be case-specific and adapted to local situations because the causes and factors behind human-bear conflicts can vary considerably across regions in respect to local ecological, cultural, and management contexts.

The majority of human-bear conflicts are associated with the opportunistic foraging behaviour of bears. It is therefore not surprising that anthropogenic food sources are among the most frequently identified drivers of conflict (Swenson et al., 2000; Herrero, 2002). Anthropogenic food sources include garbage, crops, animal carcasses, and slaughter remains, as well as food provided through some sort of intentional feeding of bears. Hence, limiting access to such food sources is often regarded as the most effective way to prevent human-bear conflicts (Herrero, 2002; Majić and Krofel, 2015). A first step to limit access to anthropogenic food is regulation and strict enforcement of direct bear feeding by people, accompanied by public education. The next step is to prevent bears from accessing anthropogenic food sources such as garbage. There are numerous approaches to effectively achieve this, including bear-proof garbage containers and compost bins, physical obstacles, and electric fences (for a review see Sowka, 2009).

A key factor causing conflict is the occurrence of potential problem individuals, i.e., bears that have changed their behaviour through habituation to human presence or conditioning to anthropogenic food (Majić and Krofel, 2015). Although such bears represent only a small portion of a bear population, they usually cause most human-bear conflicts, while the majority of the bear population comes into conflict only rarely or never. Therefore, dealing with problem bears is often the focal point in bear conflict management in Europe (Majić and Krofel, 2015) and, in particular, among neighbouring countries (Kaczensky et al., 2011). Indeed, bears may become conditioned to anthropogenic food in one country and generate conflicts in adjacent ones. For this reason, for example, there has been a proposal for common trans-boundary risk assessment protocols and management recommendations in dealing with conflict bears (Majić and Krofel, 2015).

A rapid response is required to face up to the issue of problematic bears. Non-lethal approaches typically rely on aversive conditioning (Majić and Krofel, 2015), which may be effective if applied early in the development of conflict behaviour in an individual bear, but usually fail for bears in later stages of the process of food-conditioning or habituation to human presence (Gillin et al., 1995; Herrero, 2002; Mazur, 2010). In these cases, lethal removal is the most effective short-term solution (Gunther et al., 2004). Translocations of problem bears to another area are largely considered to be ineffective, as translocated bears usually return to the capture site or start causing problems in the new area (Linnell et al., 1997). For lethal removals to be effective, it must be ensured that the correct bear is dispatched and that the removal is coupled with effective measures to prevent the development of new problem bears. The feasibility of lethal removal also depends on the conservation status of a given bear population, as well as the socio-political context of a given country. For example, in the case of threatened bear populations managers should invest more effort into non-lethal options, while for larger populations, especially where regular culling is practiced, lethal removal of problem bears may be used more liberally.

Because of the often complex considerations in handling problem bears, most policies are developed within administrative boundaries without significant trans-boundary or trans-regional coordination. Although we appreciate the difficulties related to trans-boundary coordination in terms of such policies, we note three principal issues with regard to this lack of coordination.

First, since aversive conditioning is largely dependent on the context within which the conditioning is done, a lack of coordination across borders may result in some individuals experiencing several different types of conditioning. Second, a lack of coordination in terms of lethal removal may cause different demographic pressures, in terms of increased mortality, among neighbouring zones of the same population. Such spatial variation in mortality could disrupt social structure, which is actually the reason why some bears approach human settlements, rather than being attracted by food (e.g., Elfström et al., 2014), and hence, this may have secondary demographic impacts beyond the direct removal of the animals considered as problematic. Finally, because lethal removal is only a short-term solution, a potential lack of coordinated effort in terms of limiting human habituation and food conditioning could lead to an excess removal of bears, particularly if a country that implements lethal removal coupled with actions towards limiting further habituation and food condition is neighboured by countries and regions which are doing this less efficiently.

Hunting

Legal, biological, ecological, ethical, and economic aspects need to be taken into account when bear hunting is allowed. Legal issues seem to be routinely solved through derogations from the Habitats Directive and other international agreements (e.g., the Bern Convention), together with appropriate national regulations followed by sound monitoring and management. From a biological perspective, hunting has to be sustainable and non-disruptive to the sex and age structure of the hunted population, bearing in mind the potential effects on the bear populations of neighbouring countries. Ecologically, brown bears are apex predators, i.e., highly interactive species that ultimately affect community diversity. Therefore, the hunting of apex predators may undermine their ecological role by (a) reducing predator numbers and thus their potential to affect lower trophic levels, and (b) forcing predators to adjust their own behaviour under hunting risk, which may limit a predator's contribution to the "landscape of fear" that they create for their own prey (Ordiz et al., 2013). Ethically, hunting has to satisfy the minimum requirement of not provoking unnecessary suffering (e.g., not leaving wounded animals), and local bans are often set on bait hunting and/or shooting of bear families (e.g., females with cubs are protected from hunting in Sweden and in several other countries). Economically, hunting can be designed in a way that hunters pay (a) only the trophy value, e.g., through CIC points (i.e., the scoring ascribed to a given trophy bear following the International Council for Game and Wildlife Conservation [CIC] rules), or (b) only for the right to hunt (Knott et al., 2014). Although the first option provides more income for bear managers, it often leads to the targeted hunting of the biggest bears, i.e., old males. This often results in age- and sex-biased hunting, which can have undesirable demographic effects (Swenson et al., 1997). While the biological, ecological, and ethical considerations of hunting are relatively similar among neighbouring countries and management regions, the legal and economic considerations can differ dramatically. Hence, although there are obvious benefits of coordinating the number of hunted animals in trans-boundary populations, both legal and economic aspects may be too difficult to coordinate across borders or may demand a long-term horizon.

The Dinaric-Pindos brown bear population probably best exemplifies the trans-boundary hunting of bears in a European population, and thus it is worth exploring in further detail. This bear population is shared by nine countries (Slovenia, Croatia, Bosnia and Herzegovina, Montenegro, Serbia, Kosovo, Albania, Former Yugoslav Republic of Macedonia, and Greece), and consists of ca. 3,070 bears distributed over an area of 114,100 km^2 (Kaczensky et al., 2013). These bears are hunted in the countries where they are most numerous, with Croatia having the largest number of individuals (ca. 1,000 bears). In the last 12 years (2005–2016), a total of 1,305 dead bears have been recorded, of which 1,009 (77.3%) were quota hunted, with the majority (69%) of these being adult male bears. The hunting of bears in Croatia appears to be sustainable in terms of the impact on population size, but some concerns have been raised about possible high adult male mortality. Some possible trans-boundary effects can be seen across the Croatian-Slovenian border. In the period 1998–2008, the total recorded bear mortality in Slovenia was 927 individuals, of which 724 were hunted (Krofel et al., 2012). This represents an average annual harvest mortality rate of 16% for an estimated population of 400–500 individuals (in 2007), which is one of the highest harvest rates reported for the species worldwide (Krofel et al., 2012). Until 2007, this heavy harvest was probably partially sustained by bears from Croatia, but has since become self-sustainable. In contrast to Croatia, the hunting of bears in Slovenia is concentrated on young bears to reduce side effects on the bear social system. The removal of mature males has been reported to increase infanticide, i.e., the killing of bear cubs by neighbouring males that expand their home range in response to male removal (Swenson et al., 1997; Leclerc et al., 2017).

Viewing tourism

Brown bears are a common target of wildlife-related viewing tourism (Penteriani et al., 2017). Bear viewing can take place from permanent or temporary viewing spots. Permanent viewing sites usually include camouflaged observatory posts near artificial feeding sites (e.g., Finland, Romania, and Slovenia in Europe) or some other location with a predictable occurrence of bears (e.g., Italy and Spain). Temporary viewing sites are typically opportunistic, and commonly include observation sites targeting females with cubs of the year, bear mating areas and natural bear feeding sites, such as berry-rich mountain slopes. From the perspective of the potential consequences of bear viewing on bear trans-boundary conservation and management, it is important to highlight the potential effect of viewing sites associated with artificial feeding.

An example of a potential trans-boundary management problem related to bear viewing practices comes from Finland (Penteriani et al., 2010), where about 4,000 visitors arrive annually to observe bears (Eskelinen, 2009; Kojola and Heikkinen, 2012). In this country, bear viewing is associated with an increased development of sites for the artificial feeding of brown bears. The Finnish bear population plays an important role in linking the bear populations of Russian Karelia and Scandinavia. Thus, the provision of food to bears in the Finland-Russia border region could (a) disrupt the daily and seasonal movement patterns of bears, as individuals may converge on artificial feeding areas that are outside their natural home range; (b) alter bear population density and distribution around artificial food sources; and (c) artificially increase local density, exceeding the natural carrying capacity. Moreover, bears conditioned to artificial feeding may lose their natural instinct to avoid people and also become aggressive towards humans. Under this scenario, artificial feeding sites associated with bear viewing practices at the border of two or more countries may attract bears from a neighbouring country and create food-conditioned individuals, which might later create conflicts in both countries, as discussed previously. As artificial feeding has been detected at 57% of the European bear viewing sites (Penteriani et al., 2017), management policies should not overlook the potential effects of bear viewing practices on bear management in trans-boundary populations.

Trans-regional management of brown bears in Sweden

Although most trans-boundary issues associated with bear management occur across international borders, contrasting local conditions and regional management strategies may also influence bear management on smaller administrative scales (i.e., trans-regional management). The Swedish brown bear population was widespread until the mid-18th century, but drastically declined afterwards due in part to an active policy of trying to exterminate the species. The Swedish population was estimated at approximately 130 bears by the time its protection was introduced back in 1927 (Sahlén et al., 2006). Since then, the population has recovered and the current population status is stable with 2,800–3,000 bears (Swenson et al., 2017). It is worth noting that Norway, which shares the same population, only stopped paid bounty on brown bears in 1973. This continued hunting led to a serious reduction in terms of both the size and geographic range of the Norwegian part of the bear population, which further led to an almost total population collapse in Norway.

In Sweden, large carnivores are owned and managed by the government. Central policies regarding large carnivore management are dictated by the national Environmental Protection Agency (EPA). However, Sweden has adopted the Ecosystem Approach, founded at the UN convention on biodiversity (United Nations, 1992), which, among other requisites, demands that decisions regarding environmental management should be made at scales as local as possible (Smith and Maltby, 2003). Hence, although the national policies are outlined in national management plans published by the EPA, management is carried out by regional county boards. These are localized regional governments in each of Sweden's 21 counties, which are mandated to make independent management decisions as long as they are aligned with national policies. These regional management policies are outlined in regional management plans, which are produced for all large carnivores, including brown bears. However, to facilitate some coordination of management at the regional level, the counties are grouped into three separate large carnivore management regions. Although each county still formulates its own independent management plan, counties within management regions attempt to coordinate their plans in terms of goals and management actions.

Currently, the Swedish brown bear population is based in seven counties, including all four counties in the northern region and three counties in the central region. Of these, five counties share a border with Norway and one also a border with Finland. Dispersing bears may, however, occur outside these seven counties, as well. Hence, although there are clear national goals with regard to Swedish brown bear management, these goals are intended to be met by implementing seven independent management plans, of which five are from areas that share international borders and six include annual hunting quotas. These quotas are based on minimum desirable population levels, which are set in coordination within each region, although the hunting quotas may be defined completely independently. Counties are allowed to set independent quotas as long as the management region to which they belong meets minimum population sizes defined by the EPA. Contrary to this independence, policies regarding damages are relatively centralized, particularly with respect to the removal of bears that cause problems during the calving season of semi-domesticated reindeer and compensation for livestock and crop damage. However, the implementation of these central policies still largely resides within each county, so that the removal of problematic bears, in particular, may be coordinated with annual hunting quotas.

Trans-regional management of the Cantabrian bear population in Spain

Brown bears once occupied the entire Iberian Peninsula, but during the 17th century bear distribution shrank northwards. Separation between the Cantabrian (northwest Spain) and Pyrenean bear ranges (Spanish-French border) occurred between the 17th and 18th centuries. The bear range in the Cantabrian Mountains decreased from approximately 14000 km^2 in the 19th century to only about 9000 km^2 during the 20th century, with the population separated into two areas (Nores, 1988; Nores and Naves, 1993). Cantabrian bears reached their lowest historic levels in the 1990s, after which the trend reversed, and now about 200 bears inhabit the range (Martínez Cano et al., 2016).

A ban on brown bear hunting has been in place in Spain since 1973, and bears are currently listed as "critically endangered" (Naves and Fernández-Gil, 2002). Recovery plans for each of the four Spanish Autonomous Regions in the Cantabrian Mountains (Asturias, Cantabria, Castilla-León, and Galicia) were approved between 1989 and 1992, and a National Strategy for the Conservation of the Brown Bear was established in 1999 (CNPN, 1999). Although bear conservation and management in Spain is decentralized to the autonomous regions, it is the country of Spain, as an EU member state, that is ultimately responsible for meeting its EU obligations regarding the conservation of this population. Nevertheless, due to the administrative organization of Spanish territory, even this small brown bear population represents a case of a trans-regional population shared among four different and administratively independent regions.

All of these autonomous regions have agreed to conservation goals similar to those outlined in the Action Plan for the Conservation of the Brown Bear in Europe (Swenson et al., 2000), which aimed at maintaining and restoring, in coexistence with people, viable populations of brown bears as an integral part of the ecosystems. The plan highlighted the need to identify and mitigate or remove threats to brown bear populations and their habitat (Swenson et al., 2000). Securing the viability of small, isolated bear populations, such as the Cantabrian one, by increasing population size and geographic range was also a major goal. Protective legislation and conservation initiatives for brown bears in the Cantabrian Mountains have played a role in the ongoing, partial recovery of this population. Recently, it has been suggested that changes in resource availability may be the main mechanism behind the trend observed in the Cantabrian bear population (Martínez Cano et al., 2016). However, the effect of changes due to human pressure, protective legislation, or conservation actions on the observed trend for this population deserves further evaluation. Up-to-date conservation plans confronting existing threats are essential to secure the long-term viability of the still small Cantabrian bear population. In addition, there is a need to improve monitoring techniques and coordinated assessments of conservation actions among the neighbouring Autonomous Regions.

About a decade after the four Spanish Autonomous Regions in the Cantabrian Mountains approved their bear recovery plans (between 1989 and 1992, as mentioned previously), the National Strategy for the Conservation of the Brown Bear (CNPN, 1999) argued that specific objectives, such as eliminating the loss of bears caused by people, securing a connection between the two bear nuclei in the Cantabrian Mountains, and preventing habitat loss, had not been achieved. In recent years, however, the negative trend of the brown bear population reversed and the bear population has been increasing (e.g., Martínez Cano et al., 2016). Although conservation efforts may have helped to bring about the current trend to some extent, management issues still arise. For instance, each autonomous region should have updated its respective bear recovery plan several times since the plans were approved, but this has not been the case. Such periodic revisions should provide a great opportunity to review conservation achievements and failures, outline further needs, and eventually set joint plans for the conservation of this unique bear population shared by four administrative units.

Conclusions

Differences among countries or regions may create serious challenges for addressing the trans-boundary or trans-regional character of bear conservation and management. Indeed, political and economic realities do not facilitate such an approach where administrative boundaries and bear movements need to be integrated; in most cases, each jurisdiction independently manages its own part of a shared population (Bischof and Swenson, 2012). This situation urgently calls for the adoption of coordinated population approaches in brown bear conservation and management in Europe, where individuals can routinely cross international borders. Today, trans-boundary populations represent a challenge for Europe, which would only benefit from coordinated monitoring and management between those jurisdictions that share the same carnivore populations. Because the number of individuals within a jurisdiction, or some proxy thereof, represents the basic information for guiding management policies (Bischof et al., 2015), the first step of appropriate trans-boundary management is to increase our knowledge of bear status in neighbouring countries. This is the basic premise to successfully confront, in turn, human-bear conflicts, leisure activities related to bears (hunting included), and consequently, bear conservation in Europe. Yet, effective decision making and policies concerning human dimensions in bear conservation and management need to take into account the dense web of political and administrative borders within and between countries.

Trans-boundary and trans-regional attempts to manage bear populations are not a novelty. For example, management plans published in 2005 in Croatia (Dečak et al., 2005) and in 2002 in Slovenia (Ministry of Agriculture, Forestry and Food, 2002) included concrete proposals and directives for managing local bear populations by taking into account bear movements across the borders (Huber et al., 2008), since bear management in Croatia and Slovenia is expected to influence bear populations in neighbouring countries. These directives included development of trans-boundary management strategies, maintaining habitat connectivity across national borders, harmonization of research/monitoring methodology, regular meetings, and the sharing of data/information. Similarly, even though brown bear management in Norway and Sweden differs, in March 2015 the Norwegian Environment Agency and the Swedish Environmental Protection Agency formalized coordination of carnivore monitoring between these two countries, which represents an official recognition that carnivore populations are shared across their borders. This coordination involves, for instance, a common set of regulations and definitions for how monitoring is to be carried out, as well as a shared data base that is capturing all monitoring activities. Moreover, the Scandinavian Brown Bear Project (www.bearproject. info) coordinates research activities related to brown bears in Norway and Sweden. The leadership of this project is directly shared among members of several Norwegian and Swedish institutions, and funding is sought from both countries to contribute to common research goals.

However, a thorough trans-boundary and trans-regional understanding still needs to be incorporated into most management and conservation policies across Europe. The authorities responsible for the implementation of such practices generally have a jurisdiction limited to areas that are smaller in size than those typically affected by population dynamics of large carnivores (Linnell et al., 2008; Gervasi et al., 2015). Together with the evidence that contrasting management regimes within the same population can affect population dynamics and reduce the impact of conservation and management goals on both sides of a border, incongruities in management actions also result in a waste of economic resources for conservation, which reduces the already limited resources allocated for large carnivore conservation. At the same time, different management practices can reduce some of the unwanted side effects of hunting, as in the case of Slovenia and Croatia (Krofel et al., 2012), for example. As remarked by Bischof et al. (2015), the current geopolitical reality of Europe, and the sometimes divergent interests of various local lobbies and stakeholders, might make the management of trans-boundary bear populations a utopian goal, but this is, indeed, a crucial challenge to be urgently taken into account.

Annex 15.1 Characteristics of European bear populations and related management policies by country

Country	Population size[1,2]	Percentage of bear population[3]	Bear population	Range size (1)[4]	Range size (2)[5]	% range	Conservation status	Average human density	Management agency	Conflict management	Hunting or culling	Viewing points[6]
Albania	180–200	6.3%	Dinaric-Pindos	6,600 km²	13,200 km²	1.1%	VU	123/km²	Ministry of Environment, Forestry and Water Administration	Guardian dog breeding for livestock herding	NO	N/A
Austria	3	6%	Alpine	0 km²	2,300 km²	< 1%	CR	97/km²	Hunting and natural conservation authorities of each province	Damage compensation (voluntary), damage prevention	NO	NO
Bosnia and Herzegovina	550	18.3%	Dinaric-Pindos	21,600 km²	33,200 km²	3%	VU	78/km²	Ministry of Agriculture	Supplemental feeding, damage compensation	YES	N/A
Bulgaria	100–130 & 430–460	92.6%	Eastern Balkan	16,700 km²	32,800 km²	3.3%	VU	69/km²	Ministry of Environment and Waters (MOEW) and Ministry of Agriculture and Foods	Damage compensation, prevention measures (electric fences)	YES	8
Croatia[7]	1,000	32.6%	Dinaric-Pindos	10,400 km²	12,372 km²	10.8%	VU (favourable)	75/km²	Ministry for Agriculture (Hunting Unit)	Damage compensation, hunting, emergency removals, supplemental feeding	YES	3

(Continued)

Annex 15.1 (Continued)

Country	Population size[1,2]	Percentage of bear population[3]	Bear population	Range size (1)[4]	Range size (2)[5]	% range	Conservation status	Average human density	Management agency	Conflict management	Hunting or culling	Viewing points[6]
Estonia	700	98.6%	Baltic	20,800 km²	37,000 km²	3.9%	LC	28/km²	Environmental Board under the jurisdiction of Ministry of the Environment	Damage compensation, subsidies for damage prevention, hunting	YES	3
Finland	1,600–1,800	85%	Karelian	66,800 km²	35,7900 km²	10%	LC	16/km²	Ministry of Agriculture and Forestry	Damage compensation, public funding for prevention, fast removal in villages	YES	10
France	22 including Spanish side (8 translocated from Slovenia)	100% with Spain	Pyrenean	3,400 km²	8,200 km²	< 1%	CR	111/km²	Ministries of Ecology and Agriculture	Damage compensation, financial help for prevention measures, awareness campaign for hunters	NO	N/A
Greece	50 (Eastern Balkan), plus 350 (Dinaric-Pindos)	8.2% Eastern Balkans, 11.7% Dinaric-Pindos	Eastern Balkan; Dinaric-Pindos	20,300 km²	25,000 km²	2.2%	VU	81/km²	Ministry of the Environment, Energy and Climate Change, regional administration and local forestry services	Damage compensation, prevention measures, Bear Management Team for bears approaching human settlements	NO	2

Country	Population	%	Population (region)		Area	%	IUCN	Density	Responsible authority	Management measures	Strategy	No.
Italy[8,9]	41–51 (Alpine, reintroduced), plus 35–67 (Central Apennines, autochthonous)	88% Alpine, 100% Central Apennines	Alpine; Central Apennines	10,000 km²	10,000 km²	< 1%	CR	192/km²	Ministry of Environment, regional governments and authorities of protected areas, ISPRA, Forestry and Wildlife Department of Trento Autonomous Province	Damage compensation, prevention measures	NO	5
Kosovo	N/A	N/A	Dinaric-Pindos	400	800	NA	VU	159/km²	NA	Damage compensation (failed due to economic recession)	YES	N/A
Latvia	10–15 sporadic	1.4%	Baltic	–	1,400 km²	< 1%	LC	37/km²	Ministry of Environment and Regional Development	Damage compensation by law (failed due to economic recession)	NO	N/A
Former Yugoslav Republic of Macedonia[10]	160–200	6%	Dinaric-Pindos	13,500 km²	17,200 km²	1.1%	VU	81/km²	Ministry of Agriculture	No strategy	NO	2
Montenegro	270	9%	Dinaric-Pindos	km²	km²	1.5%	VU	48/km²	Ministry of Agriculture	No strategy	YES	N/A
Norway	46 (Karelian), plus 105 (Scandinavian)	2.3 % Karelian, 3% Scandinavian	Karelian; Scandinavian	38,600 km²	19,4200 km²	< 1%	LC	13/km²	Directorate for Nature Management	Damage compensation, lethal control, few prevention and mitigation measures	YES	N/A

(*Continued*)

Annex 15.1 (Continued)

Country	Population size[1,2]	Percentage of bear population[3]	Bear population	Range size (1)[4]	Range size (2)[5]	% range	Conservation status	Average human density	Management agency	Conflict management	Hunting or culling	Viewing points[6]
Poland	80–150	1%	Carpathian	6,600 km²	10,400 km²	< 1%	LC	124/km²	Ministry of Environment along with General Directorate for Environmental Protection and its regional branches	Damage compensation, electric fences, bear-resistant garbage containers	NO	N/A
Romania	6,000	74%	Carpathian	72,000 km²	89,900 km²	33.1%	LC	80/km²	Ministry of Environment and Forestry	Plan for the prevention of damage and compensation, supplemental feeding	YES	14
Serbia	8 (Carpathian), plus 60 (Dinaric-Pindos)	0.1 % Carpathian, 2% Dinaric-Pindos	Carpathian; Dinaric-Pindos	3,100 km²	9,600 km²	< 1%	VU (Dinaric-Pindos & Eastern Balkans), LC (Carpathian)	97/km²	Ministry of Energy, Development and Environmental Protection (MEDEP)	Damage compensation, supplemental feeding	NO	1
Slovakia[11]	770–870	12%	Carpathian	21,200 km²	21,200 km²	4.8%	LC	111/km²	Environment Ministry and the Agriculture Ministry	Damage compensation and prevention, bear-proof rubbish bins, dissuasive methods, hunting, supplemental feeding	YES	8

Slovenia[12]	5–10 (Alpine), 450 (Dinaric-Pindos)	16% Alpine, 15% Dinaric-Pindos	Alpine; Dinaric-Pindos	5,800 km²	13,700 km²	2.5%	CR (Alpine), VU (Dinaric-Pindos)	95/km²	Ministry of the Environment and Spatial Planning	Damage compensation, Bear Response Team, hunting, supplemental feeding	YES	8
Spain[13]	222 (Cantabrian) + 22-27 (Pyrenean)	100%	Cantabrian; Pyrenean	7,700 + 5,100 km²	7,700 + 5,100 km²	1.4%	EN (Cantabrian mountains), CR (Pyrenees)	92/km²	Autonomous regions	Damage compensation, prevention measures	NO	18
Sweden[14]	2,800–3,000	97%	Scandinavian	14,9800 km²	31,6300 km²	18.2%	LC	18/km²	Swedish Environmental Protection Agency (Naturvårdsverket) and regional county boards	Prevention measures, damage compensation, protective hunting	YES	9

[1] Kaczensky et al., 2013; [2]Skrbinšek et al., 2015; [3]Percentage of bear population represents the % of the whole European population that inhabits a given country; [4]Range size (1) is the extent of the permanent range size only; [5]Range size (2) is the extent of the whole bear range size within a given country; [6]Viewing points refer to the presence or not of specific areas for touristic bear viewing; [7]Kocijan and Huber, 2008; [8]Gervasi et al., 2008; [9]Ciucci and Boitani, 2008; [10]Melovski and Godes, 2002; [11]Rigg and Adamec, 2007; [12]Kavčič et al., 2013; [13]Pérez et al., 2014; [14]Swenson et al., 2017

Annex 15.2 Heterogeneity in conservation status, conflict management policies and hunting practices by European bear population

Bear population[1]	Countries	Population size[2,3]	Conservation status	Conflict management[4,5]	Hunting or culling
Carpathian	Poland, Romania, Serbia, Slovakia[6]	6,858–7,028	Least concern	Damage prevention, damage compensation, hunting, dissuasive methods, supplemental feeding	Allowed in Romania and Slovakia
Scandinavian	Norway, Sweden[7]	2,905–3,105	Least concern	Damage compensation, lethal control, prevention and mitigation measures	Allowed
Dinaric-Pindos	Albania, Bosnia and Herzegovina, Croatia[8], Greece, Kosovo, Former Yugoslav Republic of Macedonia[9], Montenegro, Serbia, Slovenia[10]	2,990–3,020	Vulnerable	Damage prevention, damage compensation, supplemental feeding, hunting, emergency removals, Bear Management Team/Bear Response Team for bears approaching human settlements	Allowed in Bosnia and Herzegovina, Croatia, Kosovo, Slovenia
Karelian	Finland, Norway	1,646–1,846	Least concern	Damage compensation, public funding for prevention, fast removal in villages, lethal control, mitigation measures	Allowed

Population	Countries	Population size	Conservation status	Management measures	Hunting
Baltic	Estonia, Latvia	710–715	Least concern	Damage compensation, subsidies for damage prevention, hunting	Allowed in Estonia
Eastern Balkan	Bulgaria, Greece	150–180 & 480–510	Vulnerable	Damage compensation, damage prevention, Bear Management Team for bears approaching human settlements	Allowed in Bulgaria
Cantabrian	Spain[11]	222	Endangered	Damage compensation, damage prevention	Not allowed
Alpine	Austria, Italy[12,13], Slovenia[10]	49–64 (reintroduced in Italy)	Critically endangered	Damage compensation, damage prevention, Bear Response Team, hunting, supplemental feeding	Allowed in Slovenia
Central Apennines	Italy[12,13]	35–67 (Central Apennines, autochthonous)	Critically endangered	Damage compensation, damage prevention	Not allowed
Pyrenean	France, Spain	22–27 (8 translocated from Slovenia)	Critically endangered	Damage compensation, financial help for prevention measures, awareness campaign for hunters	Not allowed

[1] Bear populations are listed in decreasing order of population size; [2]Kaczensky et al., 2013; [3]Skrbinšek et al., 2015; [4]Damage prevention includes e.g., electric fences, bear-proof garbage and compost bins, use of guardian dogs for livestock protection; [5]Not all countries that share a single bear population apply all the listed practices; [6]Rigg and Adamec, 2007; [7]Swenson et al., 2017; [8]Kocijan and Huber, 2008; [9]Melovski and Godes, 2002; [10]Kavčič et al., 2013; [11]Pérez et al., 2014; [12]Gervasi et al., 2008; [13]Ciucci and Boitani, 2008

References

Andreasen, A. M., Stewart, K. M., Longland, W. S., Beckmann, J. P., and Forister, M. L. (2012) 'Identification of source-sink dynamics in mountain lions of the Great Basin', *Molecular Ecology*, vol 21, no 23, pp5689–5701.

Bischof, R., Brøseth, H., and Gimenez, O. (2015) 'Wildlife in a politically divided world: Insularism inflates estimates of brown bear abundance', *Conservation Letters*, vol 9, no 2, pp122–130.

Bischof, R., and Swenson, J. E. (2012) 'Linking noninvasive genetic sampling and traditional monitoring to aid management of a trans-border carnivore population', *Ecological Applications*, vol 22, no 1, pp361–373.

Blanco, J. C. (2012) '*Towards a Population Level Approach for the Management of Large Carnivores in Europe: Challenges and Opportunities*, European Commission, Brussels.

Chapron, G., Kaczensky, P., Linnell, J. D. C., von Arx, M., Huber, D., Andrén, H., López-Bao, J. V., Adamec, M., Álvares, F., Anders, O., Balčiauskas, L., Balys, V., Bedő, P., Bego, F., Blanco, J. C., Breitenmoser, U., Brøseth, H., Bufka, L., Bunikyte, R., Ciucci, P., Dutsov, A., Engleder, T., Fuxjäger, C., Groff, C., Holmala, K., Hoxha, B., Iliopoulos, Y., Ionescu, O., Jeremić, J., Jerina, K., Kluth, G., Knauer, F., Kojola, I., Kos, I., Krofel, M., Kubala, J., Kunovac, S., Kusak, J., Kutal, M., Liberg, O., Majić, A., Männil, P., Manz, R., Marboutin, E., Marucco, F., Melovski, D., Mersini, K., Mertzanis, Y., Mysłajek, R. W., Nowak, S., Odden, J., Ozolins, J., Palomero, G., Paunović, M., Persson, J., Potočnik, H., Quenette, P.-Y., Rauer, G., Reinhardt, I., Rigg, R., Ryser, A., Salvatori, V., Skrbinšek, T., Stojanov, A., Swenson, J. E., Szemethy, L., Trajçe, A., Tsingarska-Sedefcheva, E., Váňa, M., Veeroja, R., Wabakken, P., Wölfl, M., Wölfl, S., Zimmermann, F., Zlatanova, D., and Boitani, L. (2014) 'Recovery of large carnivores in Europe's modern human-dominated landscapes', *Science*, vol 346, 6216, pp1517–1519.

Ćirović, D., de Gabriel Hernando, M., Paunović, M., and Karamanlidis, A. A. (2015) 'Home range, movements, and activity patterns of a brown bear in Serbia', *Ursus*, vol 26, no 2, pp79–85.

Ciucci, P., and Boitani, L. (2008) 'The Apennine brown bear: A critical review of its status and conservation problems', *Ursus*, vol 19, no 2, pp130–145.

CNPN (1999) *Estrategia para la Conservación del Oso pardo Cantábrico (Ursus arctos) en España. Comisión Nacional para la Protección de la Naturaleza*, Ministerio de Medio Ambiente, Madrid, Spain.

Dahle, B., and Swenson, J. E. (2003) 'Seasonal range size in relation to reproductive strategies in brown bears *Ursus arctos*', *Journal of Animal Ecology*, vol 72, no 4, pp660–667.

Dečak, D., Frković, A., Grubešić, M., Huber, D., Iviček, B., Kulić, B., Sertić, D., and Štahan, Ž. (2005) *Brown Bear Management Plan for the Republic of Croatia*. Ministry of Agriculture, Forestry and Water Management, Zagreb, Croatia.

Elfström, M., Zedrosser, A., Støen, O. – G., and Swenson, J. E. (2014) 'Ultimate and proximate mechanisms underlying the occurrence of bears close to human settlements: Review and management implications', *Mammal Review*, vol 44, no 1, pp5–18.

Eskelinen, P. (2009) *Karhut elinkeinona – millaisia ovat katselupalveluja tarjoavat yritykset?*, Riista-ja kalatalouden tutkimuslaitos, Helsinki, Finland.

European Commission (1992) 'The Habitats Directive (Council Directive 92/43/EEC)', http://ec.europa.eu/environment/nature/legislation/habitatsdirective/index_en.htm, accessed 11 April 2017.

Gavrilov, G. V., Zlatanova, D. P., Spasova, V. V., Valchev, K. D., and Dutsov, A. A. (2015) 'Home range and habitat use of brown bear in Bulgaria: The first data based on GPS-telemetry', *Acta Zoologica Bulgarica*, vol 67, no 4, pp493–499.

Gervasi, V., Brøseth, H., Nilsen, E. B., Ellegren, H., Flagstad, Ø., and Linnell, J. D. C. (2015) 'Compensatory immigration counteracts contrasting conservation strategies of wolverines (*Gulo gulo*) within Scandinavia', *Biological Conservation*, vol 191, pp632–639.

Gervasi, V., Ciucci, P., Boulanger, J. B., Posillico, M., Sulli, C., Focardi, S., Randi, E., and Boitani, L. (2008) 'A preliminary estimate of the Apennine brown bear population size based on hair-snag sampling and multiple data source mark-recapture Huggins model', *Ursus*, vol 9, no 2, pp105–121.

Gillin, C. M., Hammond, F. M., and Peterson, C. M. (1995) 'Aversive conditioning of grizzly bears. Can bears be taught to stay out of trouble?', *Yellowstone Science Winter*, vol 3, no 1, pp1–7.

Gunther, K. A., Haroldson, M. A., Frey, K., Cain, S. L., Copeland, J., and Schwartz, C. C. (2004) 'Grizzly bear-human conflicts in the Greater Yellowstone ecosystem, 1992–2000', *Ursus*, vol 15, no 1, pp10–22.

Herrero, S. (2002) *Bear Attacks: Their Causes and Avoidance* (2nd ed.), Nick Lyons Books, New York.

Huber, Đ., Kusak, J., Majić-Skrbinšek, A., Majnarić, D., and Sindičić, M. (2008) 'A multidimensional approach to managing the European brown bear in Croatia', *Ursus*, vol 19, no 1, pp22–32.

Jerina, K., and Adamič, M. (2008) 'Fifty years of brown bear population expansion: Effects of six-biased dispersal on rate of expansion and population structure', *Journal of Mammalogy*, vol 89, no 6, pp1491–2501.

Kaczensky, P., Chapron, G., von Arx, M., Huber, D., Andrén, H., and Linnell, J. (2013) *Status, Management and Distribution of Large Carnivores – Bear, Lynx, Wolf & Wolverine – In Europe*, Report to the EU Commission.

Kaczensky, P., Jerina, K., Jonozovič, M., Krofel, M., Skrbinšek, T., Rauer, G., Kos, I., and Gutleb, B. (2011) 'Illegal killings may hamper brown bear recovery in the Eastern Alps', *Ursus*, vol 22, no 1, pp37–46.

Kavčič, I., Adamič, M., Kaczensky, P., Krofel, M., and Jerina, K. (2013) 'Supplemental feeding with carrion is not reducing brown bear depredations on sheep in Slovenia', *Ursus*, vol 24, no 2, pp111–119.

Knott, E. J., Bunnefeld, N., Huber, D., Reljić, S., Kerež, i V., and Milner-Gulland, E. J. (2014) 'The potential impacts of changes in bear hunting policy for hunting organisations in Croatia', *European Journal of Wildlife Research*, vol 60, no 1, pp85–97.

Kocijan, I., and Huber, Đ. (2008) *Threat to the Brown Bear in Croatia, Annex 7: Conservation Genetics of Brown Bears in Croatia: Final Report*. Report by the Ministry of Environment and Nature Protection (BBI -Matra/2006/020 through ALERTIS).

Kojola, I., and Heikkinen, S. (2012) 'Problem brown bears *Ursus arctos* in Finland in relation to bear feeding for tourism purposes and the density of bears and humans', *Wildlife Biology*, vol 18, no 3, pp258–263.

Krofel, M., Filacorda, S., and Jerina, K. (2010) 'Mating-related movements of male brown bears on the periphery of an expanding population', *Ursus*, vol 21, no 1, pp23–29.

Krofel, M., Jonozovič, M., and Jerina, K. (2012) 'Demography and mortality patterns of removed brown bears in a heavily exploited population', *Ursus*, vol 23, no 1, pp91–103.

Kryštufek, B., Flajšman, and Griffiths, H. I. (eds) (2003) *Living with Bears. A Large European Carnivore in a Shrinking World*, Ecological Forum of the Liberal Democracy of Slovenia, Ljubljana, Slovenia.

Leclerc, M., Frank, S. C., Zedrosser, A., Swenson, J. E., and Pelletier, F. (2017) 'Hunting promotes spatial reorganization and sexually selected infanticide', *Scientific Reports*, vol 7, 45222.

Linnell, J., Salvatori, V., and Boitani, L. (2008) *Guidelines for Population Level Management Plans for Large Carnivore in Europe*. A Large Carnivore Initiative for Europe, Report Prepared for the European Commission (contract 070501/2005/424162/MAR/B2).

Linnell, J. D., and Boitani, L. (2011) 'Building biological realism into wolf management policy: The development of the population approach in Europe', *Hystrix, the Italian Journal of Mammalogy*, vol 23, no 1, pp80–91.

Linnell, J. D. C., Aanes, R., Swenson, J. E., Odden, J., and Smith, M. E. (1997) 'Translocation of carnivores as a method for managing problem animals: A review', *Biodiversity and Conservation*, vol 6, no 9, pp1245–1257.

Linnell, J. D. C., Nilsen, E. B., Lande, U. S., Herfindel, I., Odden, J., Skogen, K., Andersen, R., and Breitenmoser, U. (2005) 'Zoning as a means of mitigating conflicts with large carnivores: Principles and reality', in R. Woodroffe, S. Thirgood and A. Rabinowitz (eds) *People and Wildlife: Conflict or Coexistence?* (pp. 162–175), Cambridge University Press, Cambridge.

Linnell, J. D. C., Trouwborst, A., Boitani, L., Kaczensky, P., Huber, D., Reljic, S., Kusak, J., Majic, A., Skrbinsek, T., Potocnik, H., Hayward, M. W., Milner-Gulland, E. J., Buuveibaatar, B., Olson, K. A., Badamjav, L., Bischof, R., Zuther, S., and Breitenmoser, U. (2016) 'Border security fencing and wildlife: The end of the transboundary paradigm in Eurasia?', *PLoS Biology*, vol 14, no 6, e1002483.

Majić, A., and Krofel, M. (2015) *Defining, Preventing, and Reacting to Problem Bear Behaviour in Europe*, Technical report to DG Environment, European Commission, Brussels.

Martínez Cano, I., Taboada, F. G., Naves, J., Fernández-Gil, A., and Wiegand, T. (2016) 'Decline and recovery of a large carnivore: environmental change and long-term trends in an endangered brown bear population', *Proceeding of the Royal Society of London B*, vol 283, no 1843, p20161832.

Mazur, R. L. (2010) 'Does aversive conditioning reduce human-black bear conflict?', *Journal of Wildlife Management*, vol 74, no 1, pp48–54.

Melovski, L., and Godes, C. (2002) 'Large carnivores in the "Republic of Macedonia" (recognised by Greece as: "the Former Yugoslav Republic of Macedonia")', in S. Psaroudas (ed) *Protected Areas in the Southern Balkans-Legislation, Large Carnivores, Transborder Areas* (pp. 81–93), Arcturos and Hellenistic Ministry of the Environment, Physical Planning, and Public Works, Thessaloniki, Greece.

Mertzanis, Y., Ioannis, I., Mavridis, A., Nikolaou, O., Riegler, S., Riegler, A., and Tragos, A. (2005) 'Movements, activity patterns and home range of a female brown bear (*Ursus arctos*, L.) in the Rodopi Mountain Range, Greece', *Belgian Journal of Zoology*, vol 135, no 2, p217.

Ministry of Agriculture, Forestry and Food (2002) *Brown Bear (Ursus arctos) Management Strategy in Slovenia*, Ministry for Agriculture and the Environment, Ljubljana, Slovenia, www.mop.gov.si/, accessed 21 February 2017.

Naves, J., and Fernández-Gil, A. (2002) 'Ursus arctos Linnaeus, 1758', in L. J. Palomo and J. Gisbert (eds) *Atlas de los Mamíferos terrestres de España* (Ministerio de Medioambiente, SECEM and SECEMU) (pp. 282–285), Madrid, Spain.

Nores, C. (1988) 'Reducción de areal del oso pardo en la Cordillera Cantábrica', *Acta Biologica Montana, Série Documents de Travail*, vol 2, pp7–14.

Nores, C., and Naves, J. (1993) 'Distribución histórica del oso pardo en la Península Ibérica', in J. Naves and G. Palomero (eds) *El oso pardo (Ursus arctos) en España* (ICONA) (pp. 13–33), Madrid, Spain.

Ordiz, A., Bischof, R., and Swenson, J. E. (2013) 'Saving large carnivores, but losing the apex predator?', *Biological Conservation*, vol 168, pp128–133.

Penteriani,V., Delgado, M. M., and Melletti, M. (2010) 'Don't feed the bears!', *Oryx*, vol 44, no 2, pp169–170.

Penteriani,V., López-Bao, J.V., Bettega, C., Dalerum, F., Delgado, M. M., Jerina, K., Kojola, I., Krofel, M., and Ordiz, A. (2017) 'Consequences of brown bear viewing tourism: A review', *Biological Conservation*, vol 206, pp169–180.

Pérez, T., Naves, J., Vázquez, J. F., Fernández-Gil, A., Seijas, J., Albornoz, J., Revilla, E., Delibes, M., and Domínguez, A. (2014) 'Estimating the population size of the endangered Cantabrian brown bear through genetic sampling', *Wildlife Biology*, vol 20, no 5, pp300–309.

Piédallu, B., Quenette, P. Y., Jordana, I. A., Bombillon, N., Gastineau, A., Jato, R., Miquel, C., Munoz, P., Palazón, S., Sola de la Torre, J., and Gimenez, O. (2016) 'Better together: A transboundary approach to brown bear monitoring in the Pyrenees', *bioRxiv*, pp075663.

Rigg, R., and Adamec, M. (2007) *Status, Ecology and Management of the Brown Bear (Ursus arctos) in Slovakia*, Slovak Wildlife Society, Liptovský Hrádok, Slovakia.

Sahlén, V., Swenson, J., Brunberg, S., and Kindberg, J. (2006) *Björnen i Sverige. En rapport från det Svenska björnprojektet*, Report n° 2006–4 from the Scandinavian Brown Bear Project.

Skrbinšek, T., Bragalanti, N., Calderolla, S., Groff, C., Huber, D., Kaczensky, P., Majić, A., Molinari-Jobin, A., Molinari, P., Rauer, G., Reljić, S., and Stergar, M. (2015) *Annual Population Status Report for Brown Bears in Northern Dinaric Mountains and Eastern Alps. Action C5. Population Surveillance*, LIFE DINALP BEAR, Report LIFE13 NAT/SI/000550.

Smith, R. D., and Maltby, E. (2003) *Using the Ecosystem Approach to Implement the Convention on Biological Diversity: Key Issues and Case Studies*, IUCN, Gland, Switzerland and Cambridge.

Sowka, P. (2009) *Techniques and Refuse Management Options for Residential Areas, Campgrounds, and Group-Use Area*, Living with Predators Resource Guide Series. 3rd ed. Living with Wildlife Foundation, Montana Fish, Wildlife and Parks. Swan Valley, Montana.

Swenson, J. E., Gerstl, N., Zedrosser, A., and Dahle, B. (2000) *Action Plan for the Conservation of the Brown Bear (Ursus arctos) in Europe*, Nature and Environment, No 114. Strasbourg, Council of Europe Publishing.

Swenson, J. E., Sandegren, J., Söderberg, A., Bjärvall, A., Franzén, R., and Wabakken, P. (1997) 'Infanticide caused by hunting of male bears', *Nature*, vol 386, no 6624, p450.

Swenson, J. E., Schneider, M., Zedrosser, A., Söderberg, A., Franzén, R., and Kindberg, J. (2017) 'Challenges of managing a European brown bear population; Lessons from Sweden, 1943–2013', *Wildlife Biology*, no 4, wlb.00251.

Trouwborst, A. (2010) 'Managing the carnivore comeback: International and EU species protection law and the return of lynx, wolf and bear to Western Europe', *Journal of Environmental Law*, vol 22, no 3, pp347–372.

Turgeon, K., and Kramer, D. L. (2012) 'Compensatory immigration depends on adjacent population size and habitat quality but not on landscape connectivity', *Journal of Animal Ecology*, vol 81, no 6, pp1161–1170.

United Nations (1992) 'Convention on Biological Diversity', https://www.cbd.int/doc/legal/cbd-en.pdf, accessed 11 April 2017.

16 Good practice in large carnivore conservation and management

Insights from the EU Platform on Coexistence between People and Large Carnivores

Tasos Hovardas and Katrina Marsden

Introduction

Population trends and range distribution of the four large carnivore species in Europe (brown bear [*Ursus arctos*], wolf [*Canis lupus*], Eurasian lynx [*Lynx lynx*], and wolverine [*Gulo gulo*]) indicate that they have experienced recent recovery. This implies that coexistence between people and large carnivores is possible even in Europe's heavily human-dominated landscapes (Chapron et al., 2014), where protected areas are not large enough to sustain viable populations of large carnivore species (Boitani and Ciucci, 2009; see also Trouwborst et al., 2017). However, the population increase and return of large carnivores has raised tensions between stakeholder groups whose views range between wholeheartedly supporting these developments to believing that large carnivores have no place in the existing cultural landscape of Europe. The latter often bear the brunt of the financial or social costs associated with large carnivore return; for instance, livestock depredation and fear for human safety. Both natural and socio-economic factors contribute towards the controversies and disputes around decisions on large carnivore conservation and management: large home ranges that cross physical and administrative borders, local regions with rich histories and narratives around large carnivores but in many cases little practical knowledge or financial means to deal with the adverse consequences of living with these species, and various factors related to dimensions of social identity and inter-group relations (Hovardas and Korfiatis, 2012; Lüchtrath and Schraml, 2015; von Essen, 2017).

To address these multifarious challenges, a range of organizations which represent different stakeholder groups on the EU level were supported by the European Commission in creating the EU Platform on Coexistence between People and Large Carnivores (http://ec.europa.eu/environment/nature/con servation/species/carnivores/coexistence_platform.htm). The Platform, which was established in 2014, consists of seven members: ELO – European Land-owners' Organization (which currently co-chairs the Platform, together with the European Commission); the Joint representatives of Finnish and Swedish reindeer herders; FACE – The European Federation of Associations for Hunting & Conservation; CIC – The International Council for Game and Wildlife Conservation; IUCN – International Union for Conservation of Nature, European Union Representative Office; WWF – Worldwide Fund for Nature, European Policy Office; and EUROPARC Federation – Federation of Nature and National Parks of Europe. All Platform members have signed an agreement for joint action, committing to: "promote ways and means to minimize, and wherever possible find solutions to, conflicts between human interests and the presence of large carnivore species, by exchanging knowledge and by working together in an open-ended, constructive and mutually respectful way" (http://ec.europa.eu/environment/nature/conservation/species/carnivores/pdf/EN_Agreement.pdf). This mission statement has been implemented since 2014 through jointly organized plenary meetings, regional workshops, and pieces of analysis involving consultation of and gathering information from the Platform members' own membership in the member states, as well as between the Platform members on the EU level.

The European Commission funds a Secretariat to support the Platform members, with the important task, amongst others, of collecting and analyzing case studies identifying good practice in large carnivore conservation and management in various regional contexts. Given the complex nature of human-carnivore conflicts and the inter-group tension it usually causes, in order to be considered good practice, a case must accommodate the complex inter-relationships among stakeholder positions. The objective of the activities promoted are normally to produce a change in terms of stakeholder relations – reduced disputes and disagreement and/or increased collaboration – potentially even establishing a common vision among stakeholders over time. Since damage caused by large carnivores is the main origin of human-carnivore conflicts and related disputes, good practice examples can deliver concrete results to reduce financial losses; for instance, decreased frequency of livestock depredation events or improved damage prevention methods and compensation systems. Successful implementation of these measures will require adequate funding, most probably over the long term, especially if stakeholders are to be encouraged to maintain their involvement over a significant period of time. In regard to this last point, durability of funding can also characterize good practice and provide an index of sustainability; for example, when selected actions and stakeholder collaboration continued after the initial funding of an initiative or project had been terminated, suggesting they are self-sustaining in the long-term.

This chapter aims to present the main results of the task of the Secretariat in showcasing examples of good practice in selected locations across Europe, analyzing good practice and discussing implications and recommendations for large carnivore conservation and management. We have abstained from referring to concepts or terms which would denote a supposed "ideal" situation, readily transferrable between different contexts. The context-bound character of large carnivore conservation and management involves both natural dimensions and human dimensions, and their various interactions, and these require careful consideration before any transfer is attempted (e.g., Redpath et al., 2017). Interactions between natural and human dimensions have been increasingly taken into account in recent research; for instance, in breeding site selection by large carnivores (Sazatornil et al., 2016), model simulations of habitat suitability for large carnivores (Behr et al., 2017) and identification of "ecological traps" (i.e., where large carnivore species may be attracted to due to habitat suitability but where human disturbance is relatively high) (Milanesi et al., 2017). These interactions between natural and human aspects and their effects show how place-based characteristics and practices may be decisive for large carnivore conservation and management.

Another reason for avoiding terminology suggesting "ideal" situations can be attained is a critical reading of the win–win perspective (see in this regard Pooley et al., 2017). We regard good practice as the result of ongoing interaction and compromise between stakeholder groups, reached through negotiation. Therefore, we find it very unlikely that there will be stakeholder groups either benefiting or losing entirely (see Redpath et al., 2013). Instead, our "mixed-motive" alternative to the win–win model seeks to detect the mixed impacts expected after negotiations are held and compromises reached, weighted within and across stakeholder groups (e.g., Hoffman et al., 1999; Hovardas, 2012). Indeed, stakeholders involved in large carnivore conservation and management seem to be aware of the demanding nature of negotiating solutions, and their readiness to reach a compromise should be reflected in a fair allocation of both benefits and costs in some good practice examples. The mixed-motive approach distinguishes itself from win–win conceptualizations in the sense that win–win tends to downplay the need for compromises and accepting loss and emphasizes only the possibility of all parties gaining from the process (e.g., Sunderland et al., 2008; Galuppo et al., 2014). However, when the costs which exist in reality are downplayed, largely for the purposes of encouraging convergence of viewpoints and positions amongst stakeholders in the short term, then this may have a perverse impact, since conflict may re-surface once the costs become apparent (e.g., McShane et al., 2011). The mixed-motive perspective has been used to study examples of good practice and showcase how stakeholders involved were able to negotiate tradeoffs in order to move towards the necessary compromises.

Methods

Sampling good practice

The first stage of data collection involved an extensive and non-selective sampling of case studies of good practice in large carnivore conservation and management with a concentration on stakeholder interaction. A range of experts from the Platform members and the Large Carnivore Initiative for Europe

(LCIE), as well as speakers at Platform events, were requested to suggest such examples. At this stage, no further or more detailed criteria were formulated to guide respondents and data selection aimed at gathering as many examples as possible. For each case study, a series of characteristics were described by the Secretariat using a standard template. The case characteristics were then used to assign case studies to different categories. A list was prepared with 35 case studies (Table 16.1; for a detailed description of case studies: http://ec.europa.eu/environment/nature/conservation/species/carnivores/case_studies.htm) and these were subjected to a screening procedure for further analysis.[1]

Screening good practice

Two reviewers scored each case study based on the following criteria: (1) impact on stakeholder relations (e.g., disagreements or disputes among stakeholders reduced, collaboration among stakeholders increased, common vision among stakeholders established), to examine whether pre-existing tension among stakeholder groups was addressed; (2) socio-economic benefits for residents of the target area (e.g., damage due to large carnivores decreased, compensation systems improved, income/employment opportunities created), indicating that benefits were diffused to local people; and (3) durability of funding (e.g., if all or some of the actions continued after the end of the initial funding), reflecting a mid- to long-term sustainability of the actions. For each criterion, reviewers could indicate whether changes had actually occurred in the case study under consideration or not (binary response variable). Reviewers were then encouraged to provide written comments to justify their reasoning for each response. As part of this screening, funding source and amount, geographic location and species targeted were also recorded. Background literature was consulted, including scientific papers, websites and project reports. In addition, experts and the representatives who had first suggested the case studies were contacted to provide a full overview of potential information sources. Based on this screening procedure, a ranking of case studies was performed by taking into account: (1) case studies with the highest positive score across criteria (binary responses indicating that the desired change had occurred), and (2) homogeneity in reviewer responses (matches between the two reviewers across criteria for each case study and justification in their written comments in the open-ended items). The screening also involved weighting for geographical location (Mediterranean, Balkan, Central European and Nordic dimension) and large carnivore species (wolf, bear, lynx, wolverine). The screening procedure resulted in a long list of 16 case studies. The Secretariat assessed the feedback of Platform members on the long list, from which no significant concerns or remarks emerged, and arrived at a short list of 10 case studies based again on score, divergence between reviewer ratings, geographical location, and large carnivore species. These ten case studies were further examined (Annex 16.1).

Table 16.1 Categorization of case studies depicting good practice in large carnivore conservation and management

Category	Description	Species	Member States	Number of cases
Advice/ Awareness raising[1]	Sourcing of information from individual contact points (websites, experts, volunteers) for the general public, responsible authorities or stakeholders	Bear, wolf, lynx	Austria, Germany, Lithuania, Finland	8
	Awareness raising for tourists to avoid conflict with bears	Bear	Bulgaria, Poland	2
	Avoiding infrastructure development in areas important for wolf breeding	Wolf	Portugal	1
Innovative financing	Volunteer programmes supporting livestock keepers in protecting their flocks from wolves	Wolf	France, Italy	3
	Eco-labelling schemes to increase value of farm produce coming from areas where livestock coexist with large carnivores	Bear, wolf	Austria, Croatia, France, Italy, Slovenia	3
	Eco-tourism development based on the presence of large carnivores	Wolf	Italy	1
	Payment for results scheme (number of successful young wolverines)	Wolverine	Sweden	1
Practical support	Practical measures to improve coexistence such as provision of fencing or livestock guarding dogs	Bear, wolf, lynx	Bulgaria, Greece, Italy, Slovenia	5
	Establishment of emergency teams to respond to call-outs	Bear	Greece	1
Monitoring	Good practice in involving stakeholders in monitoring of large carnivores and sharing the results with stakeholders	Bear, wolf, lynx, wolverine	Slovenia, Croatia, Italy, Finland, Sweden, Norway	4
	Good practice in cross border monitoring	Bear, wolf	Finland, Norway, Russia	1
Understanding viewpoints	Studies understanding stakeholder attitudes to different large carnivore species	Bear, wolf	Greece, Italy, Slovenia	2
	Intensive efforts to encourage stakeholders to work together	Bear, wolf, lynx	Germany, Switzerland, Spain	3

[1] No case study in this category was retained for further analysis, after screening and weighting; see "Methods," subsection "Screening good practice".

Interviews with members of stakeholder groups involved in the case studies

Selection of interviewees and data collection

For each case study, at least three informants were selected. The procedure started with experts among Platform members, the LCIE and presenters in Platform events, who had submitted the case studies, suggesting potential informants among members of stakeholder groups who had been actively and deeply involved and had gained a thorough experience of the implementations at the local level. These included local producers (e.g., stock breeders, farmers, bee keepers), hunters, landowners, authorities (e.g., regional/local and managing authorities), foresters, scientists (e.g., biologists, conservation scientists, experts in human dimensions) and members of environmental non-governmental organizations. After a first informant had been secured for each case study, he/she was asked to indicate other people engaged. When selecting interviewees, care was taken to include at least one member from all core stakeholder groups involved. By means of the snowball sampling method, a total of 34 interviewees were identified, and semi-structured interviews were conducted based on an interview matrix developed by the Secretariat. Before the interview, interviewees were informed about the aim of the study and granted their informed consent. Since the nature of the study was exploratory, the main aim was not to follow up all points raised by interviewees exhaustively but to arrive at an understanding of stakeholder positioning on the issues at hand. Therefore, one interviewee representing each interest group was regarded as sufficient and additional informants were only sought in the case that a very wide range of different groups were involved or if the complexity of issues handled indicated that additional interviews were needed. These requirements were generally met for each case after the third or fourth interview. Semi-structured interviews focused on pre-defined themes (e.g., what has worked well and what could be improved, with a concentration on stakeholder interaction, transferability and continuation of the implementation) but also gave interviewees the opportunity to expand on topics of interest to address any relevant concern. All interviews were carried out by national experts commissioned for that purpose in the national language of interviewees, and they were digitally recorded, transcribed, and translated into English.

Data analysis

Qualitative analysis of interview transcripts involved developing codes and then coding of all interview extracts. Key themes were identified in terms of: (1) perceived benefits and gains of participation, which could also refer to the added value of participation (e.g., direct or indirect financial or other benefits), and (2) costs of participation, also including unanticipated negative side effects acknowledged by respondents during or after the implementation. Perceived benefits and costs of participation were allocated across two main topics – namely, large carnivore conservation and agricultural production/ primary sector (mainly livestock raising or herding) activities – while game management and developments in the tourism sector also emerged as secondary topics in some case studies. This mixed-motive perspective was employed in each individual case study to account for aspects that had worked well and aspects that could or should be improved, and the trade-offs that should be met for effective stakeholder interaction, transferability and viability of the implementation. Particular attention was paid to discursive positioning, whereby interviewees elaborated on their viewpoints on a range of developments at the local level focusing particularly on contrasting lines of argumentation (see Hovardas, 2017a). To provide a measure of consistency in coding, 20% of transcripts have been coded by two independent raters and inter-rater reliability amounted to more than 85% across themes.

Results

Examples of good practice and stakeholder positioning

Practical support under the Slovenian rural development programme

By the end of the 19th century, habitat alterations and policies of poisoning and hunting large carnivores had driven Slovenia's brown bear and wolf populations nearly to extinction. Conservation measures introduced at the end of the 19th century allowed recovery of the populations to around 533–598 bears and around 52 wolves, according to government monitoring (Skoberne, 2017). Although estimates for both bear and wolf populations show increasing population trends, there are several concerns in terms of having reached favourable conservation status or not, which mainly refer to population dynamics (e.g., Jerina and Adamič, 2008).

The main conflict around large carnivores is depredation of livestock by both bears and wolves. A wide range of management measures have been put in place including education and awareness raising, monitoring, waste management, lethal population management and a series of compensation and prevention measures targeted specifically at the livestock sector. Measures have been trialled through a number of EU LIFE projects (e.g., "SLOWWOLF"; "DINALP"), but Slovenia has also one of the longest standing uses in the EU of their Rural Development Programme to support prevention measures, having included measures in the programme since entering the EU. The Rural Development Programmes are funded through the European Agricultural Fund for Rural Development (EAFRD), part of the EU Common Agricultural Policy, together with national co-financing. The advantage of this funding stream is that it is available across the EU, is significantly larger than the LIFE Nature and Biodiversity fund and can be accessed both by individuals and groups (Marsden et al., 2016). The activities funded under the sub-measure *Animal husbandry in central areas of appearance of large carnivores* include herd protection using mobile electric fences and nets, shepherding and support for livestock guarding dogs. In the 2007–2014 period, the total expenditure for the sub-measure was €1,304,443.28. There were 642 applications in total, meaning that payments were on average €2,090 per applicant (Slovenian Ministry of Agriculture, 2015). Numerous stakeholder groups were involved in the design of the measure, including farmer representatives, the Slovenian Forest Service and environmental non-governmental organizations. These stakeholder groups are also represented on the Programme Monitoring Committee.

Interviewees in the Slovenian case study highlighted that implementation of damage preventions methods is expected to bring about benefits such as increased tolerance of beneficiaries towards large carnivore species and a willingness to coexist with these species (Table 16.2). An increase in large carnivore populations had been also recorded within the period during which the measures were in place. The costs referred to by interviewees concentrated on the additional input required for the effective implementation of damage prevention methods – for instance, maintaining and breeding or training of livestock guarding dogs – that may not be covered by the programmes and need to be partly taken on by conservation organizations. Interviewees also underlined a series of costs of participation or unintended side effects for agricultural production and primary activities, which were related to either institutional arrangements or on-the-ground implementation of damage prevention methods. Specifically, they pointed out that when costs of participation were not fully covered for farmers, then this may force them to increase the price of agricultural products. Interviewees also criticized the programme for not allowing for a joint implementation of more than one measures (e.g., combining electric fences with livestock guarding dogs), which may have compromised the optimum potential for damage prevention methods. With regard to on-the-ground implementation, interviewees pointed out the crucial importance of installing the fences properly, since damage can be worse otherwise. Another issue to be tackled was the ability of wolves to learn how to get around fences unless they were moved frequently.

Innovative funding: conservation performance payments for wolverines, Sweden

Wolverines are rare in Europe, being found only in four northern countries: Norway, Sweden, Finland and Russia. Overall numbers are low, at around 1,200 in the three Nordic countries, with more than half of these found in Sweden (Linnell, 2014). The population decreased in the 19th century due to persecution and the fur trade. Habitat changes have also had an impact over time. The fact that the Nordic population is small and groups are often isolated means that genetic variability is low (Abramov, 2016).

Depredation of reindeer is the main cause of conflict related to wolverines. In order for wolverines to survive in an area, a certain amount of depredation must be tolerated by the indigenous Sámi reindeer herders. A compensation system for damages based on reindeer carcass documentation was put in place. Lethal control of large carnivores is also carried out based on a quota

Table 16.2 Elaboration of the benefits and costs of participation by interviewees in the Slovenian case study (Practical support under the Slovenian Rural Development Programme)

Main focus of stakeholder interest	Benefits and gains of participation; added value	Costs of participation; unanticipated side effects
Large carnivore conservation	• Adoption of damage prevention measures is usually accompanied by increased tolerance towards large carnivores and better acceptance of coexistence with people • The population of large carnivores has increased over the period the measures have been in place (though this is likely due to a range of conservation activities)	• Livestock guarding dogs need training, which is quite demanding, before they can be effective in preventing damage to livestock from large carnivores; not all of these costs for maintaining, breeding and training livestock guarding dogs are covered by the programmes and they are sometimes partly taken on by conservation organizations
Agricultural production, livestock, primary sector activities	• The implementation of damage prevention measures has decreased the number of attacks on livestock	• Where costs for farmers are not fully covered, this increases their overall costs and potentially has a knock-on impact on the price of agricultural products • Implementing multiple measures is not allowed under the programme, e.g., combining electric fences with livestock guarding dogs • If fences are incorrectly installed, damage can be worse than when it has not been there at all • Wolves can learn how to get around fences if they are not moved frequently

Note: Interviewees included one farmer, staff of the Slovenian Forest Service and a member of an environmental non-governmental organization.

scheme and reacting to problem individuals. In 1996, the Swedish government replaced compensation payments with a Conservation Performance Payments (CPP) scheme, paying reindeer herders for the number of successfully breeding wolverines in their area regardless of predation levels as well as for wolverine occurrence in districts without confirmed reproductions. A monitoring scheme is attached to the CPP in which reindeer herders are also engaged. The CPP is government funded and payments have been set at around 200,000 SEK (1 SEK = approximately €0.10) per documented wolverine reproduction (Zabel and Holm-Müller, 2007; Persson et al., 2009). This measure has been successful in reducing illegal killing, which was the primary reason for adult mortality of Swedish wolverines. The population doubled over a 10-year period (from 57 in 2002 to 125 in 2012), and expanded into previously unoccupied areas (Persson et al., 2015).

Interviewees agreed that the scheme had reduced conflict around wolverines. Since payments were received whether or not reindeer were killed, it had also encouraged better stock protection measures (Table 16.3). Additionally, herders saved time as they were not obliged to go out to search for the carcasses of reindeer in order to support compensation claims. The joint monitoring was also instrumental in reaching an agreed understanding of the background situation. While the scheme had significant benefits over the previous compensation method, it was not without its detractors and certain unanticipated side effects were experienced by those involved. The monitoring needed to back up the scheme has its weaknesses: It may disturb females with young since a close approach to the den is required in order to count young. Additionally, weather conditions may sometimes lead to false estimations of the population and potentially underpayment and local people may not always agree with the methods used. The payments are based on location of dens within administrative boundaries. If wolverines hunt within an administrative district other than the one in which they have their dens, herders may not be sufficiently compensated for their losses. Another problem is that there may be a mismatch between the CPP and other large carnivore management measures. In particular, a 10% tolerated damage threshold (10% of stock lost) has been set and in the case this is reached, lethal control may be agreed even though it may be unclear which large carnivore species is responsible for damages. Finally, some herders may take advantage of the timing of the monitoring and illegally kill young wolverines after they know it has been carried out. Since they will still receive their payment without accepting the loss of reindeer caused by the wolverines, this can be regarded as "free riding" the system.

Understanding viewpoints: Core Group Wolf, Bern, Switzerland

After the extermination of wolves in Switzerland in the 19th century, the first returning wolf was recorded in the Canton of Bern in 2006 (KORA, 2012). A pair established themselves in the area in 2016, but the female wolf was later found poisoned in the canton of Fribourg. Single wolves are present, and it is expected that there will be further pairing attempts in the area.

Depredation of livestock, particularly sheep, is the main cause of conflict. Farmers received adequate compensation, but face emotional stress due to loss of livestock and additional work to implement protection measures. Tourism is dependent on the grazed landscapes of the Alps, and there are concerns that a growing population of wolves will put this at risk through reducing the viability of livestock farming – conflict between tourism and the presence of livestock guarding dogs and potentially putting tourists off due to fear of wolves. Illegal killing of wolves still has the potential to be a major problem in the area.

Table 16.3 Elaboration of the benefits and costs of participation by interviewees in the Swedish case study (Conservation performance payments [CPP] for wolverine)

Main focus of stakeholder interest	Benefits and gains of participation; added value	Costs of participation; unanticipated side effects
Large carnivore conservation	• An increase in the wolverine population and distribution has been observed and documented through monitoring which can be directly attributed to the CPP • A corresponding reduction of human-carnivore conflict can be observed; this has set the stage for an improvement in stakeholder relations	• Pressure is put on people working with monitoring – the field personnel or the methodology are blamed for any inconsistencies and local people may not agree that the methodology used is suitable, leading to a resurfacing of conflict • The monitoring system may disturb females, since inspection needs to approach quite close to the den so that reproduction can be documented • The CPP may not always align with other large carnivore management measures; specifically, there is a 10% tolerated damage threshold for all large carnivore species – if damage is above this level, lethal action may be taken, and it may be difficult to attribute damage to a specific species • Illegal killing of wolverines does still exist as a type of "free riding", e.g., young shot after documentation of reproduction has been undertaken
Agricultural production, livestock, primary sector activities	• Payments in the form of CPP encourage efficient herding (i.e., prevention of depredation) and do not penalize with a lower compensation • They are certainly an improvement over the compensation systems based on documentation of livestock depredation for a variety of reasons (many depredated animals may be never found, significant effort is required to look for them) • Involvement in monitoring increases belief in the system and can provide an additional source of income	• Weather conditions may not always allow for a reliable assessment of monitoring indices, which may lead to an underestimation of CPP (dependence of monitoring on snow) • Reindeer herders underlined that CPP cannot fairly balance livestock losses and request an increase; they also highlight that the system has not been updated since 2002, and they would like a higher fee/subsidy for local people involved • CPP payments are dependent on the number of reproductions from a location as well as the location of the den in relation to the border of the focal district – a herder may not get compensation even though they are close to the site, if they are located on the wrong side of the focal district border

Note: Interviewees included a reindeer herder, county administration personnel and member of an environmental non-governmental organisation.

Compensation schemes have been in place since the reintroduction of the lynx in the 1970s. Originally paid by the Swiss League for the Protection of the Nature (now "Pro Natura"), they have now been taken over by Government authorities (federal and cantonal), and payments are dependent on confirmation of an attack by a game warden. In order to engage stakeholders with large carnivore management, the Department of Economic Affairs of the Canton of Bern developed an Official Strategy on Coexisting with Wolves (Kanton Bern, 2007). The Core Group Wolf was established to implement the strategy and to encourage and facilitate discussions between stakeholders in the light of an increasing wolf population (Juesy, 2015). Such groupings, which have also been established in other cantons, look at the practical implementation of damage prevention measures. Biannual meetings take place in Bern, as well as an annual excursion to look at prevention measures on the ground. Meeting venues and excursion costs are financed by the authorities, though members must attend in their own time and cover their travel costs to the meeting point.

Interviewees highlighted that by working together in the group and particularly by taking part in excursions together, relationships had been improved, meaning that on the local level, joint solutions could be found, even if the different interest groups had opposing opinions. The group has succeeded in developing a common vision and a joint understanding of their aims. Particularly important for building trust was signing and respecting an agreement not to release any press statements without prior discussion with the group members (Table 16.4). Farmer participants could learn how to better protect their flocks, and also appreciated the increased understanding of their position and the additional difficulties they faced in putting in place protection measures. Farmers, however, also faced negative side effects which were especially related to their standing with other farming colleagues and in-group dynamics. Those participating were regarded by colleagues as being "pro-wolf" and not fully representing their group's viewpoints. This also meant that the trust built up within the group was not necessarily disseminated outwards to other farmers. Additionally, concern was expressed that those farmers who did not implement protection measures would be disproportionately affected by any wolf attacks and could end up bearing substantial costs. While conservationists largely benefited from the group, participation meant that they also have to accept compromise and potentially consider lethal control in the case that damages increase beyond a threshold acceptable to farmers.

TASSU monitoring system and volunteer-based large carnivore contact network, Finland

All four European large carnivores (wolf, bear, lynx and wolverine) are present in Finland. In recent years, there has been some recovery of numbers and range. Current population estimates are as follows: wolves, 150–180 individuals, considered endangered according to IUCN criterion; bears, 1,980–2,100 individuals, near threatened; lynx, 2,355–2,495 individuals, near threatened

Table 16.4 Elaboration of the benefits and costs of participation by interviewees in the Swiss case study (Core Group Wolf)

Main focus of stakeholder interest	Benefits and gains of participation; added value	Costs of participation; unanticipated side effects
Large carnivore conservation	• Legal requirements concerning the protected status of the wolf are generally respected • Public insecurity related to wolves in the area stirred up by the media can be avoided through previously agreed upon communication flows	• The increase in wolf numbers has also led to an increase in the damage extent and to occasional illegal killing • Lethal wolf management may be considered if wolf numbers increase dramatically
Agricultural production, livestock, primary sector activities	• Farmers are recognized by stakeholder groups as possible victims of damages caused by large carnivores; the additional burden in time and investment for farmers implementing livestock protection measures is acknowledged • Excursion participants broaden their knowledge on protection measures against wolf attacks	• If damage prevention measures are not uniformly implemented at the local level, then those local producers who have not implemented the measures may suffer substantial damage • Professional colleagues (e.g., other farmers) outside of the Core Group Wolf assume that a member of this group automatically is in favour of the return of the wolf, leading to potential conflict within this interest group

Note: Interviewees included a representative of a farming organization, a nature conservation association and a game warden.

(LUKE, 2017); wolverines, 220–250 individuals, endangered (Metsähallitus, 2017a). Finnish wolf and wolverine populations are still small and suffer from lack of genetic exchange. Wolverines suffer the effects of climate change as full snow cover during winter influences the success of reproduction of this species. Lynx and bear populations are healthier (near threatened, recently down-ranked from threatened), and exchange occurs with the Russian bear population.

National management plans have been created for all large carnivore species (Metsähallitus, 2017b). The plans consist of a range of actions which aim to find solutions both to conservation issues and to conflict with other interests. Hunting quotas are allocated for all four species. In 2017, a hunting quota for wolverine was issued for the first time in 35 years. The biggest problem of coexisting with large carnivores in the case study area is damage to reindeer herding. A damage evaluation and compensation system is in place, while the compensation claimed in 2016 was €10 million (Metsähallitus, 2017c). The threat wolves pose to hunting dogs is also a significant issue, and damage caused by large carnivore species to other agricultural activities (livestock, beekeeping and other forms of agriculture) can also be a problem locally. There are also some problems with fear of large carnivores and the danger they might pose to humans.

The TASSU data collection system for large carnivores (TASSU is the Finnish word for "paw") has been in place since 1978. The aim of the system is to improve accessible scientific data and monitor population sizes though the establishment of an electronic database to which all stakeholders have access. The system is run by the Finnish State and costs between €10,000–25,000/year to run. It involves identifying and training suitable volunteers from a range of different interest groups, to collect information on the signs of the presence of large carnivores. Once confirmed by an expert, the information is entered into a publically available database. According to the interviewees, TASSU has been successful in increasing the knowledge of the engaged stakeholders about large carnivores and their trust in the monitoring system (Table 16.5). Stakeholders now at least start from a common understanding of the facts when discussing

Table 16.5 Elaboration of the benefits and costs of participation by interviewees in the Finnish case study (TASSU monitoring system and volunteer-based large carnivore contact network)

Main focus of stakeholder interest	*Benefits and gains of participation; added value*	*Costs of participation; unanticipated side effects*
Large carnivore conservation	• Better documentation of presence and numbers of large carnivores	• "Problem" wolves to be detected and removed
Agricultural production, livestock, primary sector activities	• Damage evaluation is deemed satisfactory	• Compensation after damage evaluation is not satisfactory, especially for beekeepers • Damage caused by wolves increases with number of wolves
Game management	• Hunters are the main social actor to collect data on large carnivores	• Hunters' rights and quotas may be renegotiated among stakeholders following new information gathered from monitoring.
Tourism sector	• Employment and development opportunities in the tourism sector	• Inappropriate tourism activities have been recorded, resulting in large carnivores being attracted to certain areas/food conditioned • Public opinion may be against increased tourism

Note: Interviewees were representatives from the local game management association, a research institution and a conservationist.

large carnivore management. This has supported agreement of a common vision at least on certain issues such as dealing with "problem" wolves, and conservationists have also had to accept some lethal control. Although damage caused by large carnivores is thought to be satisfactorily evaluated, there have been some complaints for compensation, especially from beekeepers. Moreover, the basic conflict related to the number of reindeer in particular lost to wolves has not yet been effectively addressed. Any increase in wolf numbers is likely to result in an increase in damages. Hunters have been the main social actors to contribute in data collection on large carnivores. However, hunters' rights and quotas may be renegotiated among stakeholders following new information gathered from monitoring. While there may be potential to carry out new tourism activities, these are often negatively perceived locally, with concerns expressed that they encourage large carnivores to return to particular areas and potentially habituate to humans.

Discussion and implications for large carnivore conservation and management

The examples of good practice in large carnivore conservation and management which have been presented in this chapter are among the most acknowledged and valued instances of successful stakeholder involvement and collaboration as identified by Platform members and national experts. The mixed-motive approach highlights how stakeholder positioning took into account trade-offs and potential points of focus and improvement in terms of stakeholder interaction, transferability of the various implementations and their viability. We have showcased how interviewees raised points related not only to the difficulties they themselves experienced in participating (time commitment needed or compromises to be made), but also how the implementation of good practice itself may have unanticipated consequences, which need to be very carefully considered in future decision-making at the local level. Unanticipated consequences may take the form of an interplay of positive or negative feedback loop effects, reinforcing unwanted outcomes or halting desired developments. In this overall frame, good practice cannot be conceptualized as a flawless trajectory of intended and fully anticipated events and outcomes, but as stakeholder joint action including adapting plans to deal with the difficulties or barriers emerging at the local level with the constellation of stakeholder interests present. This conceptualization will always incorporate some kind of compromise among stakeholders reached through negotiation and under the frame of what we have termed a mixed-motive perspective in large carnivore conservation and management.

Perhaps the most well documented positive feedback loop effect is that the increase in large carnivore numbers is highly likely to increase the odds of damage caused by large carnivores (e.g., the Swiss and Finnish case studies). In some cases, this may exceed a tolerance threshold and result in calls for lethal management of large carnivores or removal of "problem" animals. These actions will not be readily supported by all stakeholders, particularly in the light of recent research, questioning the effectiveness of lethal methods over non-lethal ones, when it comes to damage prevention (e.g., Treves et al., 2016). Disagreement in this case exemplifies how stakeholder interaction may give rise to new tension in inter-group relations (see also the Finnish case study, where hunters may need to renegotiate hunting rights and quotas with other stakeholders). However, positive outcomes for inter-group relations were also described by interviewees (e.g., the Swiss and Finnish case studies), especially that mutual recognition of the points of view of the actors involved allowed for trust building (see Young et al., 2016). A common finding in all case studies examined is that stakeholder interaction is possible even if the conflict has not been resolved and even when a fully-fledged consensus on controversial issues related to large carnivore conservation and management will never be achieved (see also Redpath et al., 2013; Young et al., 2013).

A second positive feedback was that operation of damage prevention methods was associated with additional workload (i.e., the greater the number the damage prevention methods introduced, the larger the workload needed). Stakeholders interviewed in the Slovenian case study suggested that this additional input could eventually increase the price of agricultural products. Undesirable developments after the implementation of damage prevention methods were also recorded. For instance, a negative feedback loop emerges if damage prevention methods are not taken up by all local producers uniformly in an area, which renders those not implementing them disproportionally vulnerable to large carnivore attacks (see the Swiss case study). Interviews revealed another negative feedback loop related to measures that were not put in place properly (see the Slovenian case study). For instance, incorrect installation of electric fences or the habituation of large carnivores to electric fences installed will not help decrease livestock losses and can even increase them if sheep are trapped by an incorrectly installed fence.

Conservation performance payments in Sweden reverse the burden of proof for damage caused by large carnivores: the *ex-post* compensation schemes was replaced by an *ex-ante* compensation scheme; i.e., participants were paid for the number of young wolverines born. Stakeholder input, expertise and workload is redirected from searching for livestock carcasses to prove damage and access compensation (a work-intensive and often frustrating process if the carcass is never found), towards documenting wolverine reproduction to justify payments. Indeed, this may result in a positive feedback in that engaging Sámi people in reproduction documentation may provide a means of recognizing their knowledge and expertise. The rewarding incentives cultivated by this *ex-ante* scheme may trigger a further positive feedback loop in encouraging implementation of damage prevention methods and efficient herding (e.g., Zabel et al., 2011). Namely, payments could be wiped out by damages unless damage prevention methods are put in place. Therefore, our findings support those of Skonhoft (2017) showing that *ex-ante* compensation schemes are more efficient than *ex-post* compensation.

However, even this example of good practice is not without unanticipated side effects. Indeed, aspects related to the local setting (weather conditions, which may influence monitoring systems and leave some reproductions unaccounted for) and aspects related to how monitoring is operationalized (approaching close to dens to document reproduction is likely to disturb female wolverines), as well as social norms related to monitoring (social pressure put on field personnel who perform monitoring), may result in resentment building up locally between reindeer herders. What is more, our interviewees suggested that poaching may sometimes take place after reproductions have been documented, despite the fact that previous research indicated that beneficiaries refrain from illegal killing of wolverines (Zabel et al., 2014). Such events may be characterized as "free riding" behaviour, since undiscovered poachers may enjoy the benefits of the *ex-ante* compensation scheme without losing reindeer, having to adapt their herding practices or implement any damage prevention method. Future research needs to reflect on the motives behind this "free riding" and whether it is related to an apparent lack of full correspondence between the allocation of conservation performance payments, on the one hand, and allocation of wolverine damages on the other. Payments are directed to villages in the administrative area where the reproduction is located, and not at each herder separately. Each village then is responsible for distributing the reward. Since damage caused by wolverines may not be uniformly distributed across the reindeer herding area, some herders may not receive fair allocation compared to the damage they have to bear. This is particularly problematic for herders located on the wrong side of the focal district border. This distributional issue in conservation performance payments has been underlined by previous studies (Zabel et al., 2014, p. 616; Skonhoft, 2017, p. 922).

There were two cases that pointed to the need for planning synergies between different measures carefully. In the Slovenian case study, interviewees highlighted that producers were only eligible for funding for individual measures through the rural development programmes rather than a range of complimentary measures (e.g., installing electric fences together with using livestock guarding dogs). In addition, the Swedish case study showed how the incentive structures for two different management measures may prove to be incompatible. In this case, interviewees underlined the inconsistency between conservation performance payments, on the one hand, and the 10% tolerance level set for damage caused by large carnivores, on the other. Whereas conservation performance payments involve an incentive structure in linear proportionality (i.e., the more reproductions documented the more payments directed to local communities), the 10% tolerance level introduces a change in management practices after the threshold has been surpassed (lethal control if damage is too high). These findings show the need to plan measures for large carnivore conservation and management in an integrated manner within a region. Any desirable synergies should be supported, allowing beneficiaries to take full advantage of the entire toolbox of measures offered by different funding schemes (e.g., European Rural Development funding, national compensation measures, etc.). Moreover, incentive structures in all sectors should also be addressed in a coherent manner. This should apply to tourism development and employment opportunities, as well, for which there were indications that some stakeholder hold reservations (see the Finnish case study).

Good practice in large carnivore conservation and management, as reflected in our examples, seems to imply a procedural conceptualization of social sustainability (see Hovardas, 2017b); i.e., that the procedures that promote constructive stakeholder interaction may prove more important for the transfer of good practice than concrete content or outcomes which are context bound and location dependent. Good practice can indicate the circumstances under which human-carnivore coexistence may be achievable, but needs to be accompanied by a thorough situational analysis before attempting transfer to another location. Our analysis reflects the potential of deliberation and consultation processes to nurture social learning (e.g., Keen et al., 2005; Newig, 2011; Young et al., 2013). Working together on common goals may set the stage for the mutual recognition of the positions of all involved actors, catalyzing stakeholder empowerment and collaboration further. The experience gathered and analyzed from the case studies across Europe implies that good working relationships among stakeholders are a valuable resource themselves and may contribute to empowering local networks of actors, who may continue the implementation of measures even after their initial funding has expired. The mixed-motive perspective we have employed in this research could be taken up by stakeholders in local stakeholder platforms (as demonstrated by the Swiss case study) to monitor their interactions and plan their future cooperation in a structured manner.

Acknowledgments

CALLISTO and adelphi consult GmbH provide the Secretariat of the EU Platform on Coexistence between People and Large Carnivores. This work was produced following research carried out by the Platform Secretariat for DG Environment of the European Commission, Service Contract No. 07.0202/2016/738209/SER/ENV.D.3. It does not necessarily reflect the views of the Platform or the official view of the European Commission.

Annex 16.1 Short list of the 10 case studies depicting good practice in large carnivore conservation and management

Category	Geographical location (member state)	Title	Short description	Dates
Practical support	Balkan (Slovenia)	Practical support under the Slovenian Rural Development Programme (RDP)	Payment per hectare of grassland with top ups depending on a range of protection measures adopted (livestock guarding dogs, shepherds, electric fences)	2004–present
	Balkan (Greece)	Developing a network of livestock guarding dogs (LGDs)	A network of owners of LGDs was created facilitating coordination and the exchange of puppies and adult dogs	2009–2012
	Balkan (Greece)	Damage prevention measures (e.g., fences) through the Greek RDP	Installation of electric fences around apiaries and sheepfolds for minimizing damages caused by bears	2004–2013
	Mediterranean (Italy)	Livestock protection measures through Medwolf	LIFE project encouraging collaboration among provincial administration, environmental non-governmental organizations and professional agricultural associations	2012–2017
Innovative financing	Central (France)	Labelling schemes for farm cheeses in the Haut Béarn	Marketing approach using the bear foot imprint to give value to cheese, creating socio-economic benefits for shepherds through the presence of bears	1995–ongoing
	Nordic (Sweden)	Conservation performance payments	The Swedish government replaced compensation payments with conservation performance payments for successfully breeding wolverines	1996–2011

Understanding viewpoints	Transfer and Communication Project – Baden-Württemberg	Central (Germany)	Management of conflicts and development of sound solutions, mainly by enlarging the awareness on conflict dynamics among parties through mediated discussions	2012–ongoing
	Core Group Wolf	Central (Switzerland)	Cantonal Wolf Groups established in several Swiss cantons to objectify discussions and improve relationships between stakeholders	2006–ongoing
	Cooperation of stakeholders in the Cantabrian mountains	Mediterranean (Spain)	A project facilitatating collaboration using formal agreements with hunting associations and federations to foster the social acceptance of bears, reduce poaching with illegal snares and avoid the indirect impacts of hunting activities	1993–2015
Monitoring	TASSU system and volunteer-based large carnivore contact network	Nordic (Finland)	Electronic database which tracks the presence of large carnivores based on the input from volunteers who are trained by state agencies	1978–ongoing

Note

1 It should be noted that the process of collecting cases has been continued by the Platform members. Although 35 cases had been collected when this work commenced and were considered in the screening described in this chapter, more have since been added to the Platform website.

References

Abramov, A. V. (2016) *Gulo gulo*. The IUCN Red List of Threatened Species 2016: e.T9561A45198537, http://dx.doi.org/10.2305/IUCN.UK.2016-1.RLTS.T9561A45 198537.en, accessed 11 December 2017.

Behr, D. M., Ozgul, A., and Cozzi, G. (2017) 'Combining human acceptance and habitat suitability in a unified socio-ecological suitability model: A case study of the wolf in Switzerland', *Journal of Applied Ecology*, vol 54, no 6, pp1919–1929.

Boitani, L., and Ciucci, P (2009) 'Wolf management across Europe: Species conservation without boundaries', in M. Musiani, L. Boitani, and P. C. Paquet (eds) *A New Era for Wolves and People: Wolf Recovery, Human Attitudes, and Policy*, University of Calgary Press, Calgary.

Chapron, G., Kaczensky, P., Linnell, J. D. C., von Arx, M., Huber, D., Andrén, H., López-Bao, J. V., Adamec, M., Álvares, F., Anders, O., Balčiauskas, L., Balys, V., Bedö, P., Bego, F., Blanco, J. C., Breitenmoser, U., Brøseth, H., Bufka, L., Bunikyte, R., Ciucci, P., Dutsov, A., Engleder, T., Fuxjäger, C., Groff, C., Holmala, K., Hoxha, B., Iliopoulos, Y., Ionescu, O., Jeremić, J., Jerina, K., Kluth, G., Knauer, F., Kojola, I., Kos, I., Krofel, M., Kubala, J., Kunovac, S., Kusak, J., Kutal, M., Liberg, O., Majić, A., Männil, P., Manz, R., Marboutin, E., Marucco, F., Melovski, D., Mersini, K., Mertzanis, Y., Mysłajek, R. W., Nowak, S., Odden, J., Ozolins, J., Palomero, G., Paunović, M., Persson, J., Potočnik, H., Quenette, P.-Y., Rauer, G., Reinhardt, I., Rigg, R., Ryser, A., Salvatori, V., Skrbinšek, T., Stojanov, A., Swenson, J. E., Szemethy, L., Trajçe, A., Tsingarska-Sedefcheva, E., Váňa, M., Veeroja, R., Wabakken, P., Wölfl, M., Wölfl, S., Zimmermann, F., Zlatanova, D., and Boitani, L. (2014) 'Recovery of large carnivores in Europe's modern human-dominated landscapes', *Science*, vol 346, no 6216, pp1517–1519.

Galuppo, L., Gorli, M., Scaratti, G., and Kaneklin, C. (2014) 'Building social sustainability: Multi-stakeholder processes and conflict management', *Social Responsibility Journal*, vol 10, no 4, pp685–701.

Hoffman, A. J., Gillespie, J. J., Moore, D. A., Wade-Benzoni, K. A., Thompson, L. L., and Bazerman, M. H. (1999) 'A mixed-motive perspective on the economics versus environment debate', *American Behavioural Scientist*, vol 42, no 8, pp1254–1276.

Hovardas, T. (2012) 'Can forest management produce new risk situations? A mixed-motive perspective from the Dadia-Soufli-Lefkimi Forest National Park, Greece', in J. Martin-Garcia and J. J. Diez (eds) *Sustainable Forest Management: Case Studies* (pp. 239–258), INTECH, Rijeka.

Hovardas, T. (2017a) '"Battlefields" of blue flags and seahorses: Acts of "fencing" and "defencing" place in a gold mining controversy', *Journal of Environmental Psychology*, vol 53, pp100–111.

Hovardas, T. (2017b) 'Gold mining in the Greek "Village of Gaul": Newspaper coverage of conflict and discursive positioning of opposing coalitions', *Environmental Communication*, vol 11, no 5, pp667–681.

Hovardas, T., and Korfiatis, K. J. (2012) 'Adolescents' beliefs about the wolf: Investigating the potential of human – Wolf coexistence in the European south', *Society and Natural Resources*, vol 25, no 12, pp1277–1292.

Jerina, K., and Adamič, M. (2008) 'Fifty years of brown bear population expansion: Effects of sex-biased dispersal on rate of expansion and population structure', *Journal of Mammalogy*, vol 89, no 6, pp1491–501.

Juesy, P. (2015) 'Bericht Wolf/Herdenschutz Kanton Bern (2006 bis 2015)', www.vol. be.ch/vol/de/index/natur/jagd_wildtiere/publikationen.assetref/dam/documents/ VOL/LANAT/de/Natur/Jagd_Wildtiere/PUB_LANAT_JW_Bericht_Wolf_Herdens chutz_2006-2015_Kt-Bern.pdf, accessed 27 November 2017.

Kanton Bern, Volkswirtschaftsdirektion (2007) 'Strategie Umgang mit dem Wolf im Kanton Bern' [in German], www.vol.be.ch/vol/de/index/natur/jagd_wildtiere/projekte.asse tref/content/dam/documents/VOL/LANAT/de/Natur/Jagd_Wildtiere/LANAT_JW_ Umgang_mit_dem_Wolf_de.pdf, accessed 27 November 2017.

Keen, M., Brown, V. A., and Dybal, R. (2005) *Social Learning: A New Approach to Environmental Management*, Earthscan, London.

KORA (2012) 'KORA News – Wölfe im Kanton Bern' [in German], News on KORA website, www.kora.ch/index.php?id=214&L=3&tx_ttnews%5Btt_news%5D=406&cHas h=b6fd1afad0f04e02c4b57103493e2538, accessed 27 November 2017.

Linnell, J. (2014) 'Status of wolverine in Europe', www.lcie.org/Blog/ArtMID/6987/Arti cleID/69/Status-of-wolverines-in-Europe, accessed 11 December 2017.

Lüchtrath, A., and Schraml, U. (2015) 'The missing lynx – Understanding hunters' opposition to large carnivores', *Wildlife Biology*, vol 21, no 2, pp110–119.

LUKE, Natural Resources Institute Finland (2017) 'Game and hunting', LUKE Website, www.luke.fi/en/natural-resources/game-and-hunting/, accessed 27 November 2017.

Marsden, K., Hovardas, T., Psaroudas, S., Mertzanis, Y., and Baatz, U. (2016) 'EU platform on large carnivores: Supporting good practice for coexistence – Presentation of examples and analysis of support through the EAFRD', http://ec.europa.eu/environment/nature/con servation/species/carnivores/pdf/96_LC%20Platform-case%20studies%20and%20RD. pdf, accessed 26 November 2017.

McShane, T. O., Hirsch, P. D., Trung, T. C., Songorwa, A. N., Kinzig, A., Monteferri, B., Mute-kanga, D., Van Thang, H., Dammert, J. L. Pulgar-Vidal, M., Welch-Devine, M., Brosius, J. P., Coppolillo, P., and O'Connor, S. (2011) 'Hard choices: Making trade-offs between biodiversity conservation and human well-being', *Biological Conservation*, vol 144, no 3, pp966–972.

Metsähallitus (2017a) 'Degree of endangerment', website on Finland's large carnivores, www. largecarnivores.fi/conservation-and-hunting/conservation/degree-of-endangerment. html, accessed 27 November 2017.

Metsähallitus (2017b) 'Large carnivores and reindeer herding', website on Finland's large carnivores, www.largecarnivores.fi/large-carnivores-and-us/damages/reindeer-damages. html, accessed 27 November 2017.

Metsähallitus (2017c) 'Damages caused by large carnivores', website on Finland's large carnivores, www.largecarnivores.fi/large-carnivores-and-us/damages.html, accessed 27 November 2017.

Milanesi, P., Breiner, F.T., Puopolo, F., and Holderegger, R. (2017) 'European human-dominated landscapes provide ample space for the recolonization of large carnivore populations under future land change scenarios', *Ecography*, vol 40, pp1359–1368.

Newig, J. (2011) 'Partizipation und neue Formen der Governance', in M. Gross (Hrsg) *Handbuch Umweltsoziologie* (pp. 485–502), VS Verlag, Wiesbaden.

Persson, J., Ericsson, G., and Segerström, P. (2009) 'Human caused mortality in the endangered Scandinavian wolverine population', *Biological Conservation*, vol 142, pp325–331.

Persson, J., Rauset, G. R., and Chapron, G. (2015) 'Paying for an endangered predator leads to population recovery', *Conservation Letters*, vol 8, pp345–350.

Pooley, S., Barua, M., Beinart, W., Dickman, A., Holmes, G., Lorimer, J., Loveridge, A. J., Macdonald, D. W., Marvin, G., Redpath, S., Sillero-Zubiri, C., Zimmermann, A., and Milner-Gulland, E. J. (2017) 'An interdisciplinary review of current and future approaches to improving human – Predator relations', *Conservation Biology*, vol 31, no 3, pp513–523.

Redpath, S. M., Linnell, J. D. C., Festa-Bianchet, M., Boitani, L., Bunnefeld, N., Dickman, A., Gutiérrez, R. J. Irvine, R. J., Johansson, M., Majić, A., McMahon, B. J., Pooley, S., Sandström, C., Sjölander-Lindqvist, A., Skogen, K., Swenson, J. E., Trouwborst, A., Young, J., and Milner-Gulland, E. J. (2017) 'Don't forget to look down – Collaborative approaches to predator conservation', *Biological Reviews*, vol 92, no 4, pp2157–2163.

Redpath, S. M., Young, J. Evely, A., Adams, W. M., Sutherland, W. J., Whitehouse, A., Amar, A., Lambert, R. A., Linnell, J. D. C., Watt, A., and Gutiérrez, R. J. (2013) 'Understanding and managing conservation conflicts', *Trends in Ecology & Evolution*, vol 28, no 2, pp100–109.

Sazatornil, V., Rodríguez, A., Klaczek, M., Ahmadi, M., Álvares, F., Arthur, S., Blanco, J. C., Borg, B. L., Cluff, D., Cortés, Y., García, E. J., Geffen, E., Habib, B., Iliopoulos, Y., Kaboli, M., Krofel, M., Llaneza, L., Marucco, F., Oakleaf, J. K., Person, D. K., Potočnik, H., Ražen, N., Rio-Maior, H., Sand, H., Unger, D., Wabakken, P., and López-Bao, J. V. (2016) 'The role of human-related risk in breeding site selection by wolves', *Biological Conservation*, vol 201, pp103–110.

Skoberne, P. (2017) 'The approach of Slovenia to manage the relationship between large carnivores and human presence', Slovenian Ministry of the Environment and Spatial Planning, Presentation at the 5th Regional Workshop of the EU Platform on Coexistence between People and Large Carnivores, http://ec.europa.eu/environment/nature/conservation/species/carnivores/pdf/128_Skoberne_LC%20Management%20Slovenia.pdf, accessed 26 November 2017.

Skonhoft, A. (2017) 'The silence of the lambs: Payment for carnivore conservation and livestock farming under strategic behavior', *Environmental and Resource Economics*, vol 67, no 4, pp905–923.

Slovenian Ministry of Agriculture (2015) 'Annual progress report Rural Development Programme', www.program-podezelja.si/images/SPLETNA_STRAN_PRP_NOVA/2_PRP_2007-2013/2_4_Spremljanje_in_vrednotenje/Letna_porocila_o_napredku/Letno_poro%C4%8Dilo_PRP_07-13_2015_OS_final.pdf, accessed 11 December 2017.

Sunderland, T., Ehringhaus, C., and Campbell, B. M. (2008) 'Conservation and development in tropical forest landscapes: A time to face the trade-offs?', *Environmental Conservation*, vol 34, no 4, pp276–279.

Treves, A., Krofel, M., and McManus, J. (2016) 'Predator control should not be a shot in the dark', *Frontiers in Ecology and the Environment*, vol 14, pp380–388.

Trouwborst, A., Boitani, L., and Linnell, J. D. C. (2017) 'Interpreting "favourable conservation status" for large carnivores in Europe: How many are needed and how many are wanted?', *Biodiversity and Conservation*, vol 26, no 1, pp37–61.

von Essen, E. (2017) 'Whose discourse is it anyway? Understanding resistance through the rise of "barstool biology" in nature conservation', *Environmental Communication*, vol 11, no 4, pp470–489.

Young, J. C., Jordan, A. R., Searle, K., Butler, A. S., Chapman, D., Simmons, P., and Watt, A. D. (2013) 'Does stakeholder involvement really benefit biodiversity conservation?', *Biological Conservation*, vol 158, pp359–370.

Young, J. C., Searle, K., Butler, A. S., Simmons, P., Watt, A. D., and Jordan, A. R. (2016) 'The role of trust in the resolution of conservation conflicts', *Biological Conservation*, vol 195, pp196–202.

Zabel, A., Bostedt, G., and Engel, S. (2014) 'Performance payments for groups: The case of carnivore conservation in Northern Sweden', *Environmental and Resource Economics*, vol 59, no 4, pp613–631.

Zabel, A., and Holm-Müller, K. (2007) 'Conservation performance payments for carnivore conservation in Sweden', *Conservation Biology*, vol 22, pp247–251.

Zabel, A., Pittel, K., Bostedt, G., and Engel, S. (2011) 'Comparing conventional and new policy approaches for carnivore conservation: Theoretical results and applications to tiger preservation' *Environmental and Resource Economics*, vol 48, no 2, pp287–311.

Index

abductive 64
administration: national level of 30; public 20, 29, 30, 203, 272, 273; wildlife 271, 272, 277
administrator 21, 22, 265
Africa 39, 132, 256, 257
agency: for non-humans 142, 143, 148
agriculture 22, 67, 134, 195, 276, 327
Alps 100, 155–157, 161, 292, 294, 323
Andes 58, 59, 74
anthropocentric 238, 240–241; values 37, 39, 269, 286; *see also* ecocentric; intrinsic value; value, instrumental
anthropocentrism 231–233, 237; *see also* ecocentrism; geocentrism
anthropogenic food 294–295
anti-carnivore 79, 281; *see also* pro-carnivore
anti-protectionist 206; *see also* protectionist
apex predator 99, 112, 115, 296
argumentation 4–10, 42–43, 51, 86, 320
Asia 57, 132, 148, 254, 256, 259–263, 265
assessment 65, 88, 92, 102, 211, 241, 262, 300; environmental 239; population status 294; risk 42, 103, 295; *see also* evaluation
assumption 10, 32, 63, 66, 68, 157, 197, 208–209, 214–216, 234; based on the 39, 43, 47, 138, 143, 145, 161; core 12, 91; underlying 65, 206, 242
attack: bear 73, 173, 178, 209; on children 114; coyote 213; on dogs 22, 153, 236; on humans 5–6, 22, 115, 134–135, 138, 142–144, 174, 178, 234, 236, 294; large carnivores 22, 24–25, 168, 176, 182, 209, 280, 329; leopard 135, 138, 143–144; lynx 109; 115; on pets 236; on reindeer 7; wolf 20, 22, 147, 149, 150–151, 153, 155–156, 158–159, 161–162, 173, 180–181, 325

attitude 26, 37, 40, 42–43, 62, 81–82, 89, 202; change/shift 82, 169–170, 175–176, 180, 192–200; public 25–26, 32, 168, 181, 196, 201, 203, 281; rural 168, 172, 175, 181, 193–195, 204; stakeholder 28, 46, 90, 101, 120, 133, 208; towards bears 26, 48, 178, 181, 200; towards large carnivores 3, 26–27, 100, 140, 170, 176, 178–179, 180–181, 191, 194–195, 197, 279; towards wolves 22, 26–27, 38, 43, 49, 52, 178, 180, 192, 195, 196, 200
authority 20, 21–22, 24, 31, 39, 42–43, 47, 51, 240, 270–271, 275, 281, 302, 319; competent 173, 286; local 46, 89, 279, 294; management 43–47, 53, 235, 269, 273–274, 280; public 23, 25, 203, 277; regional 12, 282–283, 286; state 283, 325
awareness 99, 134, 140, 229, 262; campaign 159, 171, 180–181; conservation 257, 264; raise 13, 106, 260, 320; self-awareness 142

bear: Andean 58–78; bear population and sub-populations in Europe 291–313; black 212; brown 4, 48, 99, 190–205, 168, 256, 269–290, 291–313, 314, 320; management 278, 292–294, 298–299, 301
beehive 209, 294
bee keeper 319
behaviour 216, 234, 240; bear 170, 294–295; change 13, 48, 132, 149; deviant 39; free riding 330; human 38, 39, 52, 82; illegal 45; intention 81–82, 210; leopard 139–140; 143; nonhuman 142; personal 228; prey 115; stakeholder 46, 51, 53, 81–82, 88, 91, 101, 212–213; towards carnivores 202;